Essential Topics in Statistics

Essential Topics in Statistics

Editor: Calanthia Wright

NY RESEARCH PRESS
New York

Published by NY Research Press
118-35 Queens Blvd., Suite 400,
Forest Hills, NY 11375, USA
www.nyresearchpress.com

Essential Topics in Statistics
Edited by Calanthia Wright

International Standard Book Number: 978-1-63238-560-4 (Hardback)

Cataloging-in-Publication Data

Essential topics in statistics / edited by Calanthia Wright.
p. cm.
Includes bibliographical references and index.
ISBN 978-1-63238-560-4
1. Statistics. 2. Mathematical statistics. I. Wright, Calanthia.
QA276 .E87 2017
519.5--dc23

Printed in the United States of America.

Contents

Preface

I am honored to present to you this unique book which encompasses the most up-to-date data in the field. I was extremely pleased to get this opportunity of editing the work of experts from across the globe. I have also written papers in this field and researched the various aspects revolving around the progress of the discipline. I have tried to unify my knowledge along with that of stalwarts from every corner of the world, to produce a text which not only benefits the readers but also facilitates the growth of the field.

Statistics is the science of compilation of numerical data that may represent a social or physical fact. This book on statistics deals with the basic methodology that is followed to derive inferences. Statistics uses the help of aggregation, averages and mean to calculate values of study. Contents included in this book discuss the various techniques through which results on the basis of data, data sets and models can be arrived at. There has been rapid progress in this field and its applications are finding their way across multiple industries. Tools such as sampling, surveys and experiments are a vital part of this field. This book on statistics will be of great help to students and researchers in the fields of economics, number theory and computational statistics.

Finally, I would like to thank all the contributing authors for their valuable time and contributions. This book would not have been possible without their efforts. I would also like to thank my friends and family for their constant support.

Editor

1

A defense against the alleged unreliability of difference scores

David Trafimow[1]*

*Corresponding author: David Trafimow, Department of Psychology, New Mexico State University, MSC 3452, P.O. Box 30001, Las Cruces, NM 88003-8001, USA

E-mail: dtrafimo@nmsu.edu

Reviewing editor: Hamdi Raïssi, IRMAR-INSA (UEB), France

Abstract: Based on a classical true score theory (classical test theory, CTT) equation, indicating that as the observed correlation between two tests increases, the reliability of the difference scores decreases, researchers have concluded that difference scores are unreliable. But CTT shows that the reliabilities of the two tests and the true correlation between them influence the observed correlation and previous analyses have not taken the true correlation sufficiently into account. In turn, the reliability of difference scores depends on the interaction of the reliabilities of the individual tests and their true correlation when the variances of the tests are equal, and on a more complicated interaction between them and the deviation ratio when the variances of the tests are not equal. The upshot is that difference scores likely are more reliable, on more occasions, than researchers have realized. I show how researchers can predict what the reliability of the difference scores is likely to be, to aid in deciding whether to carry through one's planned use of difference scores.

Subjects: Behavioral Sciences; Communication Studies; Development Studies; Education; Social Sciences; Sports and Leisure

Keywords: reliability of difference scores; classical true score theory; classical test theory; true correlation; deviation ratio

ABOUT THE AUTHOR

David Trafimow is a distinguished achievement professor of psychology at New Mexico State University, a fellow of the Association for Psychological Science, executive editor of the *Journal of General Psychology*, and also for Basic and Applied Social Psychology. He received his PhD in psychology from the University of Illinois at Urbana-Champaign in 1993. His current research interests include attribution, attitudes, cross-cultural research, ethics, morality, methodology and philosophy of science, potential performance theory, and descriptive and inferential statistical analyses.

PUBLIC INTEREST STATEMENT

There are many important issues that depend on difference scores for their elucidation. For example, to find out if an intervention worked, participants might be asked to complete a relevant measure before and after the intervention, with the hope that scores will be more favorable after it than before it. Or participants might complete two or more different measures, such as when high school students take tests in different subject areas to determine relative strengths and weaknesses. Whether the difference of interest pertains to change on a measure, or on the difference between different measures, difference scores are important. However, because of a prevalent belief that difference scores are not reliable, researchers have been discouraged from using them. The present work challenges the blanket assertion that difference scores are inherently unreliable and demonstrates the conditions under which different scores are more reliable or less reliable.

1. Introduction

Basic and applied researchers in education and psychology often have reason to be interested in the differences between people's scores on two tests. The two tests might have the same items, as in a pretest–posttest design or in other types of within-participants designs. Alternatively, the tests might have different items, such as when a school psychologist identifies that a particular student has a relative weakness in one domain compared to a relative strength in another domain. In either case, difference scores—the difference between scores on two tests—are of interest. But for many decades, psychometric experts have warned that difference scores are not reliable. Bereiter (1963, p. 3) stated the consequence of this warning with admirable clarity:

> Although it is commonplace for research to be stymied by some difficulty in experimental methodology, there are really not many instances in the behavioral sciences of promising questions going unresearched because of deficiencies in statistical methodology. Questions dealing with psychological change may well constitute the most important exceptions. It is only in relation to such questions that the writer has ever heard colleagues admit to having abandoned major research objectives solely because the statistical problems seemed to be insurmountable.

Bereiter's observation has retained its force, despite advances in the intervening decades. For example, Chiou and Spreng (1996) noted that difference scores continue to be criticized. Even very recently, Thomas and Zumbo (2012) stated, "There is such doubt in research practice about the reliability of difference scores that granting agencies, journal editors, reviewers, and committees of graduate students' theses have been known to deplore their use" (p. 37). Given the expressed negativity throughout the decades, my goal is to investigate the alleged unreliability of difference scores in a better way than has hitherto been done. The stakes are large. If previous criticisms hold up, this would constitute an important reason to avoid doing research that depends on difference scores. In contrast, if previous criticisms do not hold up, an enormous impediment to research can be removed. I will show that the reliability of difference scores depends on the interaction between the true correlation between the tests, individual test reliabilities, and the deviation ratio, which is a ratio of test variances. A close study of this three-way interaction suggests that the blanket conclusion that difference scores are inherently unreliable is too pessimistic.

1.1. The basis of the criticism

The criticism that difference scores are unreliable comes directly from a theorem based on classical true score theory or classical test theory (CTT) that can be found in many places (e.g. Cronbach, 1990; Lord, 1963; see Gulliksen, 1987 for an excellent review of CTT theorems). CTT features the notion of reliability, which is usually defined as true score variance divided by observed score variance. Equation 1 provides the general formula that expresses the reliability of difference scores ($\rho_{dd'}$) in terms of the reliabilities of the two tests ($\rho_{XX'}$ and $\rho_{YY'}$), the variances of the two tests (σ_X^2 and σ_Y^2), and the correlation between observed scores on the two tests (ρ_{XY}) (Designating a person's score on one test as X_i and on the other as Y_i implies that the difference score is $d_i = Y_i - X_i$.).

$$\rho_{dd'} = \frac{\sigma_X^2\rho_{XX'} + \sigma_Y^2\rho_{YY'} - 2\rho_{XY}\sigma_X\sigma_Y}{\sigma_X^2 + \sigma_Y^2 - 2\rho_{XY}\sigma_X\sigma_Y} \tag{1}$$

If the variances of the two tests are equal, Equation 1 reduces to Equation 2 below.

$$\rho_{dd'} = \frac{\rho_{XX'} + \rho_{YY'} - 2\rho_{XY}}{2 - 2\rho_{XY}} \tag{2}$$

Under Equation 2, it is easy to see that as the correlation between the two tests increases, keeping the reliabilities of the two tests constant, the reliability of difference scores decreases. In the case of a pretest–posttest design, for example, one would expect the correlation between pretest and posttest to be substantial, and Equation 2 seems to indicate that this should result in unreliable difference scores. Historically, researchers with psychometric expertise have made similar arguments to militate against the reliability of difference scores.

However, despite widespread criticism, some have defended difference scores. For example, one problem with unreliable difference scores is that such unreliability might render it difficult to obtain statistically significant findings. After reviewing the literature on this criticism, Thomas and Zumbo (2012) suggested that this sort of unreliability is not crucial for inferential statistical procedures, such as *t*-tests and ANOVAs (also see Gaito & Wiley, 1963). Possible limitations of this defense might be that this alleged lack of importance has not been demonstrated conclusively. Besides that, difference scores are used for many purposes other than significance tests. Suppose a researcher wishes to correlate difference scores with another variable. Or suppose a researcher wishes to base interventions on difference scores. In cases such as these, the unreliability of difference scores, if it holds up, would constitute an important problem, even if it were admitted that inferential statistics are not importantly affected.

Another defense involves a switch of focus from Equations 2 to 1. The idea here is that although difference scores are unreliable under the assumption of equal variances, if the variances are unequal, difference scores do not have to be unreliable. Chiou and Spreng (1996) provided some examples of how the unreliability of difference scores is substantially mitigated by having sufficiently unequal variances of the two tests. Thus, their analysis suggested that a potential answer to the problem is to use tests with unequal variances. A limitation, however, is that it might be difficult to arrange matters in this way, particularly if the same test is given twice.

Another defense traces back to Bereiter (1963), who pointed out that as the reliabilities of the individual tests increase, so does the reliability of the difference scores. This can be seen easily in Equation 2.

In his insightful analysis, Bereiter (1963) pointed out that increasing the reliabilities of the individual tests has contradictory effects on the reliability of the difference scores. To understand these contradictory effects, it is necessary to recall that as the reliability of the tests increase, so does the observed correlation. Therefore, on the one hand, increasing the reliabilities of the individual tests decreases the reliability of the difference scores by increasing the correlation between observed scores over what it otherwise would be. But, on the other hand, as Equation 2 shows, increasing the reliabilities of the individual tests increases the reliability of the difference scores.

Unfortunately, Bereiter did not proceed further with his analysis, and researchers have not attended to these contradictory effects. Given that Equation 1, Equation 2, and the contradictory effects of varying the reliabilities of the individual tests, all depend on CTT, this lack of attention is surprising. My goal is to take the contradictory effects seriously to investigate their consequences for difference score reliability. In the analyses that follow, I intend to return to the roots of CTT to discern exactly the extent of the deleterious vs. beneficial effects of increasing the reliabilities of the individual tests on the reliability of the difference scores. In so doing, I am not committing to an actual belief or disbelief in CTT. Rather, because the criticisms stem from CTT, an investigation from the point of view of CTT directly addresses them. It also seems worthwhile to note that although more advanced theories have been proposed and favored by many researchers, such as generalizeability theory and item response theory, these reduce down to CTT in their simplest incarnations (see Cronbach, Gleser, Nanda, & Rajaratnam, 1972; Hulin, Drasgow, & Parsons, 1983, respectively).

1.2. The importance of CTT's true correlations

Consider the attenuation formula from CTT, written below as Equation 3, where $\rho_{T_X T_Y}$ represents the true correlation between two tests (i.e. the correlation between true scores).

$$\rho_{XY} = \rho_{T_X T_Y} \sqrt{\rho_{XX'} \rho_{YY'}} \tag{3}$$

Equation 3 makes clear the importance of the reliabilities of the two tests. As the reliabilities of the tests increase, the observed correlation becomes increasingly similar to the true correlation. In the limiting case of perfectly reliable tests, the observed correlation equals the true correlation. When the tests are not perfectly reliable, the observed correlation is less than the true correlation. Thus, Equation 3 clarifies Bereiter's (1963) statement that increasing the reliabilities of the tests increases the correlation between observed scores, and hence decreases the reliability of the difference scores. To reiterate, however, Equation 2 shows a contradictory effect whereby increasing the reliability of the individual tests increases the reliability of the difference scores.

Let us again consider Equation 3, and the mathematical fact that the observed correlation depends on two things: the reliabilities of the individual tests and the true correlation coefficient. I have discussed the reliabilities of the individual tests above, but what about the true correlation coefficient? Equation 3 shows that, in general, as the true correlation increases, so does the observed correlation. An exception would be if the reliability of one or both of the individual tests equals zero, in which case, the observed correlation also would equal zero, regardless of the true correlation. Well, then, going back to Equation 2, we see that because increasing the true correlation tends to increase the observed correlation, and because increasing the observed correlation decreases the reliability of the difference scores, it follows that the reliability of the difference scores depends partly on the size of the true correlation. Specifically, the larger the true correlation, keeping the reliabilities of the individual tests constant, the less reliable the difference scores.

It is interesting that although the argument that difference scores have low reliability has its origin in CTT, and that CTT depends on the notion of a true correlation (which, in turn, depends on the notion of a true score), nobody to my knowledge has actually included true correlations in their investigation of the reliability of difference scores. In addition, my foregoing comments implicate the importance of the reliabilities of the individual tests as having complicated effects on the reliability of difference scores. Therefore, one of my main goals is to investigate how the true correlation interacts with the reliabilities of the two tests to determine the reliability of difference scores. It is not necessary to derive equations that are radically different from existing ones; rather, it is sufficient to express existing CTT equations in a form that includes true correlations and reliabilities of individual tests.

In addition, there is no law stating that the variances of the two tests have to be equal. As Chiou and Spreng (1996) suggested, the sizes of variances between the two tests also matter. Possibly, the reliability of difference scores might be influenced by the interaction between the variances of the two tests, their reliabilities, and the true correlation. Therefore, my second goal is to investigate this potential triple interaction.

2. Analyses

I will present two types of analyses. First, I will assume equal variances and explore the reliability of difference scores as a function of the interaction between the reliabilities of the individual tests and the true correlation. Second, I will take into account the possibility of unequal variances and explore the reliability of difference scores as a function of three variables. These are the reliabilities of the individual tests, the true correlation, and the ratio of variances or standard deviations between the two tests.

2.1. *Equal variances*

To perform the analyses of interest, it is necessary to render the reliability of the difference scores in Equation 2 as a function of the reliabilities of the individual tests and the true correlation. At present, however, Equation 2 includes the observed correlation rather than the true correlation. Because the observed correlation is influenced, in part, by the reliabilities of the two tests, there is no way to vary these reliabilities and the observed correlation independently of each other. This inconvenience can be remedied simply by substituting Equation 3 into Equation 2. The result is Equation 4 below that features the true correlation, as opposed to the observed correlation.

$$\rho_{dd'} = \frac{\rho_{XX'} + \rho_{YY'} - 2\rho_{T_XT_Y}\sqrt{\rho_{XX'}\rho_{YY'}}}{2 - 2\rho_{T_XT_Y}\sqrt{\rho_{XX'}\rho_{YY'}}} \tag{4}$$

Based on Equation 4, Figure 1 provides an illustration of how the reliabilities of the two tests interact with the true correlation to determine the reliability of the difference scores. In Figure 1, the reliabilities of the two tests were set equal to each other, and allowed to range from 0 to 1 along the horizontal axis. In addition, the curves represent the cases where the true correlation equals .1 (top curve), .3 (second curve), .5 (third curve), .7 (fourth curve), and .9 (bottom curve). Figure 1 illustrates three effects. Most obviously, as the true correlation increases, the reliability of the difference scores (along the vertical axis) decreases. There also is an effect of the reliabilities of the tests on the reliability of the difference scores. Figure 1 illustrates the totality of the contradictory effects I discussed earlier whereby increasing the reliabilities of the tests has both a deleterious effect on the reliability of difference scores (by increasing the true correlation) and a beneficial effect on the reliability of difference scores (by direct entry into Equation 2). The upward trend of all five curves in Figure 1, as the reliabilities of the individual tests increase, shows that the net effect of the contradictory forces is positive for the reliability of the difference scores. That is, as the reliabilities of the individual tests increase, so does the reliability of the difference scores.

Finally, the interaction between the true correlation and the reliabilities of the individual tests deserves observation. As the true correlation increases, the effect of increasing the reliabilities of the individual tests on the reliability of the difference scores becomes increasingly nonlinear. For example, when the true correlation is .9, the reliabilities of the individual tests have to be quite impressive to result in a respectable reliability of the difference scores.

Well, then, are difference scores unreliable, as we so often have been told? Figure 1 shows that the answer is, "It depends." If the true correlation is large (e.g. .9), then it takes extremely impressive reliabilities of the individual tests to result in a reasonably sized reliability of difference scores; the shape of the bottom curve is a real problem. As an example, when the true correlation is .9, even if the reliabilities of both individual tests also equal .9, the reliability of the difference scores is an unimpressive .47. The reliabilities of both individual tests must be ratcheted up to .96 for the reliability of the difference scores to exceed an arbitrary cut-off of .7.

On the other hand, the problem is substantially mitigated if the true correlation is at lesser levels. For example, for the reliability of the difference scores to pass the cut-off of .7 when the true correlation is .7, it is only necessary for the reliabilities of the two tests to exceed .89—a difficult but far from impossible requirement. And as the true correlation decreases further, it is tolerable to have lesser reliabilities of the individual tests.

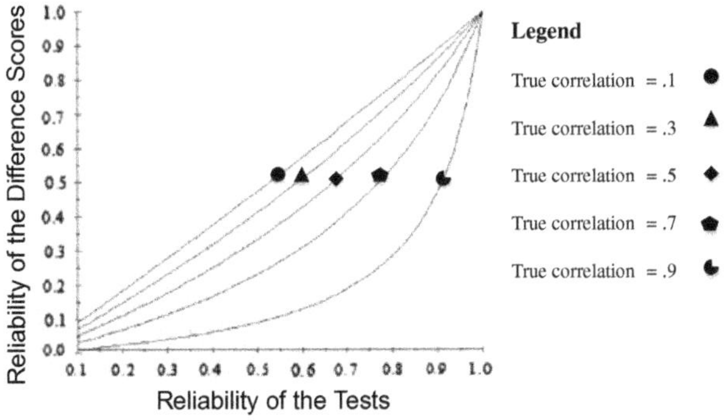

Figure 1. The reliability of difference scores is presented as a function of the reliability of the individual tests and the true correlation.

2.2. Unequal variances

Before conducting the analyses, it is again necessary to obtain an equation based on the true correlation as opposed to the observed correlation. In addition, it is necessary to include a ratio of standard deviations or variances (I will use standard deviations). The net effect of these considerations is to render a slightly more complicated derivation than in the previous section.

To commence, I define the deviation ratio (a) as the standard deviation of test X scores divided by the standard deviation of test Y scores, as is indicated in Equation 5 below. And Equation 6 is simply a rearrangement of Equation 5.

$$a = \frac{\sigma_X}{\sigma_Y} \tag{5}$$

$$\sigma_X = a\sigma_Y \tag{6}$$

Substituting Equation 6 into Equation 1 renders Equation 7.

$$\rho_{dd'} = \frac{a^2\sigma_Y^2\rho_{XX'} + \sigma_Y^2\rho_{YY'} - 2\rho_{XY}a\sigma_Y^2}{a^2\sigma_Y^2 + \sigma_Y^2 - 2\rho_{XY}a\sigma_Y^2} \tag{7}$$

Cancelling σ_Y^2 renders Equation 8.

$$\rho_{dd'} = \frac{a^2\rho_{XX'} + \rho_{YY'} - 2\rho_{XY}a}{a^2 + 1 - 2\rho_{XY}a} \tag{8}$$

Although Equation 8 is nice to have, it is not sufficient in one respect. Specifically, it still retains the observed correlation, whereas we would like it to include the true correlation. The solution is to substitute Equation 3 into Equation 8 to render Equation 9.

$$\rho_{dd'} = \frac{a^2\rho_{XX'} + \rho_{YY'} - 2\rho_{T_XT_Y}\sqrt{\rho_{XX'}\rho_{YY'}}\,a}{a^2 + 1 - 2\rho_{T_XT_Y}\sqrt{\rho_{XX'}\rho_{YY'}}\,a} \tag{9}$$

Figure 2 illustrates how Equation 9 manifests where Panels A, B, and C represent when the true correlation is .1, .5, and .9, respectively; the reliabilities of the two tests are set equal to each other and vary from 0 to 1 (as in Figure 1); and where the deviation ratio is 4 (top curve), 3, (second curve), 2 (third curve), or 1 (bottom curve). As in Figure 1, the reliability of the difference scores is indicated on the vertical axis.

Figure 2 shows that there are three trends that replicate what we saw in Figure 1. Comparing across panels in Figure 2 shows that as the true correlation increases, the reliability of the difference scores decreases. Also, the upward curves in all three panels replicate that as the reliabilities of the individual tests increase, so does the reliability of the difference scores. Finally, these two effects qualify each other, as we saw earlier.

But some new observations also emerge. First, there is a general trend, which can be seen most easily in Panel C of Figure 2, that as the deviation ratio increases, so does the reliability of the difference scores. This effect supports the argument by Chiou and Spreng (1996) in favor of high deviation ratios if one wishes to increase the reliability of difference scores. Second, there is an interesting interaction between the deviation ratio and the reliabilities of the individual tests that is particularly easy to see in Panel C of Figure 2. Hearkening back to Figure 1, recall the unfavorable shape of the curve when the true correlation was .9, which is illustrated again by the bottom curve in Panel C of Figure 2. But as the deviation ratio increases, note that the shape (and altitude) of the curve becomes increasingly less unfavorable. Thus, even in the worst-case scenario of a large true correlation, it is possible to obtain a respectable number for the reliability of the difference scores so long as the deviation ratio is high. Put

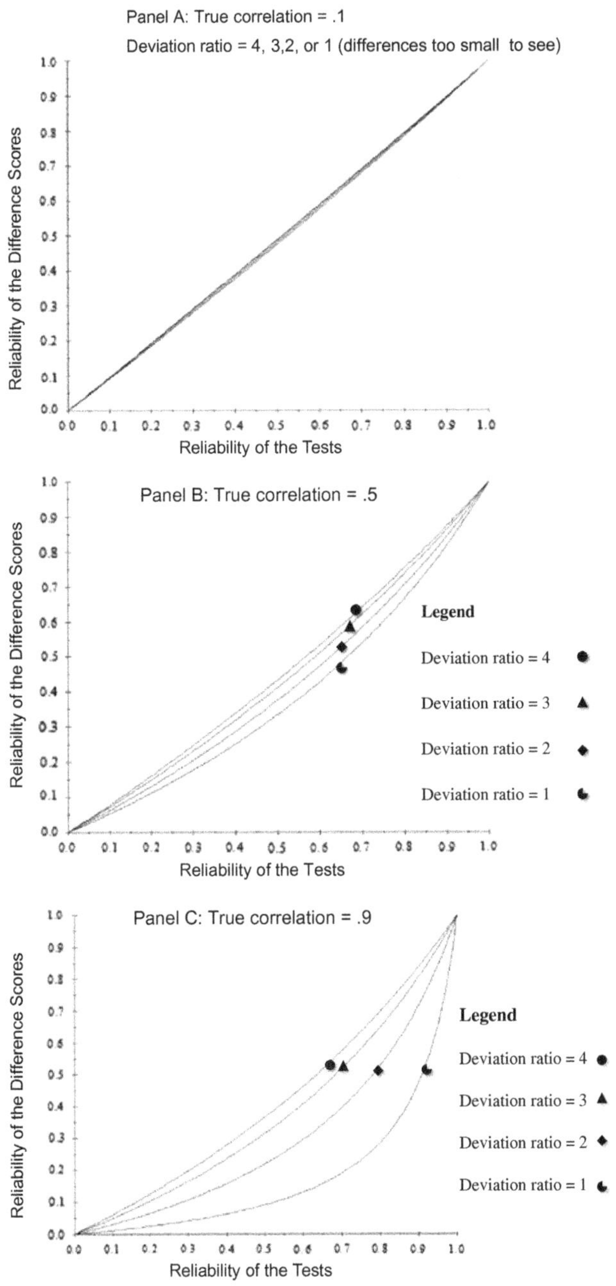

Figure 2. The reliability of difference scores is presented as a function of the reliability of the individual tests, the true correlation, and the deviation ratio.

another way, as the variances of the two tests become increasingly unequal, their reliabilities can be less impressive and still overcome the deleterious effects of a large true correlation. Finally, it is interesting to consider the three-way effect. In Panel A of Figure 2, where the true correlation is only .1, the altitude and shape of the curves is approximately the same regardless of the deviation ratio. In Panel B of Figure 2, where the true correlation is .5, the effect of the deviation ratio is perceptible, and shows that as it increases, so does the reliability of the difference scores. In Panel C, as we already have seen, the separation of the curves is impressive.

3. Discussion

Given Equations 4 and 9, as well as Figures 1 and 2, what can we conclude about the reliability of difference scores? The analyses demonstrate that, if the reliabilities of the individual tests are good, the reliability of the difference scores will be at least reasonable so long as the true correlation is not too large. The difficulty comes in when the true correlation is extremely large (e.g. .9). In this case, the reliabilities of the individual tests have to be excellent—and better than can be expected from most tests—to enable the reliability of the difference scores to be reasonable. An exception would be if the two tests have very different variances (e.g. deviation ratio is 4), in which case, the reliabilities of the individual tests can be reduced and still result in a reasonable value for the reliability of the difference scores.

This seems like a good place to pause to make sure that a simple point does not get lost in the interactions illustrated by Figure 2. Specifically, in all of the analyses, the reliability of the difference scores is limited by the reliabilities of the individual tests. The researcher can tolerate somewhat greater or lesser reliabilities of the individual tests, depending on the true correlation and the deviation ratio. However, even when the true correlation is low (as in Panel A of Figure 2), so the deviation ratio does not matter in a discernible way, the reliabilities of the individual tests still set a limit on how reliable the difference scores can be.

I commenced the present work by emphasizing the extent to which difference score research is deplored in much education and social science literature based on reliability grounds. However, the analyses show that a blanket condemnation is not justified. The analyses also provide the necessary information for sensible recommendations for answering the question, "With respect to the particular difference score research one wishes to conduct, how can one know whether or not to proceed?"

Let us recall Equation 4, which implies that with equal variances, the two important factors are the reliabilities of the individual tests and the true correlation. If one is using well-established tests, their reliabilities ought to already be known. And if not, the researcher can conduct a pilot study to find out. The true correlation is more of a problem because it is a hypothetical entity that cannot be obtained directly. However, it can be calculated indirectly. To see that this is so, consider a rearrangement of Equation 3 that is given below as Equation 10.

$$\rho_{T_X T_Y} = \frac{\rho_{XY}}{\sqrt{\rho_{XX'}\rho_{YY'}}} \tag{10}$$

If the correlation between the individual tests is known, and the reliabilities are known, or these can be determined by pilot research, Equation 10 provides an estimate of the true correlation. In turn, the true correlation and the reliabilities of the two tests can be instantiated into Equation 4 to predict the reliability of the difference scores. Using this method, anyone can determine, for the particular research under consideration, whether the difference scores will be sufficiently reliable for the purpose at hand. If so, the researcher can go ahead and perform the research without qualms about the reliability of the difference scores.

But suppose that the result of the foregoing process is not satisfactory. It still might not be necessary for the researcher to give up on his or her plans. Remember that Equation 4 assumes equal variances and this might not be so. If the variances of the two tests are known, or can be determined via pilot research, the researcher should use Equation 9 rather than Equation 4, which will result in a larger (and more accurate) predicted value for the reliability of the difference scores. If the result of using Equation 9 is satisfactory, the research can proceed without qualms about the reliability of the difference scores. However, if even the use of Equation 9 fails to result in a sufficient value for the reliability of the difference scores, then the researcher might contemplate returning to the drawing board.

Consider an example. Suppose that the reliabilities of the two tests are .85 for each, that the estimated true correlation is .5 (from Equation 10), and that the deviation ratio is 3. In that case, using Equation 4, the reliability of the difference scores would be predicted to be .74. Although this is not an outstanding value, it would take a hard-hearted reviewer to conclude that the proposed research should not be conducted. In addition, suppose we use Equation 9, rather than Equation 4, to take advantage of the deviation ratio (5). In that case, the reliability of the difference scores would be predicted to be .80, and even the hard-hearted reviewer should be satisfied.

The present analyses show that although much depends on the reliabilities of the individual tests, much also depends on the true correlation and on the deviation ratio. How likely are these to be favorable? It might depend on whether the difference score is based on the same test conducted at different times or on whether it is based on different tests. For example, if a school psychologist is interested in the difference between mathematical and verbal ability, tests of these are sufficiently different that the true correlation likely would not be particularly large. In addition, measurement in the area has advanced to the level where researchers routinely report reliabilities in excess of .9 for a variety of abilities. The combination of the low true correlation and high reliabilities of the individual tests implies that the reliability of the difference scores should be impressive, regardless of the deviation ratio. In contrast, if a researcher is interested in difference scores based on using the same test twice, the true correlation might be quite large. And this would be a problem unless the reliability of the test on both test-taking occasions is very impressive or the deviation ratio is large. Why might the deviation ratio be a large value? One reason is that the treatment might raise or lower scores so that they collect near a ceiling or floor, respectively. In that case, the variance would be much lower on the second test-taking occasion than on the first one, thereby causing a large deviation ratio. There is a risk that the reduction in variance on the second test-taking occasion might decrease the reliability of the test on that occasion, depending on the relative amounts by which true correlation variance vs. error variance decreases.

Of course, even if the same test is given before and after treatment, there are reasons why the true correlation might nevertheless not be a large value. To see why this might be so, contrast the case where the treatment has an approximately equal effect on each person against the case where the treatment has unequal effects on different persons—that is, there is a treatment × person interaction. In the former case, the true correlation will be a large number, the deviation ratio will be near unity, and so one would need extremely favorable reliabilities on both test-taking occasions to have reasonably reliable difference scores. However, if there is a treatment × person interaction, the true correlation likely will not be a large value, and so the difference scores will be reasonably reliable, provided that test is reasonably reliable on both test-taking occasions. As an additional possibility, in the absence of ceiling or floor effects, the treatment × person interaction could increase variance on the second test-taking occasion depending on the nature of the interaction. Although treatment × person interactions add error variance from the point of view of statistical tests, they can increase the reliability of difference scores by reducing the true correlation or by causing the deviation ratio to stray from unity.

Does it matter whether the difference scores are reliable? This depends on one's purpose. If the goal is simply to demonstrate that the treatment works, lack of reliability of the difference scores likely will not be fatal for the statistical test, as Thomas and Zumbo (2012) showed. But if the goal is to correlate the difference scores with another variable, then difference score reliability will matter a great deal.

In conclusion, based on Equations 1 and 2, much negativity has been directed towards difference scores. However, because these equations came out of CTT, and because CTT features the notion of true correlations, it is surprising that previous researchers who have been concerned with difference score reliability have not included the interaction between true correlations and reliabilities of individual tests in their analyses. But it is necessary to include this interaction because the reliabilities of individual tests interact with the true correlation to determine the observed correlation that plays so important a role in Equations 1 and 2. The reliabilities of the individual tests play an additional role through their direct entry in these equations. Finally, the deviation ratio also plays its part in an interaction with the reliabilities of the individual tests and the true correlation. It is only through Equations 4 and 9, derived above, that the interactions of these effects can be analyzed and understood. The analyses based on these equations suggest that difference scores may have received an undeserved bad reputation. Although there are some combinations of true correlations, reliabilities of individual tests, and deviation ratios that imply that difference scores are unreliable, there are other combinations that imply that difference scores are reasonably reliable or even extremely reliable. Rather than make a blanket statement, each person should evaluate, for herself or himself, the reliability of difference scores in the context in which they are to be used. Equation 4 (equal variances) or Equation 9 (unequal variances) provides the way for each researcher to come to his or her own conclusion with respect to the particular tests under consideration, as I demonstrated earlier. Surely, this is superior to basing research decisions on a blanket recommendation that often is wrong.

Funding
The author received no direct funding for this research.

Author details
David Trafimow[1]
E-mail: dtrafimo@nmsu.edu
[1] Department of Psychology, New Mexico State University, MSC 3452, P.O. Box 30001, Las Cruces, NM 88003-8001, USA.

References

Bereiter, C. (1963). Some persisting dilemmas in the measurement of change. In C. W. Harris (Ed.), *Problems in measuring change* (pp. 3–20). Madison: University of Wisconsin Press.

Chiou, J. S., & Spreng, R. A. (1996). The reliability of difference scores: A re-examination. *Journal of Consumer Satisfaction, Dissatisfaction, and Complaining Behavior, 9*, 158–167.

Cronbach, L. J. (1990). *Essentials of psychological testing* (5th ed.). New York, NY: Harper Collins.

Cronbach, L. J., Gleser, G. C., Nanda, H., & Rajaratnam, N. (1972). *The dependability of behavioral measurements: Theory of generalizability for scores and profiles*. New York, NY: Wiley.

Gaito, J., & Wiley, D. E. (1963). Univariate analysis of variance procedures in the measurement of change. In C. W. Harris (Ed.), *Problems in measuring change* (pp. 60–84). Madison: University of Wisconsin Press.

Gulliksen, H. (1987). *Theory of mental tests*. Hillsdale, NJ: Lawrence Erlbaum Associates.

Hulin, C., Drasgow, F., & Parsons, C. (1983). *Item response theory: Application to psychological measurement*. Homewood, IL: Dow Jones-Irwin.

Lord, F. M. (1963). Elementary models for measuring change. In C. W. Harris (Ed.), *Problems in measuring change* (pp. 21–38). Madison: University of Wisconsin Press.

Thomas, D. R., & Zumbo, B. D. (2012). Difference scores from the point of view of reliability and repeated-measures ANOVA: In defense of difference scores for data analysis. *Educational and Psychological Measurement, 72*, 37–43. http://dx.doi.org/10.1177/0013164411409929

2

Bayesian index to compare assumed survival functions of two hazards following exponential distributions

Yohei Kawasaki[1*], Masaaki Doi[2,3], Kazuki Ide[1] and Etsuo Miyaoka[4]

*Corresponding author: Yohei Kawasaki, Department of Drug Evaluation & Informatics, School of Pharmaceutical Sciences, University of Shizuoka, Shizuoka, Japan
E-mails: yk_sep10@yahoo.co.jp, ykawasaki@u-shizuoka-ken.ac.jp
Reviewing editor: Hiroshi Shiraishi, Keio University, Japan

Abstract: In clinical studies, the period before an event occurs is frequently considered as an outcome. The general term used for this analytical method in clinical studies is "survival time analysis". The method employs Cox regression to compare hazards between groups. In this paper, we propose an index defined in the Bayesian framework to compare hazards. This index compares the sizes of the hazards directly. Further, we propose two methods to calculate the sizes of the hazards. One is an approximate method, by which we can easily calculate the size, and the other is an exact method. In addition, we apply the proposed index to two actual clinical studies to assess its usability.

Subjects: Mathematics & Statistics; Medical Statistics; Medical Statistics & Computing; Medicine; Medicine, Dentistry, Nursing & Allied Health; Science; Statistics; Statistics & Probability

Keywords: Bayesian statistics; survival time analysis; superiority test; one-sided test; hypothesis test

1. Introduction

Events such as death and survival at the end of the studies are important outcomes in clinical studies. Survival time analysis is generally employed to compare groups when analyzing these outcomes.

ABOUT THE AUTHOR

Yohei Kawasaki received the BS degree in mathematics in 2005 and the MS and PhD degrees in mathematical science from Tokyo University of Science, Japan in 2007 and 2012, respectively. He joined the Mitsubishi Tanabe Pharma Corporation, the National Center for Global Health and Medicine, Tokyo University of Science, in 2005, 2013, and 2014, respectively.

He has published over 70 research articles in reputed international journals. His research interests include the classical analysis, Bayesian analysis, Biostatistics and mathematical education. He is currently a lecturer in the Department of Drug Evaluation and Informatics, School of Pharmaceutical Sciences, University of Shizuoka, Japan.

He also is a guest researcher of National Center for Global Health and Medicine and Shizuoka General Hospital as Biostatistician.

PUBLIC INTEREST STATEMENT

In clinical studies, the period before an event occurs is frequently considered as an outcome. The general term used for this analytical method in clinical studies is "survival time analysis". The method employs Cox regression to compare hazards between groups. In this paper, we propose an index defined in the Bayesian framework to compare hazards. We apply the proposed index to two actual clinical studies to assess its usability.

It is not an exaggeration to say that in the frequentist framework, the basic methodology of survival time analysis has almost been settled (Colett, 2003; Hosmer, Lemeshow, & May, 2008). Besides, recently the method of direct comparison of hazards was proposed (Heller & Mo, 2015).

In the Bayesian framework, on the other hand, a methodology has not yet been established. In this paper, Bayes introduced the idea of inverse probability, which, following the work of Laplace and others, came to be known as Bayes' Theorem. The Bayesian approach to statistical inference is useful for several reasons. Perhaps foremost among them is the formal manner in which Bayesian methods incorporate prior information, whether it be expert opinion or historical data. The sequential learning nature of the method, with prior knowledge being updated to yield posterior distributions, forms the basis for adaptive methods, in which models are continually updated with new information. Other advantages of the method include the ease of predictive model building and the avoidance of large-sample approximations in some cases. Furthermore, Bayesian analysis provides straightforward probabilistic statements about parameters of interest and inferential questions. The Bayesian influence can be found in a broad array of fields such as statistics, medicine, epidemiology, computer science, economics, engineering, physics, and philosophy of science. A number of papers on biomedical applications have suggested the use of Bayesian methods in healthcare evaluation (Spiegelhalter, 2004; Zhao, Feng, Chen, & Taylor, 2015), the pharmaceutical industry (Grieve, 2007; Racine, Grieve, Fluhler, & Smith, 1986), and clinical trials (Berry, 2006; Cornfield, 1976; Lee & Chu, 2012; Spiegelhalter, Freedman, & Parmar, 1993, 1994).

As mentioned above, Bayesian statistics are employed in clinical trials to evaluate outcomes, but no one has so far proposed an index for the easy interpretation of results. In Bayesian model selection, the most widely used tools are Bayes factors (BFs) (Kass & Raftery, 1995). However, these are unscaled when improper priors are used. In order to overcome this problem and approximate the underlying BFs, fractional and intrinsic BFs (Berger & Pericchi, 1996; O'Hagan, 1995) have been proposed in the literature. Thus, BFs are useful in Bayesian model selection. However, we cannot directly compare parameters of posterior distributions using BFs. Moreover, it is often difficult to interpret BFs for several reasons; one reason is that the coefficients of BFs do not have intuitive meaning.

Zaslavsky (2010, 2013) and Kawasaki and Miyaoka (2012b) were the first to propose a new type of index, called the "Bayesian index", for the comparison of binomial proportions and to apply it to clinical trial results. These indexes can compare parameters directly, and since the results are evaluated as a probability, it is easy to understand them intuitively. In this paper, we propose the index $\eta = P(h_1 < h_2 \,|\underline{T}_1, \underline{T}_2)$, which is developed from the index proposed by Kawasaki and Miyaoka (2012a, 2012b), for comparison of hazards, where h_i indicates the hazard of group i, with the exponential distribution assumed for the survival function. We further assume that the prior distribution is a gamma distribution with parameters α_i and β_i. $\underline{T}_i$ denotes the total observation duration of group i. This index indicates the probability that the hazard of group 1 is less than that of group 2.

The remainder of the paper is organized as follows. We propose a method for the exact calculation of the index η as well as a method for an approximate calculation, with which we can easily calculate the probability of the index, in Section 2. We compare the approximate probability with the exact probability and present the results of actual clinical trials to show the utility of η in Section 3. Finally, we conclude with a brief summary in Section 4.

2. Methods

2.1. Notation

Let C_{ij} and Y_{ij} be independent. For patient number j in group i, let Y_{ij} be the random variable representing the time to event occurrence, C_{ij} be the random variable representing the time to censoring (where $i = 1, 2$ and $j = 1, 2, \ldots, n_i$), and $T_{ij} = \min(Y_{ij}, C_{ij})$ be the variable representing the observation duration. Additionally, to judge the executing or censoring event for patient number j in group i, we define the event index as

$$\delta_{ij} = \begin{cases} 1 & \left(Y_{ij} \le C_{ij}\right) \\ 0 & \left(Y_{ij} > C_{ij}\right) \end{cases},$$

where the realized value of T_{ij} is t_{ij}. Furthermore, let the hazard function of patient number j in group i be constant as follows

$$h(t_{ij}) = h_i \quad (h_i > 0).$$

The probability density function and survival time function can be denoted, respectively, as

$$f(t_{ij}) = h_i e^{-h_i t_{ij}} \text{ and } S\left(t_{ij}\right) = e^{-h_i t_{ij}}.$$

2.2. Posterior distribution

We give the likelihood function for group i as

$$\begin{aligned} L(h_i;t_i) &= \prod_{j=1}^{n_i} [h_i e^{-h_i t_{ij}}]^{\delta_{ij}} [e^{-h_i t_{ij}}]^{1-\delta_{ij}} \\ &= h_i^{\sum_{j=1}^{n_i} \delta_{ij}} e^{\left(-h_i \sum_{j=1}^{n_i} t_{ij}\right)} \end{aligned}$$

Assuming gamma distributions with parameters α_i and β_i as prior distributions, and $i = 1, 2$, we have

$$g(h_i) = \frac{\beta_i^{\alpha_i}}{\Gamma(\alpha_i)} h_i^{\alpha_i - 1} e^{-\beta_i h_i}.$$

From Bayes' Theorem, the posterior distribution for i is

$$g(h_i|\underline{t}_i) = \frac{(\beta_i + \sum_{j=1}^{n_i} t_{ij})^{\alpha_i + \sum_{j=1}^{n_i} \delta_{ij}}}{\Gamma(\alpha_i + \sum \delta_{ij})} h_i^{\alpha_i + \sum_{j=1}^{n_i} \delta_{ij} - 1} e^{-(\beta_i + \sum_{j=1}^{n_i} t_{ij}) h_i}.$$

The posterior distribution follows a gamma distribution with parameters $a_i = \alpha_i + \sum_{j=1}^{n_i} \delta_{ij}$ and $b_i = \beta_i + \sum_{j=1}^{n_i} t_{ij}$.

2.3. Exact expression

THEOREM *The exact expression for the posterior hazard of group 1 to be smaller than that of group 2 is*

$$\begin{aligned} \eta &= P(h_1 < h_2 | \underline{T}_1, \underline{T}_2) \\ &= 1 - \frac{1}{a_2 Beta(a_1, a_2)} \left(\frac{b_2}{b_1 + b_2}\right)^{a_2} {}_2F_1\left(a_2, 1 - a_1; 1 + a_2; \frac{b_2}{b_1 + b_2}\right) \\ &= I_{\frac{b_1}{b_1 + b_2}}(a_1, a_2), \end{aligned} \tag{1}$$

where ${}_2F_1(a_1, a_2; b; x)$ denotes the hypergeometric series, $I_x(a, b)$ denotes the regularized incomplete beta function and Beta(a, b) denotes the beta function.

Proof The posterior probability is given by

$$g_i(h_i|\underline{t}_i) = \frac{b_i^{a_i}}{\Gamma(a_i)} h_i^{a_i - 1} e^{-b_i h_i}.$$

Hence, its joint probability density function is

$$f_{1,2}(h_1, h_2, |\underline{t}_1, \underline{t}_2) = \frac{b_1^{a_1} b_2^{a_2}}{\Gamma(a_1)\Gamma(a_2)} h_1^{a_1-1} e^{-b_1 h_1} h_2^{a_2-1} e^{-b_2 h_2}.$$

From the definition, η can be given as

$$\eta = P(h_1 < h_2 | \underline{T}_1, \underline{T}_2) = \frac{b_1^{a_1} b_2^{a_2}}{\Gamma(a_1)\Gamma(a_2)} \int_0^{\infty} \int_0^{h_2} h_1^{a_1-1} e^{-b_1 h_1} h_2^{a_2-1} e^{-b_2 h_2} dh_1 dh_2.$$

From Theorem 1 in Kawasaki and Miyaoka (2012a),

$$\eta = P(h_1 < h_2 | \underline{T}_1, \underline{T}_2) = 1 - \frac{1}{a_2\, Beta(a_1, a_2)} \left(\frac{b_2}{b_1 + b_2} \right)^{a_2} {}_2F_1\left(a_2, 1 - a_1; 1 + a_2; \frac{b_2}{b_1 + b_2}\right)$$

Furthermore, from Formulas 26.5.23 and 26.5.2 of Abramowitz and Stegun (2010), i.e.

$$I_x(a, b) = \frac{1}{Beta(a, b)} \cdot \frac{x^a}{a} \cdot {}_2F_1(a, 1 - b; 1 + a; x)$$

and

$$I_z(a, b) = 1 - I_{1-z}(b, a),$$

η can be given as

$$\eta = 1 - I_{\frac{b_2}{b_1 + b_2}}(a_2, a_1) = I_{\frac{b_1}{b_1 + b_2}}(a_1, a_2)$$ □

We can calculate the exact probability by Equation (1) and using expressions containing various hypergeometric series. In this paper, we use one such expression. Kawasaki and Miyaoka (2012b) give a more detailed proof in a similar problem. For the SAS programming code used for calculating the probability of η, see Appendix 1.

2.4. Approximate expression

We can easily formulate an approximate expression using the expected value of the posterior distribution and the variance. The expected value and the variance of the difference between the posterior distributions are, respectively,

$$E\left(h_{1,\text{post}} - h_{2,\text{post}}\right) = \frac{a_1}{b_1} - \frac{a_2}{b_2} \text{ and } V\left(h_{1,\text{post}} - h_{2,\text{post}}\right) = \frac{a_1}{b_1^2} + \frac{a_2}{b_2^2},$$

where $h_{i,\text{post}}$ denote the posterior hazard, and $i = 1, 2$. Therefore, we can rewrite Equation (2) as

$$\eta = P(h_1 < h_2 | \underline{T}_1, \underline{T}_2) = P\left(\frac{\left(h_{1,\text{post}} - h_{2,\text{post}}\right) - E\left(h_{1,\text{post}} - h_{2,\text{post}}\right)}{\sqrt{V\left(h_{1,\text{post}} - h_{2,\text{post}}\right)}} < \frac{-E\left(h_{1,\text{post}} - h_{2,\text{post}}\right)}{\sqrt{V\left(h_{1,\text{post}} - h_{2,\text{post}}\right)}} \right)$$

$$\approx P\left(Z < \frac{-E\left(h_{1,\text{post}} - h_{2,\text{post}}\right)}{\sqrt{V\left(h_{1,\text{post}} - h_{2,\text{post}}\right)}} \right) = \Phi\left(\frac{-\frac{a_1}{b_1} + \frac{a_2}{b_2}}{\sqrt{\frac{a_1}{b_1^2} + \frac{a_2}{b_2^2}}} \right),$$

where $\Phi(\cdot)$ is the cumulative distribution function of the standard normal distribution. Thus, η can easily be calculated if we know the cumulative distribution function of the standard normal distribution.

3. Results

3.1. Simulation

Our aim of simulation in Section 3.1 is to verify the condition under which the differences of the probability, indicated by the approximated method and that by the exact method, become greater. Hence, we compare the exact method and the approximation method using simulated data in this section. The simulation method is as follows. Assume the data occur randomly from the exponential distribution when the hazard ratio is between 0.40 and 1.50 in steps of 0.10, and that they occur randomly with binomial proportions when the censoring ratio is 10 or 40%. Further, let the sample sizes for the groups be either balanced or unbalanced. For the balanced cases, we set $n_1 = n_2 = 10$, $n_1 = n_2 = 20$, $n_1 = n_2 = 50$, or $n_1 = n_2 = 100$, and for the unbalanced cases, we set $n_1 = 15$ and $n_2 = 5$, or $n_1 = 75$ and $n_2 = 25$. In this simulation, we use the non-informative prior, that is, $\alpha_i = \beta_i = 0.001$ and $i = 1, 2$.

At the beginning, we verify the difference between the probability by the approximated method and that by the exact method with small sample size. Figure 1 compares the results under the censoring ratios 10 and 40% when $n_1 = n_2 = 10$ and $n_1 = n_2 = 20$. The box plots show the difference between the exact probability and the approximate probability, with the vertical axis representing the difference between the exact probability and the approximate probability, and the horizontal axis representing the hazard ratio. We make the following observations from Figure 1:

- When the censoring ratio is 10%, the difference between the exact probability and the approximate probability is about 4% at maximum.
- The difference is about 12% at maximum when the censoring ratio is 40% and the hazard ratio is 0.40.
- The difference shows a variance of around 0% when the hazard ratio is 1.

Next, we verify that with large sample size. Figure 2 compares the results under the censoring ratios 10 and 40% when $n_1 = n_2 = 50$ and $n_1 = n_2 = 100$. Figure 2 shows the following:

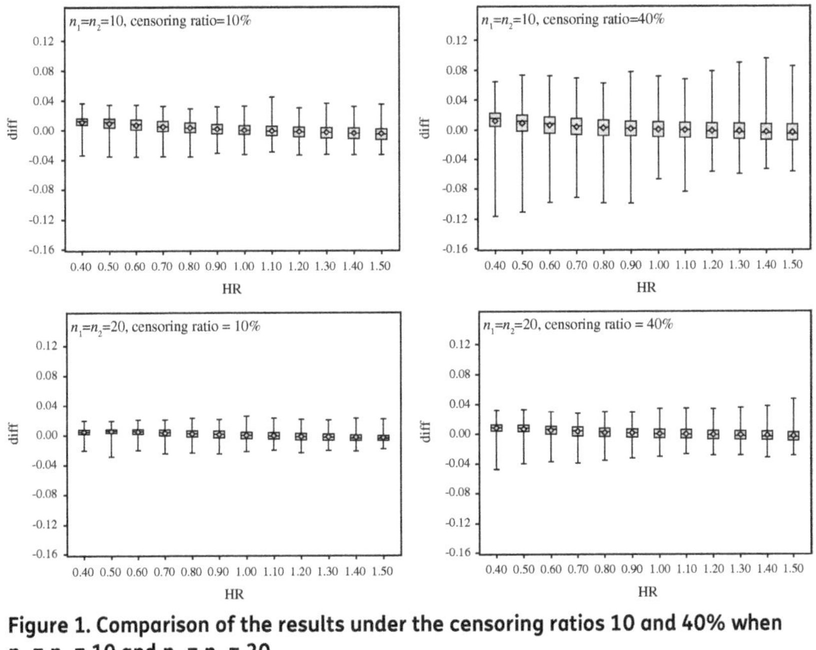

Figure 1. Comparison of the results under the censoring ratios 10 and 40% when $n_1 = n_2 = 10$ and $n_1 = n_2 = 20$.

Notes: The vertical axis shows the difference between the exact probability and the approximate probability, and the horizontal axis indicates the hazard ratio.

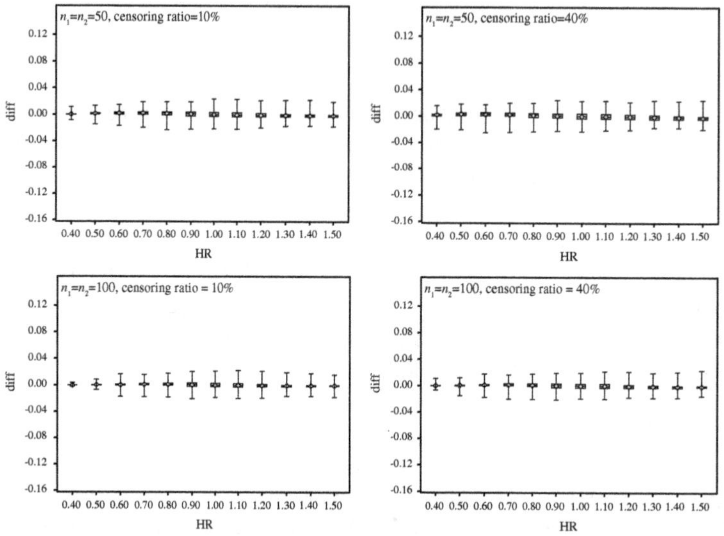

Figure 2. Comparison of the results under the censoring ratios 10 and 40% when $n_1 = n_2 = 50$ and $n_1 = n_2 = 100$.

Notes: The vertical axis shows the difference between the exact probability and the approximate probability, and the horizontal axis indicates the hazard ratio.

- The results are not affected by censoring, and the difference between the exact probability and the approximate probability goes up to about 2.5%.
- The interquartile range is very narrow.

At the end, we verify that with unbalanced sample size. Figure 3 compares the results under the censoring ratios 10 and 40% for the combinations $n_1 = 15$, $n_2 = 5$ and $n_1 = 75$, $n_2 = 25$. Figure 3 shows the following:

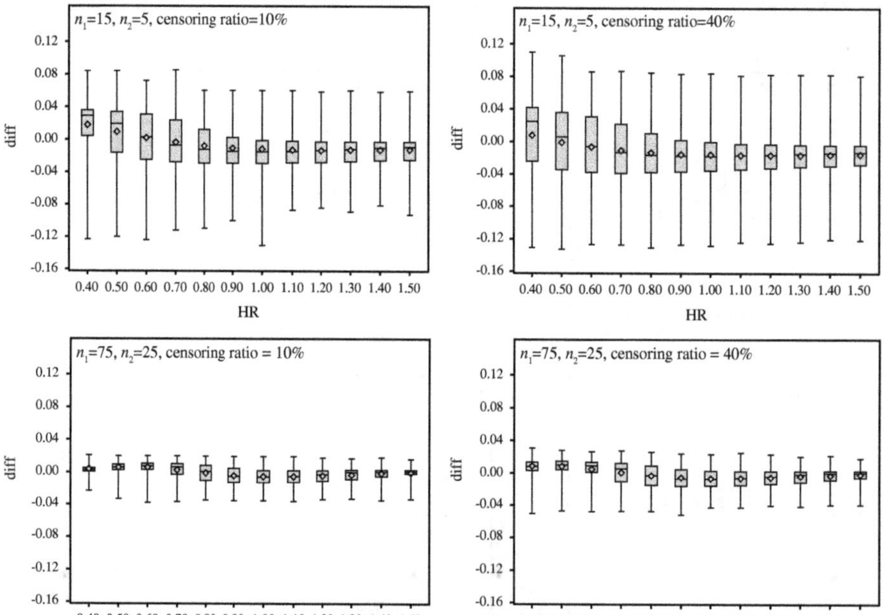

Figure 3. Comparison of the results under the censoring ratios 10 and 40% when $n_1 = 15$, $n_2 = 5$ and $n_1 = 75$, $n_2 = 25$.

Notes: The vertical axis shows the difference between the exact probability and the approximate probability, and the horizontal axis indicates the hazard ratio.

- Boxes are wide, and the difference between the exact probability and the approximate probability is large.
- The difference is even larger relative to the case where both groups have the same sample size in total.

3.2. Examples

3.2.1. Chronic active hepatitis cases

In this section, we apply the proposed index to clinical trial data used in Kirk, Jain, Pocock, Thomas, and Sherlock (1980). In this trial, 44 chronic active hepatitis cases were assigned randomly to either the prednisolone group (treatment group) or the control group with no treatment (control group). From the treatment group's 22 cases, 11 were observed and the remaining 11 were censored, and from the control group's 22 cases, 16 were observed and the remaining 6 were censored. The most interesting outcome in the trial was the survival time of patients after the trial commenced.

The *p*-value was 0.0309 by the log-rank test and 0.0105 by the Wilcoxon test. This indicates a longer survival time for the treatment group at the significance level of 5%. The proposed index indicates 98.35% as the approximate probability and 99.01% as the exact probability, with the hazard of the treatment group lower than that of the control group. We assume the prior distribution to be non-informative for both groups.

3.2.2. Lung cancer data analysis

In this section, we apply the proposed index to clinical trial data used in Chen et al. (2007).

We obtained the non-small-cell lung cancer data used in Chen et al. (2007) from http://www.ncbi.nlm.nih.gov/projects/geo/ with accession number GSE4882. The data contained 672 gene profiles of 125 lung cancer patients. From among them, 38 patients died, and the remainder were censored.

The median follow-up of 101 patients was 20 months. The patients with a high-risk gene signature showed a shorter median overall survival than those with a low-risk gene signature. The p-value was 0.001 by the log-rank test; the proposed index gave 99.99% probability as calculated approximately as well as exactly. A high-risk gene signature was associated with a median relapse-free survival period of 13 months, whereas a low-risk gene signature was associated with a median relapse-free survival period of 29 months. Here, the *p*-value was 0.002 by the log-rank test, and the proposed index gave 99.53% as the approximate probability and 99.72% as the exact probability.

4. Discussion and conclusions

In this paper, we have proposed an index to compare the hazards defined in the Bayesian framework. Further, we have presented two probability functions: the approximation function, which provides for easy calculation, and the exact function.

Our simulation experiment shows a difference between the exact probability and the approximate probability. Specifically, when we censor more, the difference becomes larger. By contrast, we found that the difference becomes smaller with larger sample sizes; the effect of censoring too becomes less. With unbalanced sample sizes for the groups, the difference becomes larger. Thus, we recommend using the exact probability calculation for cases with small samples, unbalanced sample sizes, or numerous censorings.

We applied the proposed index to actual clinical trial data and confirmed that it shows the difference between groups very clearly. Because, we indicated it in example 1 that the hazard of the treatment group becomes lower than that of the control group with about 99% probability. Thus, as this index provides comparison of the hazards with probability, the result can be indicated within the range of 0–100%. So, the index can be very easy to understand intuitively.

Thus, the index proposed in this paper to compare hazards is easy to understand and intuitive and should be beneficial for use in research studies.

Acknowledgements
The authors also wish to thank the anonymous reviewers for their suggestions that helped to improve the manuscript.

Funding
The authors received no direct funding for this research.

Author details
Yohei Kawasaki[1]
E-mails: yk_sep10@yahoo.co.jp, ykawasaki@u-shizuoka-ken.ac.jp
Masaaki Doi[2,3]
E-mail: Masaaki_Doi@nts.toray.co.jp
Kazuki Ide[1]
E-mail: s13403@u-shizuoka-ken.ac.jp
Etsuo Miyaoka[4]
E-mail: miyaoka@rs.kagu.tus.ac.jp
[1] Department of Drug Evaluation & Informatics, School of Pharmaceutical Sciences, University of Shizuoka, Shizuoka, Japan.
[2] Clinical Data Science & Quality Management Department, Toray Industries, Inc., Tokyo, Japan.
[3] Graduate School of Science and Engineering, Chuo University, Tokyo, Japan.
[4] Department of Mathematics, Tokyo University of Science, Tokyo, Japan.

References

Abramowitz, M., & Stegun, I. A. (2010). *Handbook of mathematical functions: With formulas, graphs, and mathematical tables*. United States: Dover Publications.

Berger, J. O., & Pericchi, L. R. (1996). The intrinsic Bayes factor for model selection and prediction. *Journal of the American Statistical Association, 91*, 109–122.

Berry, D. A. (2006). Bayesian clinical trials. *Nature Reviews Drug Discovery, 5*, 27–36. http://dx.doi.org/10.1038/nrd1927

Chen, H.-Y., Yu, S.-L., Chen, C.-H., Chang, G.-C., Chen, C.-Y., Yuan, A., ... Yang, P.-C. (2007). A five-gene signature and clinical outcome in non-small-cell lung cancer. *New England Journal of Medicine, 356*, 11–20. http://dx.doi.org/10.1056/NEJMoa060096

Colett, D. (2003). *Modeling survival data in medical research* (2nd ed.). UK: Chapman and Hall/CRC.

Cornfield, J. (1976). Recent methodological contributions to clinical trials. *American Journal of Epidemiology, 104*, 408–421.

Grieve, A. P. (2007). 25 years of Bayesian methods in the pharmaceutical industry: A personal, statistical bummel. *Pharmaceutical Statistics, 6*, 261–281. http://dx.doi.org/10.1002/(ISSN)1539-1612

Heller, G., & Mo, Q. (2015). Estimating the concordance probability in a survival analysis with a discrete number of risk groups. *Lifetime Data Analysis, 2*, 1–17.

Hosmer, D. W., Lemeshow, S., & May, S. (2008). *Applied survival analysis: Regression modeling of time-to-event data* (2nd ed.). United States: Wiley-Interscience. http://dx.doi.org/10.1002/9780470258019

Kass, R., & Raftery, A. (1995). Bayes factors. *Journal of the American Statistical Association, 90*, 773–795.

Kawasaki, Y., & Miyaoka, E. (2012a). A Bayesian inference of $P(\lambda_1<\lambda_2)$ for two Poisson parameters. *Journal of Applied Statistics, 39*, 2141–2152. http://dx.doi.org/10.1080/02664763.2012.702264

Kawasaki, Y., & Miyaoka, E. (2012b). A Bayesian inference of P $(\pi_1 > \pi_2)$ for two proportions. *Journal of Biopharmaceutical Statistics, 22*, 425–437. http://dx.doi.org/10.1080/10543406.2010.544438

Kirk, A. P., Jain, S., Pocock, S., Thomas, H. C., & Sherlock, S. (1980). Late results of the Royal Free Hospital prospective controlled trial of prednisolone therapy in hepatitis B surface antigen negative chronic active hepatitis. *Gut, 21*, 78–83. http://dx.doi.org/10.1136/gut.21.1.78

Lee, J. J., & Chu, C. T. (2012). Bayesian clinical trials in action. *Statistics in Medicine, 31*, 2955–2972.

O'Hagan, A. (1995). Fractional Bayes factors for model comparison. *Journal of the Royal Statistical Society. Series B (Method-ological), 57*, 99–138.

Racine, A., Grieve, A. P., Fluhler, H., & Smith, A. F. (1986). Bayesian methods in practices: Experiences in the pharmaceutical industry (with discussion). *Applied Statistics, 35*, 93–150. http://dx.doi.org/10.2307/2347264

Spiegelhalter, D. J. (2004). Incorporating bayesian ideas into health-care evaluation. *Statistical Science, 19*, 156–174. http://dx.doi.org/10.1214/088342304000000080

Spiegelhalter, D. J., Freedman, L. S., & Parmar, M. K. (1993). Applying Bayesian ideas in drug development and clinical trials. *Statistics in Medicine, 12*, 1501–1511. http://dx.doi.org/10.1002/(ISSN)1097-0258

Spiegelhalter, D. J., Freedman, L. S., & Parmar, M. K. (1994). Bayesian approaches to randomised trials (with discussion). *Journal of the Royal Statistical Society. Series A (Statistics in Society), 157*, 357–416. http://dx.doi.org/10.2307/2983527

Zaslavsky, B. G. (2010). Bayesian versus frequentist hypotheses testing in clinical trials with dichotomous and countable outcomes. *Journal of Biopharmaceutical Statistics, 20*, 985–997. http://dx.doi.org/10.1080/10543401003619023

Zaslavsky, B. G. (2013). Bayesian hypothesis testing in two-arm trials with dichotomous outcomes. *Biometrics, 69*, 157–163. http://dx.doi.org/10.1111/j.1541-0420.2012.01806.x

Zhao, L., Feng, D., Chen, G., & Taylor, J. M. (2015). A unified Bayesian semiparametric approach to assess discrimination ability in survival analysis. *Biometrics*, 554–562. doi:10.1111/biom.12453

Appendix 1

SAS programming code

The following SAS macro can be used to calculate $\eta = P(h_1 < h_2 | \underline{T}_1, \underline{T}_2)$.

The parameters are defined as follows: N1 = sample size of group 1; N2 = sample size of group 2; X1 = total survival time of group 1; X2 = total survival time of group 2; e1 = total event count for group 1; e2 = total event count for group 2; alpha1 and beta1 = hyperparameters for group 1; alpha2 and beta2 = hyperparameters for group 2.

Subject: Calculation of $\eta = P(h_1 < h_2 | \underline{T}_1, \underline{T}_2)$

Output variables: Prob_E (Exact method) and Prob_A (Approximate method)

```
%MACRO PROBMK(N1=,N2=,X1=,X2=,e1=,e2=,alpha1=,beta1=,alpha2=,beta2=);
data BASE01;
  /* sample sizes */
  n1 = &N1; n2 = &N2;
  /* all survival times */
  x1 = &X1; x2 = &X2;
  /* all event counts */
  e1 = &e1; e2 = &e2;

  XT1 = X1/N1; XT2 = X2/N2;
  /* Input variables */

  /* Hyperparameters */
  al1 = &alpha1; be1 = &beta1;
  al2 = &alpha2; be2 = &beta2;

  a1 = e1 + al1; b1=(n1*XT1)+be1;
  a2 = e2 + al2; b2=(n2*XT2)+be2;
run;

  %let N = 1000;
  /* Hypergeometric work */

data work01;
  set BASE01;
  o = a2; p = 1-a1;
  q = 1 + a2;
  u = b2/(b1 + b2);
  retain K1 - K&N 1;
  array K(&N);
  do i = 1 to &N;
    do j = 1 to i;
```

```
    o1 = o + j-1;
    p1 = p + j-1;
    q1 = q + j-1;
    K(j)=(o1*p1)/(q1);
  end;
  %macro kkk;
    SS + %do i = 1%to %eval(&N-1);
      K&i*
      %end;
  K&N*(u**i)/FACT(i);
      %mend kkk;
      %kkk
  end;
  F = sum(1,SS);
  /* Output variables */
  /* Exact */
  Prob_E = 1-round(1/(a2*Beta(a1,a2))*((b2/(b1 + b2))**a2)*F,0.0001);

  /* Approximate */
  FFF1=((-a1/b1)+(a2/b2));
  FFF2 = Sqrt(a1/b1**2 + a2/b2**2);
  FFF = FFF1/FFF2;
  Prob_A = round(CDF('Normal',FFF),0.0001);

  keep n1 n2 x1 x2 a1 b1 a2 b2 FFF1 FFF2 Prob_E Prob_A;
run;

  proc print data = work01; var Prob_E Prob_A; run;
  %MEND PROBMK;
  * Program End;
```

3

MANOVA, LDA, and FA criteria in clusters parameter estimation

Stan Lipovetsky[1]*

*Corresponding author: Stan Lipovetsky, GfK North America, 8401 Golden Valley Road, Minneapolis, MN 55427, USA
E-mail: stan.lipovetsky@gfk.com

Reviewing editor: Heung Wong, Hong Kong Polytechnic University, Hong Kong

Abstract: Multivariate analysis of variance (MANOVA) and linear discriminant analysis (LDA) apply such well-known criteria as the Wilks' lambda, Lawley–Hotelling trace, and Pillai's trace test for checking quality of the solutions. The current paper suggests using these criteria for building objectives for finding clusters parameters because optimizing such objectives corresponds to the best distinguishing between the clusters. Relation to Joreskog's classification for factor analysis (FA) techniques is also considered. The problem can be reduced to the multinomial parameterization, and solution can be found in a nonlinear optimization procedure which yields the estimates for the cluster centers and sizes. This approach for clustering works with data compressed into covariance matrix so can be especially useful for big data.

Subjects: Mathematics & Statistics; Multivariate Statistics; Science; Statistics; Statistics & Computing; Statistics & Probability; Statistics for Business, Finance & Economics

Keywords: MANOVA; LDA; Wilks' lambda; Lawley–Hotelling trace; Pillai's trace; Joreskog's classification for FA; cluster analysis; multinomial optimization

1. Introduction

Multivariate analysis of variance (MANOVA) is a well-known generalization of the analysis of variance (ANOVA) extended from one to many dependent variables, and the multivariate analysis of

ABOUT THE AUTHOR

Stan Lipovetsky, PhD, senior research director, GfK, Marketing Sciences. Stan has numerous publications in multivariate statistics, multiple criteria decision-making, econometrics, microeconomics, and marketing research.

The methods of multivariate statistical analysis are widely applied in marketing research. Clustering technique described in the current paper can be especially useful for big data with possible hundreds of thousands or millions of observations when regular clustering algorithms presents a hard computational burden. The suggested method operates with the data already compressed into covariance matrix. When the cluster centers and sizes are estimated, the actual clustering, or assignment of each observation to one or another cluster can be performed by allocating them to the closest cluster.

PUBLIC INTEREST STATEMENT

The paper describes several main multivariate statistical techniques, such as multivariate analysis of variance, linear discriminant analysis, and factor analysis in relation to cluster analysis. It shows that known in those techniques criteria of quality of solutions can be used for data clustering as well. These criteria are employed to find cluster centers and sizes, because optimizing such objectives corresponds to the best distinguishing between clusters. The problem is expressed via multinomial parameterization of the clusters characteristics, and solution can be found in optimization procedure yielding estimates for clusters. This approach uses only sample covariance matrix and not the observations themselves, so it can be especially valuable in difficult clustering tasks on big data from data bases, data warehouses, and data- mining problems.

numerical covariates (MANCOVA) is a similar extension of ANCOVA. Linear discriminant analysis (LDA) and MANOVA can be considered as dual techniques—in LDA the independent variables are the predictors (or attributes) and the dependent variables are the groups, while in MANOVA vice versa—the independent variables are the groups (dummy variables identifying the clusters belonging) and the dependent variables are the attributes. Both these techniques can be presented via the canonical correlation analysis (CCA) between two sets of variables, with an additional step of prediction of one set by another one. When the canonical aggregates of the variables are obtained, all the tests for multivariate variables can be reduced to the known tests for one variable. LDA consists in testing significance of the discriminant functions of the attributes (with additional classification), and MANOVA—in testing significance of differences between groups' vectors of means (with additional identification which of the attributes have different means across the groups). There are various criteria known for testing quality of LDA and MANOVA solutions (see, for instance, Dillon & Goldstein, 1984; Härdle & Simar, 2012; Izenman, 2008; Timm, 1975).

The current paper considers possibilities to apply these criteria for clustering purposes. Indeed, if such criteria permit testing quality of LDA and MANOVA solutions, they can be optimized to obtain the best distinguishing between the clustering groups as well. As it is well-known in multivariate statistical analysis the total (T) variance-covariance matrix can be presented as the sum $T = B + W$ of the so-called "Between" (B) and "Within" (W) matrices defined by the variance-covariance between and within the groups, respectively. The loadings for aggregation of the attributes in LDA or MANOVA are commonly found by maximizing a criterion of the quotient of the between-to-within quadratic forms which can be presented via the generalized eigenproblem with these matrices. Using its eigenvalues, it is possible to test the quality of LDA and MANOVA solutions. For this aim, the so-called Wilks' lambda criterion Λ (a multivariate generalization of F-statistics) compares the determinants (generalized variances) calculated by within and total variances. Wilks' Λ varies from 0 to 1, and $\Lambda = 0$ indicates that the groups' mean vectors differ, while $\Lambda = 1$ shows that the groups' means are the same. Thus, minimizing an objective of the Wilks' criterion by the parameters of the cluster centers permits to find them, as well as the group base sizes, which define the best distinguishing between groups. Wilks' lambda has the so called U-distribution which is tabulated, for instance, in Timm (1975). It also can be numerically approximated and presented as a regular F-test.

There are other overall tests on significance known in MANOVA, for instance, Hotelling T^2 (a multivariate generalization of the t-test for comparing vectors of means for two groups), and its generalization to Lawley–Hotelling trace criterion. The maximization of this criterion can be used for clustering problem. A modification of this criterion is given by another criterion widely used in MANOVA—the so-called Pillai's trace. It also can be used for clustering aims via the corresponding maximized objective. Such criteria are very convenient for clustering problem because they do not require calculation of each eigenvalue but only their total, which coincides with the trace, or sum of the diagonal elements of the matrix used in maximization. Practical application of all these tests in LDA and MANOVA are demonstrated, for instance, throughout the monograph (Timm, 1975). There are other tests, like Roy's largest eigenvalue λ_{max}, and more specific ones like Levene's test to define whether the variances between groups are equal, partial eta-square (similar to partial F-statistics) to see the variance explained by individual independent variable, and other extensions to the multivariate statistics. However, for clustering aims, we are interested in the criteria operating with the totals of the eigenvalues which can be reduced to some functions of the main variance-covariance T, B, and W matrices.

Contemporary cluster analysis includes a large spectrum of methods developed in the areas of segmentation, pattern recognition, machine learning, data mining, and others (for instance, see Bishop, 2006; Eldén, 2007; Frey & Dueck, 2007; Gan, Ma, & Wu, 2007; Hastie, Friedman, & Tibshirani, 2001; Lipovetsky, 2012; Lipovetsky, Tishler, & Conklin, 2002; Liu & Motoda, 2008; Nowakowska, Koronacki, & Lipovetsky, 2014; Ripley, 1996). Various works suggest different ways to divide data into groups of observations more closely related within each group in comparison with the relations between the groups by variance-covariance matrices (Brusco & Steinley, 2007; DeSarbo, Carroll,

Clark, & Green, 1984; Friedman & Meulman, 2004; Heiser & Groenen, 1997; Szekely & Rizzo, 2005). The current paper suggests a new approach based on the special objectives corresponding to the MANOVA and LDA criteria which optimization guarantees the best quality of the distinguishing among the groups estimated using the same criteria. This problem can be reduced to the optimizing procedure for nonlinear approximation of the covariance matrix by the total of the outer products of the distances from cluster centers to total center. The means and sizes of the clusters can be found using parameterization via multinomial shares—a technique developed and successfully applied for solving various problems in regression, principal components, singular value decomposition, and clustering (Lipovetsky, 2009a, 2009b, 2013a). The relation with factor analysis by least squares (LS) and generalized least squares (GLS) objectives (Joreskog, 1977; Jöreskog, 1967; Jöreskog & Goldberger, 1972; Lawley & Maxwell, 1971; Maxwell, 1983) are discussed as well. This approach can be especially useful for large data-sets with thousands and millions of observations because it operates not with original multiple observations but with the data compressed into covariance matrix.

2. Criteria for finding cluster centers and sizes

Let X denote N by n matrix with elements x_{ij} (rows $i = 1, 2, \ldots, N$ of observations by the variables in n columns $x_1, x_2, \ldots, x_n$). The elements of the total matrix S_{tot} of second moments are defined as:

$$(S_{tot})_{jk} = \sum_{i=1}^{N} (x_{ij} - M_j)(x_{ik} - M_k), \tag{1}$$

where M_j denotes the mean value of each x_j. Let observations be divided into K subsets, and these clusters are numbered as $q = 1,2, \ldots, K$, and each q-th cluster have N_q observations, so their total equals the sample base:

$$N_1 + N_2 + \ldots + N_K = N. \tag{2}$$

Consider decomposition of the cross-product (1) into the items related to the data subsets with sizes (2). Such a transformation is known in ANOVA (Ladd, 1966; Lipovetsky & Conklin, 2005) and can be presented as:

$$\sum_{i=1}^{N} (x_{ij} - M_j)(x_{ik} - M_k) = \sum_{q=1}^{K} \sum_{i=1}^{N_q} (x_{ij}^q - m_j^q)(x_{ik}^q - m_k^q) + \sum_{q=1}^{K} N_q (m_j^q - M_j)(m_k^q - M_k) \tag{3}$$

where x_{ij}^q indicates that the i-th observation by the j-th variable belongs to the q-th cluster, and m_j^q is the mean value of each j-th variable within q-th cluster. The relation between the subsets and total means for each j-th variable is as follows:

$$N_1 m_j^1 + N_2 m_j^2 + \ldots + N_K m_j^K = N M_j, \quad j = 1, 2, \ldots, n, \tag{4}$$

where both sides express simply the total by each x_j.

The double sum in (3) equals the pooled second moment within each cluster:

$$(S_{within})_{jk} = \sum_{q=1}^{K} (S_{within}^q)_{jk} = \sum_{q=1}^{K} \sum_{i=1}^{N_q} (x_{ij}^q - m_j^q)(x_{ik}^q - m_k^q). \tag{5}$$

The last sum in (3) corresponds to the weighted by group sizes second moment between the cluster means centered by the total means:

$$(S_{between})_{jk} = \sum_{q=1}^{K} N_q (m_j^q - M_j)(m_k^q - M_k). \tag{6}$$

So (3) can be presented as the matrix sum:

$$S_{tot} = S_{within} + S_{between} = S_{within} + \sum_{q=1}^{K} N_q (m^q - M)(m^q - M)', \tag{7}$$

with the outer product of vectors of distances from the centers m^q for each q-th cluster to the total center M, where each vector m^q consists of the means m_j^q by all the variables, and the vector M contains the total means M_j. For a given matrix S_{tot} (7), the data clustering corresponds to maximizing the distances between the groups and minimizing them within the groups. If observations within each q-th cluster collapse to one point of its center, the elements of the matrix S_{within} (5) reach zero. Thus, to find clusters, we can minimize the total of squared elements of matrix S_{within}, or in other words—the total of differences between elements of known matrix S_{tot} and unknown matrix $S_{between}$ in (7):

$$F = \left\| S_{tot} - S_{between} \right\|^2 = \left\| S_{tot} - \sum_{q=1}^{K} N_q (m^q - M)(m^q - M)' \right\|^2 \rightarrow \min. \tag{8}$$

The objective (8) presents the squared Frobenius norm for a matrix (also known as Hilbert–Schmidt, or Schur norm). This formulation corresponds to the LS objective for the nonlinear regression model of fitting the values in S_{tot} by the known vector M and the sets of unknown constants N_q and unknown vectors m^q. Estimation of these parameters in the approach of multinomial parameterization is considered in Lipovetsky (2013a, 2013b) where the problem is reduced to nonlinear regression modeling. A brief description of this technique is given in Appendix.

The basic relation (7) is also the fundamental equation for MANOVA and LDA where it is usually presented in one of the following known notations:

$$S_{tot} = S_{within} + S_{between} = W + B = E + H, \tag{9}$$

where $S_{within} = W = E$ denotes the Within (W), or Error (E) matrix, and $S_{between} = B = H$ denotes the Between (B), or Hypothesis (H) matrix of second moments. Finding vectors of loadings, or discriminant functions α in LDA or MANOVA is commonly performed by maximizing the criterion of the Rayleigh quotient of the between-to-within quadratic forms:

$$F = \alpha' H \alpha / \alpha' E \alpha, \tag{10}$$

which can be presented as the generalized eigenproblem:

$$H\alpha = \lambda E \alpha, \tag{11}$$

with the eigenvalues λ of the matrix $E^{-1}H$.

The so-called Wilks' lambda criterion (a multivariate generalization of F-statistics) compares the generalized variances within the groups and in the whole data-set:

$$\Lambda = \frac{|S_{within}|}{|S_{tot}|} = \frac{|E|}{|E+H|} = \frac{|E|}{|E| \cdot |1 + E^{-1}H|} = \frac{1}{|1 + E^{-1}H|} = \prod_{j=1}^{n} \frac{1}{1+\lambda_j}. \tag{12}$$

This criterion (12) can be used for finding the groups' centers. For this aim it can be rewritten via the means of $S_{between}$ (6) as the unknown parameters of interest and minimized:

$$\Lambda = \frac{|S_{tot} - S_{between}|}{|S_{tot}|} = |S_{tot}^{-1}| \cdot |S_{tot} - S_{between}| = |I - S_{tot}^{-1} S_{between}|, \tag{13}$$

where in the numerator we have minimization similar to used in (8), and in the denominator is the constant of the determinant of the total variance-covariance matrix. The difference in LS (8) and Wilks' (13) criteria consists in using Euclidean norm squared or the generalized variance in determinants, respectively.

Another overall test in MANOVA is Hotelling T^2-statistic, a multivariate generalization of the t-test for comparing vectors of means for two groups, and its generalization to Lawley-Hotelling trace (Tr, the total of diagonal elements) criterion:

$$T^2 = Tr\left(E^{-1}H\right) = \sum_{j=1}^{n} \lambda_j. \tag{14}$$

The maximized criterion (14) can be used for clusters parameter estimation by trace of the matrix:

$$\begin{aligned} T^2 &= Tr\left(E^{-1}H\right) = Tr\left((S_{tot} - S_{between})^{-1} S_{between}\right) \\ &= Tr\left((S_{tot} - S_{between})^{-1}(S_{between} - S_{tot} + S_{tot})\right) = Tr\left((I - S_{tot}^{-1} S_{between})^{-1} - I\right) \end{aligned} \tag{15}$$

with the same $S_{between}$ (6).

A modification of (14) widely used in MANOVA is the so-called Pillai's trace:

$$V = Tr\left((E+H)^{-1}H\right) = Tr\left((1 + E^{-1}H)^{-1}E^{-1}H\right) = \sum_{j=1}^{n} \frac{\lambda_j}{1+\lambda_j} \tag{16}$$

For estimation of cluster centers this test corresponds to the objective for maximization:

$$V = Tr\left((E+H)^{-1}H\right) = Tr\left(S_{tot}^{-1} S_{between}\right). \tag{17}$$

Both (15) and (17) criteria use optimization by the same matrix $S_{tot}^{-1} S_{between}$ of fitting used in (13) as well. All these criteria are convenient for clusters parameter estimations because they do not require calculation of each eigenvalue but work with the total matrices. The meaning of all these objectives, including LS (8), is similar—to identify the parameters of cluster centers by closeness of $S_{between}$ to S_{tot} (7).

The LS (8) and MANOVA (12)–(17) objectives for clusters parameter estimations correspond to the criteria in Joreskog's classification for methods of factor analysis (Joreskog, 1977; Jöreskog, 1967; Jöreskog & Goldberger, 1972; Lawley & Maxwell, 1971; Maxwell, 1983) based on a given covariance matrix S approximated by a matrix Σ of lower rank (built as the product of a matrix of loadings and its transposed). Joreskog (1977) distinguishes the unweighted least-squares (ULS)

$$\text{ULS} = \frac{1}{2} Tr(S - \Sigma)^2, \tag{18}$$

the generalized least-squares

$$\text{GLS} = \frac{1}{2} Tr(I_n - S^{-1}\Sigma)^2, \tag{19}$$

and the maximum likelihood (ML)

$$ML = Tr(\Sigma^{-1}S) - \ln(\Sigma^{-1}S) - n. \tag{20}$$

Our notations for the matrices S_{tot} and $S_{between}$ in (8) correspond to the matrices S and Σ in Joreskog's notations. The total of all elements squared, or squared Frobenius norm in (8), can be equally presented as the trace of a matrix multiplied by its transposition. Thus, we see that up to the constant ½, the expression (8) equals ULS (18).

The objective (8) can be transformed as follows:

$$F = \|S_{tot} - S_{between}\|^2 = \left\|S_{tot}(I_n - S_{tot}^{-1}S_{between})\right\|^2 \le \|S_{tot}\|^2 \cdot \left\|I_n - S_{tot}^{-1}S_{between}\right\|^2. \quad (21)$$

Besides of (8), skipping the constant term $\|S_{tot}\|^2$ in (21), it is possible to use the objective of total residual sum of squares in minimizing the following deviations:

$$\tilde{F} = \left\|I_n - S_{tot}^{-1}S_{between}\right\|^2 = \left\|I_n - S_{tot}^{-1}\sum_{q=1}^{K} N_q(m^q - M)(m^q - M)'\right\|^2 \rightarrow \min. \quad (22)$$

This objective coincides with GLS (19), up to the term ½. MANOVA (13), (15), (17) objectives also use the matrix $S_{tot}^{-1}S_{between}$ which corresponds to $\Sigma^{-1}S$ in (19) and (20).

Quality of data fit for the objective (8) can be estimated via pseudo-R^2 similar to the coefficient of multiple determination in nonlinear regression, defined as:

$$R^2 = 1 - F_{\min}/Tr(S_{tot}^2), \quad (23)$$

where $F_{\min}$ is the residual sum of squares in the minimum of the objective (8), divided by the total sum of squares of all the elements in the fitted matrix expressed via the trace of the squared matrix S_{tot}. This measure will be in favor of the ULS results obtained by the objective (18). Similarly for the objective (22), the pseudo-R^2 can be defined as one minus $\tilde{F}_{\min}$ divided by the original sum of squares equal n for the identity matrix I_n. This measure would correspond to the quality of fit for the GLS results (19).

The objectives (8) and (22), and the presentation in (18) and (19), are theoretically meaningful and correspond to MANOVA relations (12)–(17). In implementation for the numerical estimations, the totals of the squared deviations of the numerical covariance matrix' elements and their parametric counterparts are used, and there the first objective (8), or ULS (18), is preferable because it does not include the inversion of the covariance matrix which could be prone to multicollinearity in the data.

3. Numerical examples

Consider the iris data (Fisher, 1936) on the measured sepal and petal length and width of fifty iris specimens for each of three species, *Iris setosa* (SE), *Iris versicolor* (VE), and *Iris virginica* (VI). This data can be found in the *Iris* file available in the software package (S-PLUS'2000, 1999), or in R data-sets. The variables are highly correlated: except the sepal width, the correlations range from 0.81 to 0.96.

In Table 1, the first three numerical columns show the means of the variables for each kind of iris, the next three columns show the groups centers and sizes estimated by the ULS (18), then the next three columns show the results by the GLS (19), and the last three columns present the regular K-means clustering for comparison. The last row presents the pseudo-R^2 (23), which is, of course, favorable to the criterion (8), but important is that the quality of the ULS is the same as of K-means. The vectors of cluster centers for the ULS outperform the GLS—they are noticeably closer to the original centers of the iris specimens. ULS results are very similar to K-means as well, but in contrast to K-means the ULS cluster centers and base sizes are obtained using only covariance matrix, without the data-set itself.

Estimation by the objective (8), or ULS (18), does not require inversion of the covariance matrix, thus, the clustering results are more robust and less prone to multicollinearity within the data. It is the reason why ULS regularly outperforms the GLS technique (19) or (22), which employs the inverted covariance matrix with possible inflated values of elements and leads to worse clustering results that was observed by various data-sets.

Table 1. Cluster centers and sizes estimation

	Mean values			ULS			GLS			*K*-means		
Irises	**SE**	**VE**	**VI**	**SE**	**VE**	**VI**	**SE**	**VE**	**VI**	**SE**	**VE**	**VI**
Sepal length	5.01	5.94	6.59	5.32	5.27	7.01	5.39	5.71	6.19	5.01	5.90	6.85
Sepal width	3.43	2.77	2.97	3.74	2.73	3.02	2.46	3.05	3.38	3.43	2.75	3.07
Petal length	1.46	4.26	5.55	1.41	3.30	6.07	3.82	3.88	3.63	1.46	4.39	5.74
Petal width	0.25	1.33	2.03	0.22	1.01	2.17	1.36	1.04	1.23	0.25	1.43	2.07
N	50	50	50	35	67	48	35	49	66	50	62	38
R^2				0.99			0.89			0.99		

4. Summary

This work considers the problem of finding cluster centers and sizes by fitting covariance matrix with the *between-cluster* matrix of a lower rank constructed by outer products of the parameters of cluster centers weighted by cluster sizes. Relation of this approach to the criteria from multivariate analysis of variance, MANOVA, and linear discriminant analysis, LDA, in the objectives for optimization in cluster analysis is discussed. Such criteria as Wilks' lambda, Lawley–Hotelling trace, and Pillai's trace for building objectives for finding clusters parameters produces the best distinguishing between the clusters. Solutions can be found in a nonlinear optimization procedure with the multinomial parameterization which yields estimates for the cluster centers and sizes.

This approach can be especially useful for big data-sets. Indeed, for a big data with possible hundreds of thousands or millions of observations any regular clustering algorithm presents a hard computational burden, while the suggested method operates with the data already compressed into covariance matrix. When the cluster centers and sizes are estimated, the actual clustering, or assignment of each observation to one or another cluster can be performed by allocating them to the closest cluster due to the shortest distance to the centers. The described approach employs only the sample covariance matrix and not the observations themselves, so it can be valuable in difficult clustering tasks on huge data-sets from data bases, data warehouses, and data mining problems. It can also be useful for finding cluster structure when the data itself is already unavailable for any reason and only the covariance matrix can be used.

Acknowledgments
Stan Lipovetsky would like to thank two reviewers for the comments which improved the paper.

Funding
The author received no direct funding for this research.

Author details
Stan Lipovetsky[1]
E-mail: stan.lipovetsky@gfk.com
[1] GfK North America, 8401 Golden Valley Road, Minneapolis, MN 55427, USA.

References

Bishop, C. M. (2006). *Pattern recognition and machine learning*. New York, NY: Springer.
Brusco, M. J., & Steinley, D. (2007). A comparison of heuristic procedures for minimum within-cluster sums of squares partitioning. *Psychometrika, 73*, 125–144.
DeSarbo, W. S., Carroll, J. D., Clark, L. A., & Green, P. E. (1984). Synthesized clustering: A method for amalgamating alternative clustering bases with differential weighting of variables. *Psychometrika, 49*, 57–78. http://dx.doi.org/10.1007/BF02294206
Dillon, W. R., & Goldstein, M. (1984). *Multivariate analysis: Methods and applications*. New York, NY: Wiley.
Eldén, L. (2007). *Matrix methods in data mining and pattern recognition*. Philadelphia, PA: SIAM. http://dx.doi.org/10.1137/1.9780898718867
Fisher, R. A. (1936). The use of multiple measurements in taxonomic problems. *Annals of Eugenics, 7*, 179–188 http://dx.doi.org/10.1111/j.1469-1809.1936.tb02137.x
Frey, B. J., & Dueck, D. (2007). Clustering by passing messages between data points. *Science, 315*, 972–976. http://dx.doi.org/10.1126/science.1136800
Friedman, J. H., & Meulman, J. J. (2004). Clustering objects on subsets of attributes (with discussion). *Journal of the Royal Statistical Society: Series B (Statistical Methodology), 66*, 815–849. http://dx.doi.org/10.1111/rssb.2004.66.issue-4
Gan, G., Ma, C., & Wu, J. (2007). *Data clustering: Theory, algorithms, and applications*. Philadelphia, PA: SIAM. http://dx.doi.org/10.1137/1.9780898718348
Härdle, W. K., & Simar, L. (2012). *Applied multivariate statistical analysis*. New York, NY: Springer. http://dx.doi.org/10.1007/978-3-642-17229-8
Hastie, T., Friedman, J., & Tibshirani, R. (2001). *The elements of statistical learning: Data mining, inference, and prediction*. New York, NY: Springer. http://dx.doi.org/10.1007/978-0-387-21606-5

Heiser, W. J., & Groenen, P. J. F. (1997). Cluster differences scaling with a within-clusters loss component and a fuzzy successive approximation strategy to avoid local minima. *Psychometrika, 62*, 63–83. http://dx.doi.org/10.1007/BF02294781
Izenman, A. J. (2008). *Modern multivariate statistical techniques*. New York, NY: Springer. http://dx.doi.org/10.1007/978-0-387-78189-1
Joreskog, K. G. (1977). Factor analysis by least-squares and maximum-likelihood methods. In K. Enslein, A. Ralston, & H. S. Wilf (Eds.), *Statistical methods for digital computers* (pp. 125–153). New York, NY: Wiley.
Jöreskog, K. G. (1967). Some contributions to maximum likelihood factor analysis. *Psychometrika, 32*, 443–482. http://dx.doi.org/10.1007/BF02289658
Jöreskog, K. G., & Goldberger, A. S. (1972). Factor analysis by generalized least squares. *Psychometrika, 37*, 243–260. http://dx.doi.org/10.1007/BF02306782
Ladd, G. W. (1966). Linear probability functions and discriminant functions. *Econometrica, 34*, 873–885. http://dx.doi.org/10.2307/1910106
Lawley, D. N., & Maxwell, A. E. (1971). *Factor analysis as a statistical method*. New York, NY: American Elsevier.
Lipovetsky, S. (2009a). Linear regression with special coefficient features attained via parameterization in exponential, logistic, and multinomial-logit forms. *Mathematical and Computer Modelling, 49*, 1427–1435. http://dx.doi.org/10.1016/j.mcm.2008.11.013
Lipovetsky, S. (2009b). PCA and SVD with nonnegative loadings. *Pattern Recognition, 42*, 68–76. http://dx.doi.org/10.1016/j.patcog.2008.06.025
Lipovetsky, S. (2012). Total odds and other objectives for clustering via multinomial-logit model. *Advances in Adaptive Data Analysis, 4*, doi:10.1142/S1793536912500197
Lipovetsky, S. (2013a). Additive and multiplicative mixed normal distributions and finding cluster centers. *International Journal of Machine Learning and Cybernetics, 4*(1), 1-11. doi:10.1007/s13042-012-0070-3
Lipovetsky, S. (2013b). Finding cluster centers and sizes via multinomial parameterization. *Applied Mathematics and Computation, 221*, 571–580. http://dx.doi.org/10.1016/j.amc.2013.06.098
Lipovetsky, S., & Conklin, M. (2005). Regression by data segments via discriminant analysis. *Journal of Modern Applied Statistical Methods, 4*, 63–74.
Lipovetsky, S., Tishler, A., & Conklin, W. M. (2002). Multivariate least squares and its relation to other multivariate techniques. *Applied Stochastic Models in Business and Industry, 18*, 347–356. http://dx.doi.org/10.1002/(ISSN)1526-4025
Liu, H., & Motoda, H. (Eds.). (2008). *Computational methods of feature selection*. Boca Raton, FL: Chapman & Hall/CRC.
Maxwell, A. E. (1983). Factor analysis. In S. Kotz & N. L. Johnson (Eds.), *Encyclopedia of Statistical Sciences* (Vol. 3, pp. 2–8). New York, NY: Wiley.
Nowakowska, E., Koronacki, J., & Lipovetsky, S. (2014). Clusterability assessment for Gaussian mixture models. *Applied Mathematics and Computation, 256*, 591–601. doi:10.1016/j.amc.2014.12.038
Ripley, B. D. (1996). *Pattern recognition and neural networks*. Cambridge: Cambridge University Press. http://dx.doi.org/10.1017/CBO9780511812651
S-PLUS'2000. (1999). Seattle, WA: MathSoft.
Szekely, G. J., & Rizzo, M. L. (2005). Hierarchical clustering via joint between-within distances: Extending ward's minimum variance method. *Journal of Classification, 22*, 151–183. http://dx.doi.org/10.1007/s00357-005-0012-9
Timm, N. H. (1975). *Multivariate analysis with applications in education and psychology*. Monterey, CA: Brooks/Cole.

Appendix

Estimation via multinomial parameterization

Dividing relation (7) by the total number of observations N and denoting the sample variance–covariance matrix as $C = S_{tot}/N$, let us present (8) in a more convenient form:

$$F = \left\| C - \sum_{q=1}^{K} \frac{N_q}{N}(m^q - M)(m^q - M)' \right\|^2 \rightarrow \min, \tag{A1}$$

where N_q/N are the q-th cluster's size shares in the total base. For a data with n variables x_j, and for a chosen number of clusters K, with the restrictions (2) and (4), there are $K - 1$ free parameters N_q, and $K - 1$ vectors m^q with $(K - 1)n$ parameters of means m_j^q, so the total number of parameters is $(K - 1)(n + 1)$. Taking (4) into account we represent the total outer products of distances in (A1) as follows:

$$\sum_{q=1}^{K} \frac{N_q}{N}(m^q - M)(m^q - M)' = \sum_{q=1}^{K} \frac{N_q}{N}(m^q)(m^q)' - MM'$$
$$= M\left\{ \sum_{q=1}^{K} diag\left(\frac{m^q N_q}{MN}\right)\left(\frac{N}{N_q}\right) diag\left(\frac{m^q N_q}{MN}\right) - 1 \right\} M'. \tag{A2}$$

In (A2) we have share parameters: N_q/N with their total equal to one due to (2), and the means $m^q N_q/(MN)$ with the total equal to one due to (4). The multinomial parameterizations for these shares can be defined by the new sets of unknown parameters as following. Instead of the shares N_q/N, let us use the multinomial parameterization

$$\gamma_q = \frac{\exp(\alpha_q)}{1+\sum_{p=2}^{K}\exp(\alpha_p)}, \quad \alpha_1 = 0, \tag{A3}$$

with the first parameter put to zero and needed $K-1$ parameters α_q. Similarly, in place of the shares $m_j^q N_q/(M_j N)$ in (A2), for each variable x_j we define a new multinomial parameterization:

$$g_j^q = \frac{\exp(\beta_j^q)}{1+\sum_{p=2}^{K}\exp(\beta_j^p)}, \quad \beta_j^1 = 0, \quad j = 1,2,\ldots,n, \tag{A4}$$

with the needed $(K-1)n$ parameters β_j^q. Using parameterization (A3) and (A4) in place of the shares in the diagonal matrices (A2) and substituting the expression (A2) as the outer product into objective (A1) yields:

$$F = \left\| C - \sum_{q=1}^{K} (g^q M)\left(\frac{1}{\gamma_q}\right)(g^q M)' + MM' \right\|^2 \rightarrow \min, \tag{A5}$$

with g^q denoting a vector of n-th order of multinomial shares (A4) for each q-th cluster, and the expression $g^q M$ corresponds to the element-wise product of two vectors.

The objective (A5) corresponds to the nonlinear regression of the dependent variable presented by the values of elements in the matrix C by the values of the total means within the complex structure of the unknown parameters. Minimization (A5) by the parameters α_q and β_j^q of the multinomial shares (A3) and (A4) can be performed by any software for nonlinear optimization, or attained directly by the Newton–Raphson procedure which is described in more detail in Lipovetsky (2013b). When the parameters α_q and β_j^q are found, they are used for calculating the share values (A3) and (A4). With γ_q and g_j^q estimates, the quotients N_q/N are known, so the cluster sizes are

$$N_q = \gamma_q N, \tag{A6}$$

and the cluster centers equal to

$$m_j^q = N M_j g_j^q / N_q = M_j g_j^q / \gamma_q. \tag{A7}$$

Having the cluster centers and sizes, the actual clustering can be performed well.

4

Analysis of a redundant system with priority and Weibull distribution for failure and repair

Ashish Kumar[1*], Monika Saini[1] and Kuntal Devi[1]

*Corresponding author: Ashish Kumar, Department of Mathematics, Manipal University Jaipur, Jaipur 303007, Rajasthan, India
E-mail: ashishbarak2020@gmail.com

Reviewing editor: Zudi Lu, University of Southampton, UK

Abstract: The main objective of the present study is to analyse a redundant system under the concept of priority and Weibull distribution for all random variables with different scale parameter and common shape parameter. For this purpose, by using semi-Markovian approach and regenerative point technique a reliability model is developed for a two non-identical unit system. A single repair facility is provided to the system to perform all repair activities. After a pre-specified time to enhance the performance and efficiency of the system the unit goes for preventive maintenance (PM). Priority to the repair of original unit is given over the PM of duplicate unit. Numerical results for the mean time of system failure, availability and profit function have been derived for a particular case to highlight the importance of the study.

Subjects: Mathematics & Statistics; Science; Statistics & Probability

Keywords: redundant system; non-identical units; Weibull failure and repair laws; priority; preventive maintenance and maximum operation time

1. Introduction

In the current scenario, the configuration and design of industrial systems such as air crafts, textile manufacturing systems, computer systems, communication systems, carbon recovery systems in fertiliser plants and satellite systems become more and more complex. The reliability, availability and efficiency of any system depend more or less on the design of the system. So, system designers continuously make efforts to design high-reliability systems using various techniques. Redundancy, i.e. the provision of spare unit is always considered as an effective technique to enhance the reliability of the system. Redundancy is classified into three categories: cold standby, hot standby and warm standby. In cold standby redundancy, the probability of failure of the standby unit is zero. Many researchers, such as Gopalan and Nagarwalla (1985), Goel and Sharma (1989), Cao and Wu (1989) and Gopalan and Bhanu (1995) analysed standby systems using the concept of cold standby redundancy. Chandrasekhar, Natarajan, and Yadavalli (2004) analysed a two-unit cold standby

ABOUT THE AUTHOR

Ashish Kumar, PhD, works as an assistant professor in Department of Mathematics & Statistics, School of Basic Sciences, Manipal University Jaipur, Jaipur, Rajasthan, India. He published more than 50 research papers in the field of reliability modelling and sampling theory. He participated in various workshops/seminar/conferences and presented his research findings within India and abroad.

PUBLIC INTEREST STATEMENT

Reliability modelling and analysis play a very important role in the lifecycle management of a product. Reliability describes the ability of a system or component to function under stated conditions for a specified period of time. Reliability may also describe the ability to function at a specified moment or interval of time. Reliability plays a key role in the cost-effectiveness of systems. A lot of research work has already been carried out in the direction of cold standby redundant systems of identical units. But, for non-identical systems not much work has been carried out. So, this work helps in the analysis of non-identical units systems.

system with Erlangian repair time. Wu and Wu (2011) developed a reliability model with two cold standby units, one repairman and a switch under Poisson shocks. Moghaddass, Zuo, and Qu (2011) analysed the reliability and availability of repairable system with repairman subject to shut-off rules.

A lot of research papers such as Malik and Nandal (2010), Mahmoud and Moshref (2010) and Kumar, Malik, and Barak (2012) appear in the literature to strengthen the idea that by conducting preventive maintenance (PM) after a pre-specified time period, the system can be restored to a younger state. Kumar and Malik (2012) developed many stochastic models for a computer system using the concept of PM after maximum operation time and independent h/w and s/w failure. Priority in repair disciplines is also an effective technique for reliability improvement. Zhang and Wang (2009) suggested a geometric model for a repairable cold standby system with priority in use and repair. Chhillar, Barak, and Malik (2014) developed a reliability model for a cold standby system with priority to repair over corrective maintenance subject to random shocks.

Many researchers such as Malik and Barak (2013) and Kumar, Baweja, and Barak (2015) developed reliability models under the assumption of constant failure and repair rates. But, many mechanical and electrical system's failure and repair rates behave arbitrarily. Osaki and Asakura (1970), Kapur and Kapoor (1974), Gupta, Kumar, and Gupta (2013) and Kishan and Jain (2014) suggested some reliability models for cold standby redundant systems in which all random variables are arbitrarily distributed.

In all studies discussed above, stochastic models for cold standby systems having identical units under different set of assumptions are developed. But, it is not always possible to keep an identical unit in standby due to economic reasons. However, a duplicate unit can be kept in standby to improve the reliability and availability of the system. In most of the studies, all the researchers made the assumption that all the random variables related to failure time of the unit distributed exponentially and repair times are either arbitrary or constantly distributed. But, the performance of most of the mechanical, industrial and electrical systems varies with respect to passes of time. So, their repair and failure are not necessarily constantly distributed but may behave as any arbitrary distribution. There are many distributions such as Weibull, normal and lognormal distributions that are useful in analysing failure processes of standby systems. These distributions have hazard rate functions that are not constant over time, thus providing a necessary alternative to the exponential failure law. The most important probability distribution in reliability modelling is the Weibull distribution. The Weibull failure distribution may be used to model both increasing and decreasing failure rates. Suppose random variable V denotes the maximum operation time of an item/device having Weibull distribution, then its PDF is denoted by $f_1(t) = \theta\eta t^{\eta-1}\exp(-\theta t^{\eta})\ t \geq 0$ and $\theta,\ \eta > 0$. It is characterised by a hazard rate function of the form $h(t) = \theta\eta t^{\eta-1}$, $t \geq 0$ and $\theta, \eta > 0$ which is a power function. The function $\lambda(t)$ is increasing for $\eta > 0, \theta > 0$ and is decreasing for $\eta < 0, \theta < 0$. The reliability function is given by $R(t) = \exp(-\theta t^{\eta})$. Thus the failure-free operating time of the system has a Weibull distribution with parameters θ and η. Here η is referred to as the shape parameter. Its effect on the distribution can be seen for several different values. For $\eta < 1$, the PDF is similar in shape to the exponential, and for large values of η ($\eta \geq 3$), the PDF is somewhat symmetrical, like the normal distribution. For $1 < \eta < 3$, the density is skewed. If we put $\eta = 1$ in PDF, Weibull distribution reduces to Exponential distribution and if $\eta = 2$, it reduces to Rayleigh distribution. Kumar and Saini (2014) analysed the cost-benefit of a single-unit system under PM and Weibull distribution for random variables.

In the present paper, we develop a reliability model for a non-identical cold standby system for the evaluation of various reliability measures of the system by considering all time random variables as Weibull distributed. Priority to repair of the original unit is given over the PM of the duplicate unit. The possible states of the proposed problem have been given in section entitled system description. A single repair facility has been provided to do repair and maintenance activities of original and duplicate units. After a pre-specified time, the unit goes for PM. All random variables are statistically independent. Switch devices and repairs are perfect. Semi-Markov process and regenerative point techniques are used to draw recurrence relations for various reliability characteristics. All time random variables are Weibull distributed. The probability density function of maximum operation time of original and duplicate units are denoted by $g(t) = \alpha\eta t^{\eta-1}\exp(-\alpha t^{\eta})$. The PDF of failure times of the original and duplicate unit are denoted by $f(t) = \beta\eta t^{\eta-1}\exp(-\beta t^{\eta})$ and $f_2(t) = h\eta t^{\eta-1}\exp(-ht^{\eta})$, respectively. The PM rate of the original and duplicate units is denoted by the probability density function $g_1(t) = \gamma\eta t^{\eta-1}\exp(-\gamma t^{\eta})$. The random variables corresponding to repair rate of the original and duplicate units have the probability density function $f_1(t) = k\eta t^{\eta-1}\exp(-kt^{\eta})$ and $f_3(t) = l\eta t^{\eta-1}\exp(-lt^{\eta})$, respectively with $t \geq 0$ and $\theta, \eta, \alpha, \beta, h, k, l > 0$. The probability/cumulative density functions of direct transition time from regenerative state i to a regenerative state j or to a failed state j visiting state k, r once in (0, t] have been denoted by $q_{ij.kr}(t)/Q_{ij.kr}(t)$. To improve the importance of the study, graphs are drawn for a particular case for mean time to system failure (MTSF), availability and profit function.

2. System description

In this section, a reliability model for two non-identical unit's system using the concept of priority and arbitrary distribution is developed. The system may be any of the following states described as follows:

State 0 Original unit is operative, duplicate unit in cold standby and system is in upstate. The service facility at S_0 remains idle.

State 1 Original unit is under PM after completion of maximum operation time, duplicate unit is operative and system is in upstate. The service facility at S_1 is busy under PM of the original unit.

State 2 Original unit is under repair after failure, duplicate unit is operative and system is in upstate. The service facility at S_2 is busy in repair activity of the failed original unit.

State 3 Original unit is operative, duplicate unit is under repair after failure and system is in upstate. The service facility at S_3 is busy in repair activity of the failed duplicate unit.

State 4 Original unit is operative, duplicate unit is under PM after completion of maximum operation time and system is in upstate. The service facility at S_4 is busy in PM of the duplicate unit.

State 5 It is the priority state. Here priority is given to repair of failed original unit over PM of duplicate unit. The service facility at S_5 is busy in repair of the original unit.

State 6 Original unit has failed and continuously under repair from past state, duplicate unit has failed and waiting for repair and system is in downstate. The service facility at S_6 is busy in repair of the original unit.

State 7 Original unit has failed and continuously under repair from past state, duplicate unit is under waiting for PM and system is in downstate. The service facility at S_7 is busy in repair of the original unit.

State 8 Original unit is waiting for PM after completion of maximum operation time, duplicate unit is under PM continuously after completion of maximum operation time from previous state and system is in downstate. The service facility at S_8 is busy in PM of the duplicate unit.

State 9 Original unit is continuously under PM after completion of maximum operation time from previous state, duplicate unit is waiting for PM after completion of maximum operation time and system is in downstate. The service facility at S_9 is busy in PM of the original unit.

State 10 Original unit is continuously under PM after completion of maximum operation time from previous state, duplicate failed unit is waiting for repair and system is in downstate. The service facility at S_{10} is busy in PM of the original unit.

State 11 Original unit is waiting for repair, duplicate failed unit is continuously under repair from previous state and system is in downstate. The service facility at S_{11} is busy in repair of the failed duplicate unit.

State 12 Original unit is waiting for PM after completion of maximum operation time, duplicate failed unit is continuously under repair from previous state and system is in downstate. The service facility at S_{12} is busy in repair of the failed duplicate unit.

Out of these, states S_0, S_1, S_2, S_3, S_4 and $S_5$are regenerative states, while all other are non-regenerative and failed states.

3. Transition probabilities and mean sojourn times

Simple probabilistic considerations yield the following expressions for the non-zero elements:

$$p_{ij} = Q_{ij}(\infty) = \int q_{ij}(t)dt \text{ as} \tag{1}$$

$$\begin{aligned} p_{01} &= \frac{\alpha}{\alpha+\beta}, p_{02} = \frac{\beta}{\alpha+\beta}, p_{10} = \frac{\gamma}{\alpha+h+\gamma}, p_{1.10} = \frac{h}{\alpha+\gamma+h} = p_{13.10}, p_{19} = \frac{\alpha}{\alpha+\gamma+h} \\ &= p_{14.9}, p_{20} = \frac{k}{\alpha+k+h}, p_{26} = \frac{h}{\alpha+h+k} = p_{23.6}, p_{27} = \frac{\alpha}{\alpha+h+k} = p_{24.7}, p_{30} \\ &= \frac{l}{l+\alpha+\beta}, p_{3.12} = \frac{\alpha}{\alpha+\beta+l} = p_{31.12}, p_{3.11} = \frac{\beta}{\alpha+\beta+l} = p_{32.11}, p_{40} = \frac{\gamma}{\alpha+\beta+\gamma}, p_{45} \\ &= \frac{\beta}{\alpha+\beta+\gamma} = p_{42.5}, p_{48} = \frac{\alpha}{\alpha+\beta+\gamma} = p_{41.8}, p_{52} = p_{63} = p_{74} = p_{81} = p_{94} = p_{10.3} = p_{11.2} \\ &= p_{12.1} = 1 \end{aligned} \tag{2}$$

It can be easily verified that

$$\begin{aligned} &p_{01} + p_{02} = p_{10} + p_{19} + p_{1.10} = p_{10} + p_{14.9} + p_{10} = p_{20} + p_{27} + p_{26} \\ &= p_{20} + p_{23.6} + p_{24.7} = p_{30} + p_{3.12} + p_{3.11} = p_{30} + p_{31.12} + p_{32.11} = p_{40} + p_{45} + p_{48} = p_{40} + p_{418} \\ &\quad + p_{425} = p_{52} = p_{63} = p_{74} = p_{81} = p_{94} = p_{10.3} = p_{11.2} = p_{12.1} = 1 \end{aligned} \tag{3}$$

The mean sojourn times (μ_i) in the state S_i are

$$\begin{aligned} &\mu_0 = \frac{\Gamma(1+1/\eta)}{(\alpha+\beta)^{1/\eta}}, \quad \mu_3 = \frac{\Gamma(1+1/\eta)}{(\alpha+\beta+l)^{1/\eta}}, && \mu_4 = \frac{\Gamma(1+1/\eta)}{(\alpha+\beta+\gamma)^{1/\eta}}, \\ &\mu_1' = \Gamma(1/\eta+1)\left[\frac{1}{(\alpha+h+\gamma)^{\frac{1}{\eta}}} + \frac{1}{(\alpha+\gamma+h)(\gamma)^{\frac{1}{\eta}}}(\alpha+h)\right], && \\ &\mu_2' = \Gamma(1/\eta+1)\left[\frac{1}{(\alpha+h+k)^{\frac{1}{\eta}}} + \frac{1}{(\alpha+k+h)(k)^{\frac{1}{\eta}}}(\alpha+h)\right], && \mu_2 = \frac{\Gamma(1+1/\eta)}{(\alpha+k+h)^{1/\eta}} \\ &\mu_3' = \Gamma(1/\eta+1)\left[\frac{1}{(\alpha+\beta+l)^{\frac{1}{\eta}}} + \frac{1}{(\alpha+\beta+l)(l)^{\frac{1}{\eta}}}(\alpha+\beta)\right], && \mu_1 = \frac{\Gamma(1+1/\eta)}{(\alpha+\gamma+h)^{1/\eta}} \\ &\mu_4^1 = \Gamma(1/\eta+1)\left[\frac{1}{(\alpha+\beta+\gamma)^{\frac{1}{\eta}}} + \frac{\alpha}{(\alpha+\gamma+\beta)(\gamma)^{\frac{1}{\eta}}}\right], && \mu_5 = \frac{\Gamma(1+\frac{1}{\eta})}{(k)^{1/\eta}}, \end{aligned} \tag{4}$$

4. Reliability and MTSF

Let $\varphi_i(t)$ be the CDF of first passage time from the regenerative state *i* to a failed state. Regarding the failed state as absorbing state, we have the following recursive relations for $\varphi_i(t)$:

$$\varphi_i(t) = \sum_j Q_{i,j}(t)®\varphi_j(t) + \sum_k Q_{i,k}(t) \tag{5}$$

where *j* is an unfailed regenerative state to which the given regenerative state *i* can transit and *k* is a failed state to which the state *i* can transit directly. Taking LST of above relation (5) and solving for $\tilde{\varphi}_0(s)$, we have

$$R^*(s) = \frac{1-\tilde{\varphi}_0(s)}{s} \tag{6}$$

The reliability of the system model can be obtained by taking Laplace inverse transform of (6).

The MTSF is given by

$$\begin{aligned} &\text{MTSF} = \lim_{s\to 0} \frac{1-\tilde{\varphi}_0(s)}{s} = \frac{N}{D} \quad \text{where} \\ &N = \mu_0 + p_{01}\mu_1 + p_{02}\mu_2 \quad \text{and } D = 1 - p_{01}p_{10} - p_{02}p_{20} \end{aligned} \tag{7}$$

5. Steady state availability

Let $A_i(t)$ be the probability that the system is in upstate at instant "t" given that the system entered regenerative state i at $t = 0$. The recursive relations for $A_i(t)$ are given as

$$A_i(t) = M_i(t) + \sum_j q_{i,j}^{(n)}(t)©A_j(t) \tag{8}$$

where j is any successive regenerative state to which the regenerative state i can transit through n transitions. $M_i(t)$ is the probability that the system is initially in state $S_i \in E$ up at time t without visiting any other regenerative state, we have

$$M_0(t) = e^{-(\alpha+\beta)t''},\ M_1(t) = e^{-(\alpha+h+\gamma)t''},\ M_2(t) = e^{-(\alpha+k+h)t''},\ M_3(t) = e^{-(\alpha+\beta+l)t''},\ M_4(t) = e^{-(\alpha+\beta+\gamma)t''} \tag{9}$$

Taking LT of above relations (8) and solving for $A_0^*(s)$, the steady-state availability is given by

$$A_0(\infty) = \lim_{s\to 0} sA_0^*(s) = \frac{N_2}{D_2}, \text{where} \tag{10}$$

$$N_2 = (M_0(t)((-p_{41.8})(p_{24.7}p_{32.11}p_{13.10} + p_{14.9}(1 - p_{32.11}p_{23.6})) + (1 - p_{45}p_{54})((1 - p_{31.12}p_{13.10})$$
$$(1 - p_{32.11}p_{23.6}) - p_{32.11}p_{13.10}p_{31.12}p_{23.6})) + (M_1(t) + M_3(t)p_{13.10})(p_{01}(1 - p_{32.11}p_{23.6})$$
$$(1 - p_{54}p_{45}) + p_{02}(p_{31.12}p_{23.6}(1 - p_{45}p_{54}) + p_{24.7}p_{41.8})) + (M_2(t) + M_3(t)p_{23.6})(p_{01}(p_{32.11}p_{13.10}$$
$$(1 - p_{45}p_{54})) + p_{02}((1 - p_{31.12}p_{13.10})(1 - p_{45}p_{54}) - p_{14.9}p_{41.8})) - M_4(t)(-p_{01}(p_{24.7}p_{32.11}p_{13.10}$$
$$+ p_{14.9}(1 - p_{32.11}p_{23.6})) + p_{02}(-p_{24.7}(1 - p_{31.12}p_{13.10}) - p_{14.9}p_{31.12}p_{23.6}))$$

$$D_2 = \mu_0((-p_{41.8})(p_{24.7}p_{32.11}p_{13.10} + p_{14.9}(1 - p_{32.11}p_{23.6})) + (1 - p_{45}p_{54})((1 - p_{31.12}p_{13.10})$$
$$(1 - p_{32.11}p_{23.6}) - p_{32.11}p_{13.10}p_{31.12}p_{23.6})) + (\mu_1' + \mu_3'p_{13.10})(p_{01}(1 - p_{32.11}p_{23.6})(1 - p_{54}p_{45})$$
$$+ p_{02}(p_{31.12}p_{23.6}(1 - p_{45}p_{54}) + p_{24.7}p_{41.8})) + (\mu_2' + \mu_3'p_{23.6})(p_{01}(p_{32.11}p_{13.10}(1 - p_{45}p_{54}))$$
$$+ p_{02}((1 - p_{31.12}p_{13.10})(1 - p_{45}p_{54}) - p_{14.9}p_{41.8})) - \mu_4'(-p_{01}(p_{24.7}p_{32.11}p_{13.10} + p_{14.9}$$
$$(1 - p_{32.11}p_{23.6})) + p_{02}(-p_{24.7}(1 - p_{31.12}p_{13.10}) - p_{14.9}p_{31.12}p_{23.6}))$$

6. Busy period analysis for server

6.1. Due to repair

Let $B_i^R(t)$ be the probability that the server is busy in repairing the unit at an instant "t" given that the system entered state i at $t = 0$. The recursive relations for $B_i^R(t)$ are as follows:

$$B_i^R(t) = W_i(t) + \sum_j q_{i,j}^{(n)}(t)©B_j^R(t) \tag{11}$$

where j is any successive regenerative state to which the regenerative state i can transit through n transitions. $W_i(t)$ be the probability that the server is busy in state S_i up to time t without making any transition to any other regenerative state or returning to the same via one or more non-regenerative states and so $W_2(t) = e^{-(\alpha+k+h)t''}, W_3(t) = e^{-(\alpha+\beta+l)t''}$

By taking LT of (11) and solving for $B_0^{*R}(s)$, the busy period of the server due to repair is given by $B_0^R = \lim_{s\to 0} sB_0^{*R}(s) = \frac{N_3^R}{D_2}$

$$N_3^r = (W_3(t)p_{13.10})(p_{01}(1 - p_{32.11}p_{23.6})(1 - p_{54}p_{45}) + p_{02}(p_{31.12}p_{23.6}(1 - p_{45}p_{54}) + p_{24.7}p_{41.8}))$$
$$+(W_2(t) + W_3(t)p_{23.6})(p_{01}(p_{32.11}p_{13.10}(1 - p_{45}p_{54})) + p_{02}((1 - p_{31.12}p_{13.10})(1 - p_{45}p_{54}) - p_{14.9}p_{41.8}))$$
$$-(W_5(t)p_{45})(-p_{01}(p_{24.7}p_{32.11}p_{13.10} + p_{14.9}(1 - p_{32.11}p_{23.6})) + p_{02}(-p_{24.7}(1 - p_{31.12}p_{13.10})$$
$$-p_{14.9}p_{31.12}p_{23.6}))$$

And D_2 is already mentioned in the previous section.

6.2. Due to PM

Let $B_i^{Pm}(t)$ be the probability that the server is busy in preventive maintenance of the system (unit) at an instant "t" given that the system entered state i at $t = 0$. The recursive relations for $B_i^{Pm}(t)$ are as follows:

$$B_i^{pm}(t) = W_i(t) + \sum_j q_{i,j}^{(n)}(t) © B_j^{pm}(t) \tag{12}$$

where j is any successive regenerative state to which the regenerative state i can transit through n transitions. $W_i(t)$ be the probability that the server is busy in state S_i due to PM up to time t without making any transition to any other regenerative state or returning to the same via one or more non-regenerative states and so $W_1(t) = e^{-(\alpha+k+\gamma)t^{\eta}}, W_4(t) = e^{-(\alpha+\beta+\gamma)t^{\eta}}$

By taking LT of (12) and solving for $B_0^{*Pm}(s)$, the busy period of the server due to PM is given by

$$B_0^{Pm} = \lim_{s\to 0} sB_0^{*Pm}(s) = \frac{N_4^{Pm}}{D_2} \tag{13}$$

$$\begin{aligned}N_4^{Pm} &= (W_1(0))(p_{01}(1 - p_{32.11}p_{23.6})(1 - p_{54}p_{45}) + p_{02}(p_{31.12}p_{23.6}(1 - p_{45}p_{54}) + p_{24.7}p_{41.8}))\\ &-W_4(0)(-p_{01}(p_{24.7}p_{32.11}p_{13.10} + p_{14.9}(1 - p_{32.11}p_{23.6})) + p_{02}(-p_{24.7}(1 - p_{31.12}p_{13.10})\\ &-p_{14.9}p_{31.12}p_{23.6}))\end{aligned}$$

And D_2 is already mentioned in previous section.

7. Expected number of repairs

Let $E_i^R(t)$ be the expected number of repairs by the server in (0, t] given that the system entered the regenerative state i at $t = 0$. The recursive relations for $E_i^R(t)$ are given as

$$E_i^R(t) = \sum_j Q_{i,j}^{(n)}(t) ® \left[\delta_j + E_j^R(t)\right] \tag{14}$$

where j is any regenerative state to which the given regenerative state i transits and $\delta j = 1$ if j is the regenerative state where the server does the job afresh, otherwise $\delta j = 0$. Taking LST of relations (14) and solving for $\tilde{E}_0^{R}(s)$, the expected number of repairs per unit time is given by

$$E_0^R(\infty) = \lim_{s\to 0} s\tilde{E}_0^{R}(s) = \frac{N_5^R}{D_2} \tag{15}$$

$$\begin{aligned}N_5^R &= ((p_{30} + p_{31.12} + p_{32.11})p_{13.10})(p_{01}(1 - p_{32.11}p_{23.6})(1 - p_{54}p_{45}) + p_{02}(p_{31.12}p_{23.6}(1 - p_{45}p_{54})\\ &+ p_{24.7}p_{41.8})) + ((p_{20} + p_{23.6} + p_{24.7}) + (p_{30} + p_{31.12} + p_{32.11})p_{23.6})(p_{01}(p_{32.11}p_{13.10}(1 - p_{45}p_{54}))\\ &+ p_{02}((1 - p_{31.12}p_{13.10})(1 - p_{45}p_{54}) - p_{14.9}p_{41.8})) - (p_{45}p_{54})(-p_{01}(p_{24.7}p_{32.11}p_{13.10}\\ &+ p_{14.9}(1 - p_{32.11}p_{23.6})) + p_{02}(-p_{24.7}(1 - p_{31.12}p_{13.10}) - p_{14.9}p_{31.12}p_{23.6}))\end{aligned}$$

And D_2 is already mentioned in the previous section.

8. Expected number of PM (PM)

Let $E_i^{Pm}(t)$ be the expected number of PM by the server in (0, t] given that the system entered the regenerative state i at $t = 0$. The recursive relations for $E_i^{Pm}(t)$ are given as

$$E_i^{Pm}(t) = \sum_j Q_{i,j}^{(n)}(t) ® \left[\delta_j + E_j^{Pm}(t)\right] \tag{16}$$

where j is any regenerative state to which the given regenerative state i transits and $\delta j = 1$ if j is the regenerative state where the server does the job afresh, otherwise $\delta j = 0$. Taking LST of relations (16) and solving for $\tilde{E}_0^{Pm}(s)$, the expected number of PM per unit time is given by

$$E_0^{Pm}(\infty) = \lim_{s\to 0} s\tilde{E}_0^{Pm}(s) = \frac{N_6^{Pm}}{D_2} \tag{17}$$

$$N_6^{Pm} = (p_{10} + p_{13.10} + p_{14.9})(p_{01}(1 - p_{32.11}p_{23.6})(1 - p_{54}p_{45}) + p_{02}(p_{31.12}p_{23.6}(1 - p_{45}p_{54}) + p_{24.7}p_{41.8})) - (p_{40} + p_{41.8} + p_{45})(-p_{01}(p_{24.7}p_{32.11}p_{13.10} + p_{14.9}(1 - p_{32.11}p_{23.6})) + p_{02}(-p_{24.7}(1 - p_{31.12}p_{13.10}) - p_{14.9}p_{31.12}p_{23.6}))$$

And D_2 is already mentioned in the previous section.

9. Expected number of visits by the server

Let $N_i(t)$ be the expected number of visits by the server in (0, t] given that the system entered the regenerative state i at $t = 0$. The recursive relations for $N_i(t)$ are given as

$$N_i(t) = \sum_j Q_{i,j}^{(n)}(t)\circledR\left[\delta_j + N_j(t)\right] \tag{18}$$

where j is any regenerative state to which the given regenerative state i transits and $\delta j = 1$ if j is the regenerative state where the server does the job afresh, otherwise $\delta j = 0$. Taking LST of relation (18) and solving for $\tilde{N}_0(s)$, the expected number of visits per unit time by the server is given by

$$N_0'(\infty) = \lim_{s\to 0} s\tilde{N}_0(s) = \frac{N_7}{D_2} \tag{19}$$

$$N_7 = (p_{01} + p_{02})((-p_{41.8})(p_{24.7}p_{32.11}p_{13.10} + p_{14.9}(1 - p_{32.11}p_{23.6})) + (1 - p_{45}p_{54})((1 - p_{31.12}p_{13.10})(1 - p_{32.11}p_{23.6}) - p_{32.11}p_{13.10}p_{31.12}p_{23.6}))$$

And D_2 is already mentioned in the previous section.

10. Profit analysis

The profit incurred by the system model in steady state can be obtained as

$$P = K_0A_0 - K_1B_0^{Pm} - K_2B_0^R - K_3E_0^{Pm} - K_4E_0^R - K_5N_0 \tag{20}$$

K_0 is the revenue per unit up-time of the system; K_1 is the cost per unit time for which the server is busy due PM; K_2 is the cost per unit time for which the server is busy due to repair; K_3 is the cost per unit time for which the server is busy due to expected number of PM; K_4 is the cost per unit time due to expected number of repairs; K_5 is the cost per unit time visit by the server.

10.1. Case studies with discussions

(1) When shape parameter $\eta = 0.5$ then maximum operation/failure of original unit/failure of duplicate unit/PM/repair of original/repair of duplicate unit time distributions reduce to:

$$g(t) = \frac{\alpha}{2\sqrt{t}}e^{-\alpha\sqrt{t}},\ f(t) = \frac{\beta}{2\sqrt{t}}e^{-\beta\sqrt{t}},\ g_1(t) = \frac{\gamma_1}{2\sqrt{t}}e^{-\gamma_1\sqrt{t}},\ f_1(t) = \frac{k}{2\sqrt{t}}e^{-k\sqrt{t}}$$
$$f_3(t) = \frac{l}{2\sqrt{t}}e^{-l\sqrt{t}},\ f_2(t) = \frac{h}{2\sqrt{t}}e^{-h\sqrt{t}};\ \text{where } t \geq 0 \text{ and } \eta,\ \alpha\,\beta,,\gamma_1,\ h,k,l > 0 \tag{21}$$

(2) When shape parameter $\eta = 1.0$, then failure/PM/arrival time of the server/replacement/transition rate/repair time distributions reduce to exponentials, then

$$g(t) = \alpha e^{-\alpha t},\ f(t) = \beta e^{-\beta t},\ g_1(t) = \gamma_1 e^{-\gamma_1 t},\ f_1(t) = ke^{-kt},\ f_3(t) = le^{-lt},\ f_2(t) = he^{-ht};$$
$$\text{where } t \geq 0 \textit{ and } \eta,\ \alpha\,\beta,\ \gamma_1,\ h,k,l > 0 \tag{22}$$

(3) When shape parameter $\eta = 2.0$, then failure/PM/arrival time of the server/replacement/transition rate/repair time distributions reduce to Rayleigh having the pdf-

$$g(t) = 2\alpha e^{-\alpha t^2},\ f(t) = 2\beta e^{-\beta t^2},\ g_1(t) = 2\gamma_1 e^{-\gamma_1 t^2},\ f_1(t) = 2ke^{-kt^2},\ f_3(t) = 2le^{-lt^2},$$
$$f_2(t) = 2he^{-ht^2};\quad \text{where } t \geq 0 \text{ and } \eta,\ \alpha\,\beta, \gamma_1,\ h, k, l > 0 \tag{23}$$

Table 1. Values of MTSF for different values of α and η with respect to β

β	$\alpha = 2$, $\eta = 0.5$, $\gamma = 5$, $h = 0.009$, $k = 1.5$, $l = 1.4$	$\alpha = 2.4$, $\eta = 0.5$, $\gamma = 5$, $h = 0.009$, $k = 1.5$, $l = 1.4$	$\alpha = 2$, $\eta = 1, \gamma = 5$, $h = 0.009$, $k = 1.5$, $l = 1.4$	$\alpha = 2.4$, $\eta = 1, \gamma = 5$, $h = 0.009$, $k = 1.5$, $l = 1.4$	$\alpha = 2$, $\eta = 2, \gamma = 5$, $h = 0.009$, $k = 1.5$, $l = 1.4$	$\alpha = 2.4$, $\eta = 2, \gamma = 5$, $h = 0.009$, $k = 1.5$, $l = 1.4$
0.01	4.9918	2.1095	5.9647	3.0460	8.9395	4.9639
0.02	4.7950	2.0642	5.7599	2.9933	8.6486	4.8851
0.03	4.6123	2.0207	5.5696	2.9426	8.3781	4.8092
0.04	4.4423	1.9788	5.3921	2.8937	8.1260	4.7362
0.05	4.2836	1.9385	5.2264	2.8466	7.8904	4.6657
0.06	4.1351	1.8998	5.0712	2.8012	7.6698	4.5978
0.07	3.9960	1.8624	4.9255	2.7573	7.4627	4.5323
0.08	3.8654	1.8264	4.7886	2.7150	7.2680	4.4689
0.09	3.7426	1.7916	4.6596	2.6741	7.0846	4.4078
0.1	3.6268	1.7581	4.5379	2.6346	6.9116	4.3486

Table 2. Values of availability for different values of α and η with respect to β

β	$\alpha = 2$, $\eta = 0.5$, $\gamma = 5$, $h = 0.009$, $k = 1.5$, $l = 1.4$	$\alpha = 2.4$, $\eta = 0.5$, $\gamma = 5$, $h = 0.009$, $k = 1.5$, $l = 1.4$	$\alpha = 2$, $\eta = 1, \gamma = 5$, $h = 0.009$, $k = 1.5$, $l = 1.4$	$\alpha = 2.4$, $\eta = 1, \gamma = 5$, $h = 0.009$, $k = 1.5$, $l = 1.4$	$\alpha = 2$, $\eta = 2, \gamma = 5$, $h = 0.009$, $k = 1.5$, $l = 1.4$	$\alpha = 2.4$, $\eta = 2, \gamma = 5$, $h = 0.009$, $k = 1.5$, $l = 1.4$
0.01	0.9405	0.9066	0.8947	0.8627	0.8719	0.8457
0.02	0.9368	0.9026	0.8929	0.8611	0.8710	0.8449
0.03	0.9332	0.8986	0.8912	0.8596	0.8701	0.8442
0.04	0.9296	0.8946	0.8895	0.8580	0.8692	0.8435
0.05	0.9260	0.8907	0.8878	0.8565	0.8683	0.8427
0.06	0.9225	0.8869	0.8861	0.8550	0.8674	0.8420
0.07	0.9189	0.8831	0.8844	0.8535	0.8666	0.8413
0.08	0.9155	0.8793	0.8828	0.8521	0.8657	0.8407
0.09	0.9120	0.8756	0.8812	0.8506	0.8649	0.8400
0.1	0.9086	0.8719	0.8796	0.8492	0.8641	0.8393

Table 3. Values of profit for different values of α and η with respect to β

β	$\alpha = 2$, $\eta = 0.5$, $\gamma = 5$, $h = 0.009$, $k = 1.5$, $l = 1.4$	$\alpha = 2.4$, $\eta = 0.5$, $\gamma = 5$, $h = 0.009$, $k = 1.5$, $l = 1.4$	$\alpha = 2$, $\eta = 1, \gamma = 5$, $h = 0.009$, $k = 1.5$, $l = 1.4$	$\alpha = 2.4$, $\eta = 1, \gamma = 5$, $h = 0.009$, $k = 1.5$, $l = 1.4$	$\alpha = 2$, $\eta = 2, \gamma = 5$, $h = 0.009$, $k = 1.5$, $l = 1.4$	$\alpha = 2.4$, $\eta = 2, \gamma = 5$, $h = 0.009$, $k = 1.5$, $l = 1.4$
0.01	4362.9	4064.9	4181.3	3982.3	4136.5	3993.4
0.02	4341.7	4042.3	4171.3	3973.4	4131.2	3989.1
0.03	4320.7	4020.0	4161.3	3964.6	4126.0	3984.9
0.04	4299.8	3998.0	4151.6	3956.0	4120.9	3980.7
0.05	4279.2	3976.2	4141.9	3947.4	4115.8	3976.6
0.06	4258.7	3954.7	4132.4	3939.0	4110.9	3972.6
0.07	4238.3	3933.4	4123.0	3930.7	4106.0	3968.6
0.08	4218.2	3912.4	4113.7	3922.5	4101.2	3964.7
0.09	4198.2	3891.6	4104.6	3914.4	4096.5	3960.9
0.1	4178.4	3871.1	4095.6	3906.4	4091.9	3957.1

11. Conclusion

For a particular case, having values $\alpha = 2, \gamma = 5, h = 0.009, k = 1.5, l = 1.4$ the behaviour of various reliability measures such as MTSF, availability and net expected steady-state profit of the system discussed here for a two-unit cold standby system under priority and Weibull distribution. The values of K_i for profit function are assumed as $K_1 = 200$, $K_0 = 5000$, $K_2 = 150$, $K_3 = 100$, $K_4 = 75$, $K_5 = 80$. From the numerical results depicted in Tables 1–3 show that the MTSF, availability and profit of the system decline with the increase of failure rate (β). While with respect to shape parameter (η) the value of MTSF increases, the availability and profit decrease. With the increment of the repair rate and PM rate, availability and profit show increasing behaviour. Finally, we conclude that by increasing the repair rate of the original and duplicate units, the system can be made more profitable.

Nomenclature

O	operative unit
DCs	duplicate cold standby unit
Do	duplicate unit is operative
~/*	symbol for Laplace–Steiltjes transform (LST)/Laplace transform (LT)
Ⓢ/©	symbol for Laplace–Stieltjes convolution/Laplace convolution
Fur/FUR	failed original unit under repair/continuously under repair
DFur/DFUR	failed duplicate unit under repair/continuously under repair
DPm/DPM	duplicate unit under preventive maintenance/continuously under preventive maintenance
Pm/PM	original unit under preventive maintenance/continuously under preventive maintenance
WPm/WPM	original unit waiting for preventive maintenance/continuously waiting for preventive maintenance
DWPm/DWPM	duplicate unit waiting for preventive maintenance/continuously waiting for preventive maintenance
Fwr/FWR	original unit after failure waiting for repair/continuously waiting for repair
DFwr/DFWR	duplicate unit after failure waiting for repair/continuously waiting for repair
MTSF	mean time to system failure

Funding
The authors received no direct funding for this research.

Author details
Ashish Kumar[1]
E-mail: ashishbarak2020@gmail.com
Monika Saini[1]
E-mail: drmnksaini4@gmail.com
Kuntal Devi[1]
E-mail: akbrk@rediffmail.com
[1] Department of Mathematics, Manipal University Jaipur, Jaipur 303007, Rajasthan, India.

References

Cao, J., & Wu, Y. (1989). Reliability analysis of a two-unit cold standby system with a replaceable repair facility. *Microelectronics Reliability, 29*, 145–150. http://dx.doi.org/10.1016/0026-2714(89)90561-1

Chandrasekhar, P., Natarajan, R., & Yadavalli, V. S. S. (2004). A study on a two unit standby system with Erlangian repair time. *Asia-Pacific Journal of Operational Research, 21*, 271–277. http://dx.doi.org/10.1142/S0217595904000242

Chhillar, S. K., Barak, A. K., & Malik, S. C. (2014). Reliability measures of a cold standby system with priority to repair over corrective maintenance subject to random shocks. *International Journal of Statistics & Economics, 13*, 79–89.

Goel, L. R., & Sharma, S. C. (1989). Stochastic analysis of a 2-unit standby system with two failure modes and slow switch. *Microelectronics Reliability, 29*, 493–498. http://dx.doi.org/10.1016/0026-2714(89)90332-6

Gopalan, M. N., & Bhanu, K. S. (1995). Cost analysis of a two unit repairable system subject to on-line preventive maintenance and/or repair. *Microelectronics Reliability, 35*, 251–258. http://dx.doi.org/10.1016/0026-2714(95)90090-D

Gopalan, M. N., & Nagarwalla, H. E. (1985). Cost-benefit analysis of a one-server two-unit cold standby system with repair and preventive maintenance. *Microelectronics Reliability, 25*, 267–269. http://dx.doi.org/10.1016/0026-2714(85)90011-3

Gupta, R., Kumar, P., & Gupta, A. (2013). Cost benefit analysis of a two dissimilar unit cold standby system with Weibull failure and repair laws. *International Journal of System Assurance Engineering and Management, 4*, 327–334. http://dx.doi.org/10.1007/s13198-012-0091-z

Kapur, P. K., & Kapoor, K. R. (1974). A two-unit warm standby system with repair and preventive maintenance. *Indian Journal of Pure and Applied Mathematics, 7*, 13–27.

Kishan, R., & Jain, D. (2014). Classical and Bayesian analysis of reliability characteristics of a two-unit parallel system with Weibull failure and repair laws. *International Journal of System Assurance Engineering and Management, 5*, 252–261. http://dx.doi.org/10.1007/s13198-013-0154-9

Kumar, A., Baweja, S., & Barak, M. (2015). Stochastic behavior of a cold standby system with maximum repair time. *Decision Science Letters, 4*, 569–578. http://dx.doi.org/10.5267/j.dsl.2015.5.002

Kumar, A., & Malik, S. C. (2012). Reliability modeling of a computer system with priority to s/w replacement over h/w replacement subject to MOT and MRT. *International Journal of Pure and Applied Mathematics, 80*, 693–709.

Kumar, A., Malik, S. C., & Barak, M. S. (2012). Reliability modelling of a computer system with independent H/W and S/W failures subject to maximum operation and repair times. *International Journal of Mathematical Archive (IJMA), 3*(7). ISSN 2229-5046.

Kumar, A., & Saini, M. (2014). Cost-benefit analysis of a single-unit system with preventive maintenance and Weibull distribution for failure and repair activities. *Journal of Applied Mathematics, Statistics and Informatics, 10*, 5–19.

Kumar, A., Malik, S. C., & Barak, M. S. (2012). Reliability modelling of a computer system with independent H/W and S/W failures subject to maximum operation and repair times. *International Journal of Mathematical Archive (IJMA), 3*(7). ISSN 2229-5046.

Mahmoud, M. A. W., & Moshref, M. E. (2010). On a two-unit cold standby system considering hardware, human error failures and preventive maintenance. *Mathematical and Computer Modelling, 51*, 736–745. http://dx.doi.org/10.1016/j.mcm.2009.10.019

Malik, S. C., & Barak, S. K. (2013). Reliability measures of a cold standby system with preventive maintenance and repair. *International Journal of Reliability, Quality and Safety Engineering, 20*(06), 1350022. http://dx.doi.org/10.1142/S0218539313500228

Malik, S. C., & Nandal, P. (2010). Cost-analysis of stochastic models with priority to repair over preventive maintenance subject to maximum operation time, edited book, learning manual on modeling, optimization and their applications. In *Optimization and their applications* (pp. 165–178). Excel India Publishers.

Moghaddass, R., Zuo, M. J., & Qu, J. (2011). Reliability and availability analysis of a repairable-out-of-system with repairmen subject to shut-off rules. *IEEE Transactions on Reliability, 60*, 658–666. http://dx.doi.org/10.1109/TR.2011.2161703

Osaki, S., & Asakura, T. (1970). A two-unit standby redundant system with repair and preventive maintenance. *Journal of Applied Probability*, 641–648. http://dx.doi.org/10.2307/3211943

Wu, Q., & Wu, S. (2011). Reliability analysis of two-unit cold standby repairable systems under Poisson shocks. *Applied Mathematics and Computation, 218*, 171–182. http://dx.doi.org/10.1016/j.amc.2011.05.089

Zhang, Y. L., & Wang, G. J. (2009). A geometric process repair model for a repairable cold standby system with priority in use and repair. *Reliability Engineering & System Safety, 94*, 1782–1787.

5

Phase-I design structure of Bayesian variance chart

Aamir Saghir[1*]

*Corresponding author: Aamir Saghir, Department of Mathematics, Mirpur University of Science and Technology (MUST), Mirpur, Pakistan
E-mail: aamirstat@yahoo.com

Reviewing editor: Zudi Lu, University of Southampton, UK

Abstract: This article develops a new design structure for S^2-Chart, namely Bayesian variance chart, in Phase-I analysis assuming the normality of the quality characteristic to incorporate the parameter uncertainty. Our approach consists of two stages: (i) construction of the control limits for S^2-Chart and (ii) performance evaluation of the proposed control limits. The comparison of the proposed design structure with the frequentist design structure of S^2-Chart is examined in terms of (i) width of control region and (ii) OC curves when the process variance goes out of control. It is observed that the proposed Phase-I S^2-Chart is more efficient than the frequentist S^2-Chart in discriminatory power of detecting a shift in the process dispersion. When the process variance is in-control (after implementation of Bayesian variance chart), then the control limits for $\bar{X}$-Chart using in-control standard deviation are also given here for monitoring unknown mean under unknown standard deviation case.

Subjects: Mathematics & Statistics; Science; Statistics; Statistics & Computing; Statistics & Probability

Keywords: process control; variance Chart; posterior distribution; OC function; predictive limits

1. Introduction

There is a large literature on the process variability control charts. To develop a variability control chart, a basic assumption is that the underlying distribution of the quality characteristics should be normal. Here, we assume that the lot-to-lot quality (process standard) observed after a fixed time interval remains constant throughout. The constant environmental stress on the operating conditions of the process over a long period leads to an unduly restrictive and unrealistic assumption about the constant standards of the process. The situation becomes alarming when one is going for quality control of the process of the same nature accomplishing the same task in varying conditions. Obviously, for overcoming the situation, it seems logical to assume variations in process standard represented by known suitable prior distribution. More so, the process control (PC) is a continuous quality valuation process and, as such, in all PC techniques, a strong prior information representing variations in quality is available as discussed by Sharma, Singh, and Geol (2007).

ABOUT THE AUTHOR

Aamir Saghir obtained his PhD degree in Statistics from the Department of Mathematics, Zhejiang University, Hangzhou, China, under CSC program. He is currently working in Mirpur University of Science and Technology (MUST), Mirpur, Pakistan, as a senior lecturer. His research interests include Statistical Quality Control, Mathematical Statistics, and Applied Statistics.

PUBLIC INTEREST STATEMENT

The products are produced every day in industry to fulfill the requirement of the common people. The most important issue of any produced products is its quality that is the major concern of any buyer. The quality control is an important field in industrial engineering. This work helps quality engineers to meet the challenges of buyers in the market. This work gives a very useful method to control the quality of outgoing item after meeting the specification.

Chhikara and Guttman (1982) developed a procedure for prediction limits for the inverse Gaussian distribution for both frequentists and a Bayesian viewpoint. Sharma and Bhutani (1992) extended the concept of modified classical consumer's risk and Bayes consumer's risk. In all these studies, the main emphasis has been to update the prior distribution with experimental data to get posterior distribution. Menzefricke (2002, 2007) developed control limits for $\bar{X}$-Chart and generalized variance chart based on posterior predictive distribution. Menzefricke (2010) developed a control chart for the variance of the normal distribution and equivalently, the coefficient of variation of a log-normal distribution. These control limits are referring to as prospective or Phase-II control limits because these limits are based on the future outcome (predictive inference) of the process. Sharma et al. (2007) discussed the performance of $\bar{X}$-Chart and R-Chart when standards vary randomly using OC function and *ARL* as performance measures. Saghir (2007) evaluated the performance of $\bar{X}$-Chart and S-Chart when standards vary randomly based on power function as performance measure. Recently, Saghir (2015) proposed a Phase-I control limits of $\bar{X}$-Chart based on posterior distribution of the statistic to monitor the mean of normal distribution. He concluded that when the process mean is statistically in-control for Phase-I samples based on these Bayesian limits, the Phase-II monitoring of the sample mean using Menzefricke (2002) control limits could be used.

This study proposes a new design structure for variance control chart namely S^2-Chart based on Bayesian approach for monitoring the Phase-I data following Menzefricke (2002, 2007, 2010), Chen, Morris, and Martin (2005), and Saghir (2007, 2015). The performance evaluation of S^2-Chart has been studied following Chhikara and Guttman (1982) and Saghir (2007, 2015). The rest of the article is summarized as follows. In Section 2, the control limits of S^2-Chart for Phase-I analysis based on posterior distribution for informative and non-informative priors are constructed. In Section 3, the evaluation performance of the constructed control limits is made. In Section 4, the comparison between the Bayesian variance chart and the frequentist S^2-Chart for Phase-I analysis is made through simulation. Section 5 developed the design structure of $\bar{X}$-Chart for simultaneous monitoring of process mean and standard deviation. Section 6 offers some concluding remarks and recommendations for further study. The necessary derivations of the distribution used in this study are provided in Appendix A.

2. The proposed design structure

The Phase-I design structure of S^2-Chart based on posterior distribution for monitoring process variation assuming the parameter uncertainty following Woodall and Montgomery (1999), Menzefricke (2002, 2007, 2010), and Saghir (2007, 2015) will be proposed in this section. In the following subsections, the control limits of S^2-Chart, using (i) informative prior and (ii) non-informative prior distributions, are constructed.

2.1. Construction of the control limits based on informative prior

Let $x_1, x_2, \ldots, x_n$ be a random sample of size n drawn from normal distribution with unknown mean μ and unknown variance σ^2. The probability density function and the likelihood function of random variable X are defined, respectively, as:

$$f(X|\mu,\sigma^2) = \frac{1}{\sqrt{2\pi\sigma}} e^{-\frac{1}{2\sigma^2}(x_i-\mu)^2}; \quad -\infty < x_i < \infty, \sigma > 0. \tag{1}$$

and

$$L(X|\mu,\sigma^2) = \prod_{i=1}^{n} f(x_i|\mu,\sigma^2) \tag{2}$$

There are two sources of information regarding unknown mean μ and variance σ^2, the prior information and results from a calibration sample, which help update the information about unknown parameters. The parameter μ has simple family of conjugate prior distribution (normal distribution). The parameter σ^2 does not have any simple family of conjugate prior distributions because its marginal likelihood depends in a complex way on the data (see Hill, 1965; Tiao & Tan, 1965). However, the inverse-gamma family is conditionally conjugate for σ^2 (Gelman, 2006). This conditionally conjugacy allows σ^2 to be updated easily using the Gibbs sampler (Gelfand & Smith, 1990) and also allows the prior distribution to be interpreted in terms of equivalent data (see Box & Tiao, 1973).

The usual priors used for unknown mean and precision (see Menzefricke, 2002, 2010) are:

$$\left.\begin{aligned} p(\mu|\sigma^2) &= N\left(m_0, \frac{\sigma^2}{n_0}\right) \\ p(\sigma^{-2}) &= Ga\left(\frac{v_0}{2}, \frac{v_0}{2}s_0^2\right) \end{aligned}\right\}, \tag{3}$$

where "N" and "Ga" denote normal and gamma distributions, respectively, and $\left\{m_0, \frac{\sigma^2}{n_0}\right\}, \left\{\frac{v_0}{2}, \frac{v_0}{2}s_0^2\right\}$ are hyper-parameters of the prior distributions as defined in Menzefricke (2002) and similar to hyper-parameter $\{1, \sigma_x^{-2}\}$ used for unknown precision, r, of Gaussian mixture model in Chen et al. (2005).

The prior distribution for variance can be derived from precision distribution as;

$$p\left(\sigma^2\right) = \frac{\left(\frac{v_0}{2}s_0^2\right)^{\frac{v_0}{2}}}{\Gamma\frac{v_0}{2}}\left(\frac{1}{\sigma^2}\right)^{\frac{v_0}{2}+1} e^{-\frac{\left(\frac{v_0}{2}s_0\right)^2}{\sigma^2}};0 < \sigma^2 < \infty \tag{4}$$

This density is known as inverse-gamma distribution as discussed in Bernardo and Smith (2004, p. 119), Spiegelhalter, Thomas, Best, Gilks, and Lunn (1994/2003) and Gelman (2006). Therefore, the prior for unknown mean and variance becomes:

$$\left.\begin{aligned} p(\mu|\sigma^2) &= N\left(m_0, \frac{\sigma^2}{n_0}\right) \\ p(\sigma^2) &= IG\left(\frac{v_0}{2}, \frac{v_0}{2}s_0^2\right) \end{aligned}\right\} \tag{5}$$

If the data $x_1, x_2, \ldots, x_n$ are from stable process, the sufficient statistics are the sample mean $\bar{x} = \sum_{i=1}^{n} x_i/n$ and variance $S_x^2 = \sum_{i=1}^{n}(x_i - \bar{x})^2/(n-1)$ for a calibration sample of size n (see Menzefricke, 2010). The joint distribution of sample mean and variance, i.e. $f(\bar{x}, s_x^2|\mu, \sigma^2)$, is thus:

$$f(\bar{x}, s_x^2|\mu, \sigma^2) = f(\bar{x}|\mu, \sigma^2)f(s_x^2|\sigma^2) \tag{6}$$

where $f(\bar{x}|\mu, \sigma^2) = N\left(\mu, \frac{\sigma^2}{n}\right)$ and $f(s_x^2|\sigma^2) = Ga\left(\frac{n-1}{2}, \frac{n-1}{2\sigma^2}\right)$ as defined in Menzefricke (2002, 2010).

Following the methodology of Aitchison and Dunsmore (1975), the posterior distribution of $f(\mu, \sigma^2)$ given $(\bar{x}, s_x^2)$ is:

$$\begin{aligned} p(\mu, \sigma^2|\bar{x}, S_x^2) &= \frac{\sqrt{n_c}}{\sigma\sqrt{2\pi}} e^{-\frac{n_c}{2\sigma^2}(\mu - m_1)^2}\left(\frac{\beta}{\Gamma\alpha}\right)^{\alpha}\left(\frac{1}{\sigma^2}\right)^{\alpha+1} e^{-\left(\frac{\beta}{\sigma^2}\right)} \quad -\infty < \mu < \infty,\ 0 < \sigma^2 < \infty. \\ &= p(\mu|\bar{x}, \sigma^2)p(\sigma^2|s_x^2) \end{aligned} \tag{7}$$

The derivation of (7) is given in Appendix A. The random variable $\mu|(\bar{x}, \sigma^2)$ is normally distributed random variable with mean $E(\mu|(\bar{x}, \sigma^2)) = m_1$ (posterior mean) and variance $V(\mu|(\bar{x}, \sigma^2)) = \frac{\sigma^2}{n_c}$

(posterior variance); $\sigma^2 \left| S_x^2 \right.$ is inverse-gamma distributed random variable with parameters $\alpha = \frac{v_1}{2}$, $\beta = \left(\frac{v_0}{2} S_0^2 + \frac{n-1}{2} s_x^2 + \frac{n_0 n}{n_0+n} (m_0 - \bar{x})^2 \right)$ and the values of v_1, m_1, and n_c are defined in Appendix A.

The lower control limit, central line, and upper control limit are the three parameters of a Shewhart-type control chart. Assuming the measurable quality characteristic, X, to be normally distributed with unknown mean μ and unknown variance σ^2, the control limits for the usual S^2-Chart for retrospective analysis are defined as:

$$\left.\begin{array}{l} LCL = \frac{\overline{S_x^2}}{n-1} \chi^2_{(1-\frac{\alpha}{2})} \\ CL = \overline{S_x^2} \\ UCL = \frac{\overline{S_x^2}}{n-1} \chi^2_{(\frac{\alpha}{2})} \end{array}\right\}, \tag{8}$$

where α is the level of significance or false alarm rate in quality terminology, $\overline{S_x^2}$ is the average of variance of k-subgroups each of size n, $\chi^2_{(1-\frac{\alpha}{2})}$ and $\chi^2_{(\frac{\alpha}{2})}$ are the quantile points of the chi-square distribution as defined in Montgomery (2004, p. 249).

Now in a situation where variations in process variance σ^2 are assumed to be represented by a prior distribution given in (5) and using the sample information variation, the process variance can be updated in the form of posterior distribution as given in (7). The usual three-sigma control limits, for S^2-Chart using the updated posterior distribution, are given by:

$$\begin{array}{l} LCL = \frac{\beta}{\alpha - 1} - 3 \frac{\beta}{(\alpha-1)(\alpha-2)^{\frac{1}{2}}} \\ CL = \frac{\beta}{\alpha - 1} \\ UCL = \frac{\beta}{\alpha - 1} + 3 \frac{\beta}{(\alpha-1)(\alpha-2)^{\frac{1}{2}}} \end{array} \tag{9}$$

where the parameters $\alpha = \frac{v_1}{2}$, $\beta = \left(\frac{v_0}{2} S_0^2 + \frac{n-1}{2} s_x^2 + \frac{n_0 n}{n_0+n} (m_0 - \bar{x})^2 \right)$ and the values of v_1, m_1, and n_c are defined in Appendix A.

The distribution of sufficient statistic S_x^2 is not a symmetric even for moderate to large samples. Therefore, the true probability limits for S^2-Chart using posterior distribution following Chhikara and Guttman (1982) are given by:

$$\left.\begin{array}{l} \int_{-\infty}^{LCL} p(\sigma^2 \left| S_x^2 \right.)\, d\sigma^2 = \frac{\alpha}{2} \\ \int_{UCL}^{\infty} p(\sigma^2 \left| S_x^2 \right.)\, d\sigma^2 = \frac{\alpha}{2} \end{array}\right\}, \tag{10}$$

where $p(\sigma^2 \left| S_x^2 \right.)$ is as given in (7) and α is the level of significance, which is 0.0027 in Shewhart-type control charts. We are updating the control limits of S^2-Chart for monitoring the unknown process variance. So, the plotted statistic is the sample variance S_x^2 because it is an unbiased estimator of σ^2 which is a variable of interest.

2.2. Control limits for $\bar{X}$-Chart based on non-informative prior

In this subsection, we have proposed control limits for S^2-Chart based on non-informative prior distribution. We have considered uniform prior and Jeffrey's prior as non-informative prior.

2.2.1. Control limits based on Jeffrey's prior

In this subsection, we have proposed control limits for S^2-Chart based on Jeffrey's prior for unknown mean and variance of the normal distribution, so that we are then relying primarily on the likelihood involved for our inference following the work of Box (1980), Chhikara and Guttman (1982), and Gelman (2006).

Consider the sampling distribution of X, likelihood function of X, and sampling distribution of $(\bar{X}, S^2)$ as defined in Equations (1), (2), and (6). The Jeffrey's prior for unknown mean and variance as discussed in Banerjee and Bhattacharyya (1979) and Gelman (2006) is:

$$p(\mu, \sigma^2) \infty \sigma^{-}2 \quad -\infty < \mu < \infty, 0 < \sigma^2 < \infty. \tag{11}$$

where σ^{-2} is the precision of normal distribution. Then, following the methodology of Menzefricke (2002, 2007, 2010), the posterior distribution of $f(\mu, \sigma^2)$ given $(\bar{x}, s_x^2)$is defined as:

$$\begin{aligned} p(\mu, \sigma^2 \middle| \bar{x}, S_x^2) &= \frac{\sqrt{n}}{\sigma\sqrt{2\pi}} e^{-\frac{n}{2\sigma^2}(\mu-\bar{x})^2} \left(\frac{\beta}{\Gamma\alpha}\right)^{\alpha} \left(\frac{1}{\sigma^2}\right)^{\alpha+1} e^{-\left(\frac{\beta}{\sigma^2}\right)}; -\infty < \mu < \infty, 0 < \sigma^2 < \infty. \\ &= p(\mu \middle| \bar{x}, \sigma^2) p(\sigma^2 \middle| s_x^2) \end{aligned} \tag{12}$$

where $\mu|(\bar{X}, \sigma^2)$ is normally distributed random variable with mean $E(\mu \middle| (\bar{X}, \sigma^2)) = \bar{x}$ and variance $V(\mu \middle| (\bar{X}, \sigma^2)) = \frac{\sigma^2}{n}$; $\sigma^2 \middle| S_x^2$ is inverse-gamma distributed random variable with parameters$\alpha = \frac{n-1}{2}$, $\beta = (\frac{n-1}{2} S_x^2)$. The derivation of (12) is given in Appendix A.

Now in a situation where variation in process variance σ^2 is assumed to be represented by a prior distribution given in (11) and using the sample information variation, the process variance can be updated in the form of posterior distribution as given in (12). The true probability control limits for S^2-Chart using posterior distribution using Jeffrey's non-informative prior are given by:

$$\left. \begin{aligned} \int_{-\infty}^{LCL} p(\sigma^2 \middle| S_x^2)\, d\sigma^2 = \frac{\alpha}{2} \\ \int_{UCL}^{\infty} p(\sigma^2 \middle| S_x^2)\, d\sigma^2 = \frac{\alpha}{2} \end{aligned} \right\}, \tag{13}$$

where $p(\sigma^2 \middle| S_x^2)$ is as given in (12) and α is the level of significance, which is 0.0027 in Shewhart-type control charts.

2.2.2. Control limits based on uniform prior

In this subsection, we have proposed control limits for S^2-Chart based on uniform prior distribution for unknown mean and variance of the normal distribution, so that we are then relying primarily on the likelihood involved for our inference following the work of Box (1980), Chhikara and Guttman (1982), and Gelman (2006).

The uniform prior for unknown mean and variance is $p(\mu, \sigma^2) \infty 1$. The posterior distribution of $f(\mu, \sigma^2)$ given $(\bar{x}, s_x^2)$is defined as:

$$p(\mu,\sigma^2|\bar{x},S_x^2) = \frac{n}{\sigma\sqrt{2\pi}}e^{-\frac{n}{2\sigma^2}(\mu-\bar{x})^2}\left(\frac{\beta}{\Gamma\alpha}\right)^{\alpha}\left(\frac{1}{\sigma^2}\right)^{\alpha+1}e^{-\left(\frac{\beta}{\sigma^2}\right)} \quad -\infty<\mu<\infty,\ 0<\sigma^2<\infty. \tag{14}$$
$$= p(\mu|\bar{x},\sigma^2)p(\sigma^2|s_x^2)$$

where $\mu|(\bar{x},\sigma^2)$ is normally distributed random variable with mean $E(\mu|(\bar{x},\sigma^2))=\bar{x}$ and variance $V(\mu|(\bar{x},\sigma^2))=\frac{\sigma^2}{n}$; $\sigma^2|S_x^2$ is inverse-gamma distributed random variable with parameters $\alpha=\frac{n-3}{2}$,$\beta=(\frac{n-1}{2}S_x^2)$.

The control limits for S^2-Chart using Bayesian inference following Chhikara and Guttman (1982) based on uniform prior distribution are given by:

$$\left.\begin{aligned}\int_{-\infty}^{LCL} p(\sigma^2|S_x^2)\,d\sigma^2 &= \frac{\alpha}{2}\\ \int_{UCL}^{\infty} p(\sigma^2|S_x^2)\,d\sigma^2 &= \frac{\alpha}{2}\end{aligned}\right\}, \tag{15}$$

where $p(\sigma^2|S_x^2)$ is as given in (14) and α is the level of significance, which is 0.0027 in Shewhart-type control charts.

The limits defined in Equations (10), (13), and (15) are based on updated information and can be used to monitor the unknown normal process variance. The successive sample of a given size n are observed if a sample variance is outside the control limits defined in Equations (10), (13), and (15) chart signals. If a sample variance or more sample variances falls outside the control limits, the trial control limits will be revised until all the sample variances lie within the control region. When the process is in-control, the future sample variances are generated from the posterior predictive distribution as proposed by Menzefricke (2010); otherwise, the use of these will be misleading as discussed by many authors including Montgomery (2004) and Saghir (2015). This work is related to the Phase-I monitoring of S^2-Chart while the work of Menzefricke (2010) is for Phase-II monitoring. The Phase-I monitoring is necessary before applying the Phase-II study, see Montgomery (Box, 1980).

3. Evaluation of control limits

For the evaluation of the control limits obtained by frequentist (sampling) and Bayesian methods, we use OC function under the hypothetical situation that the variance of the normal distribution does not remain at level σ^2 following Sharma et al. (2007), Saghir (2007), and Menzefricke (2010). At this stage, we use sampling theoretical considerations approach following Menzefricke, (2002, 2007, 2010) and Saghir (2007, 2015). Let the sampling distribution for the next sample variance be $IG(\alpha, \frac{\beta}{\alpha_2\beta_0^2})$ where α_2 is an amount of shift in the process variance σ_0^2.

The OC function in the corresponding situations provides a measure of the sensitivity of the control limits, i.e. their ability to detect a shift in the mean of the process quality characteristic following Menzefricke (2002, 2010) and Sharma et al. (2007). The relative distribution of $\sigma^2|S_x^2$ to shifted variance $\alpha_2\sigma_0^2$ is derived in Appendix A. The OC function for S^2-Chart based on the control limits defined in (10) and relative distribution is defined as:

$$\begin{aligned}B_I &= \Pr[\text{not detecting a shift in the variance of data generating process}\\ &\quad \text{(using informative prior) on the first sample following the shift}]\\ &= \Pr[LCL \le \sigma_0^2 \le UCL|\alpha_2\sigma_0^2]\\ &= \int_{LCL}^{UCL} \alpha_2\sigma_0^2 * IG(\alpha, \frac{\beta}{\alpha_2\sigma_0^2})d\sigma^2\end{aligned} \tag{16}$$

where $\alpha = \frac{v_1}{2}, \beta = (\frac{v_0}{2}S_0^2 + \frac{n-1}{2}S_x^2 + \frac{n_0 n}{n_0+n}(m_0 - \bar{x})^2)$ and "IG" denotes inverse-gamma distribution and a_2 denotes the amount of shift in variance σ_0^2.

The OC function for S^2-Chart based on the control limits defined in Equation (13) and relative distribution is defined as:

$$B_{JNI} = \Pr \text{[not detecting a shift in the variance of data generating process (using Jeffrey's non-informative prior) on the first sample following the shift]}$$

$$= \Pr [LCL \leq S_x^2 \leq UCL | \alpha_2 \sigma_0^2]$$

$$B_{JNI} = \int_{LCL}^{UCL} IG(\frac{n-1}{2}, \frac{(n-1)S_x^2}{2\,\alpha_2\sigma_0^2}) d\sigma^2 \quad (17)$$

The OC function for S^2-Chart based on the control limits defined in Equation (15) and relative distribution is defined as:

$$B_{UNI} = \Pr \text{[not detecting a shift in the variance of data generating process (using Uniform non- informative prior) on the first sample following the shift]}$$

$$= \Pr [LCL \leq S_x^2 \leq UCL | \alpha_2 \sigma_0^2]$$

$$B_{UNI} = \int_{LCL}^{UCL} IG(\frac{n-3}{2}, \frac{(n-1)S_x^2}{2\,\alpha_2\sigma_0^2}) d\sigma^2 \quad (18)$$

The corresponding OC function for usual control limits of S^2-Chart is:

$$B_F = \Pr \text{[not detecting a shift in the variance of data generating process on the first sample following the shift]}$$

$$= \Pr \left[LCL \leq S_x^2 \leq UCL \middle| a_2\sigma_0^2\right]$$

$$\int_{LCL}^{UCL} Ga(\frac{n-1}{2}, \frac{n-1}{2a_2\sigma_0^2}) dS_x^2. \quad (19)$$

where "Ga" denotes gamma distribution with parameters $\frac{n-1}{2}$ and $\frac{n-1}{2a_2\sigma_0^2}$.

4. Simulation study and comparison

In this section, we have made a comparison among the proposed (informative and non-informative) control limits and usual control limits of S^2-Chart. First, we have made comparison based on control region obtained by the proposed and usual limits following Chhikara and Guttman (Sharma et al., 2007) and then the comparison based on OC S_x^2 function following Menzefricke (2002, 2007, 2010), Sharma et al. (2007), and Saghir (2007, 2015) is provided.

4.1. *Control region-based comparison*

In this subsection, a comparison of the proposed limits with the frequentist control limits of S^2-Chart for the initials samples is made. The comparison based on width of the control limits is made for Phase-I process monitoring. Using Monte Carlo simulation technique, 10,000 random samples are drawn from standard normal process of size nine and sample variance along standard error is calculated and given in Tables 1–3. Let us assume that (i) α = 0.0027 and (ii) $\bar{\bar{x}} = m_0$ (without loss of generality). We have considered three situations when (i) $S_x^2 = S_0^2$, (ii) $S_x^2 < S_0^2$, and (iii) $S_x^2 > S_0^2$. The critical region defined in Equation (10) changes with the posterior degree of freedom v_1 which is a measure of the amount of uncertainty regarding the unknown process variance, σ^2 (see Menzefricke, 2002, 2010). The control limits $v_1$for S^2-Chart for different values of v_1 based on these random samples and updated posterior distribution are calculated and given here in Tables 1–3 for comparison purposes.

Tables 1–3 give the following considerable points:

(1) The width of the control region for the proposed limits decreases as posterior degree of freedom v_1 increases (see last column of Tables 1–3).

(2) Comparison across the informative prior distribution-based limits reveals that the control limits are more contract in terms of minimum width of the control region when $S_x^2 = S_0^2$ than other choices (see Tables1–3).

(3) The control limits obtained by non-informative prior distributions, which do not incorporate the parameter uncertainty (Menzefricke, 2002, 2010), are wider than frequentist control limits (see first and last two rows of Tables 1–3).

(4) As the posterior degree of freedom v_1 (which is a quantitative assessment of how much certain we are about the accuracy of the prior variance "S_0^2", prior estimate of σ^2, lower the variance,

Table 1. Control limits of S^2-Chart for S_x^2 = 0.92884 (0.0034) and S_0^2 = 0.92884

Control limits of S^2-Chart		(LCL, UCL)	Width
Frequentist		(0.13100, 2.73710)	2.60610
IN prior based	$v_1 = 10$	(0.34524, 4.98952)	4.64428
	$v_1 = 15$	(0.37851, 3.80397)	3.42546
	$v_1 = 20$	(0.41885, 3.01139)	2.59254
	$v_1 = 40$	(0.51452, 2.02075)	1.50623
JNI prior based		(0.29299, 7.89839)	7.60540
UNI prior based		(0.34179, 17.54890)	17.20711

Table 2. Control limits of S^2-Chart for S_x^2 = 0.92884 (S.E = 0.0034), S_0^2 = 0.5000

Control limits of S^2-Chart		(LCL, UCL)	Width
Frequentist		(0.13100, 2.73710)	2.60610
IN prior based	$v_1 = 10$	(0.29289, 5.32285)	5.02996
	$v_1 = 15$	(0.29696, 2.98429)	2.68733
	$v_1 = 20$	(0.30282, 2.17721)	1.87439
	$v_1 = 40$	(0.32448, 1.27438)	0.94990
JNI prior based		(0.29299, 7.89839)	7.60540
UNI prior based		(0.34179, 17.54890)	17.20711

Table 3. Control limits of S^2-Chart for S_x^2 = 0.92884 (S.E = 0.0034), S_0^2 = 1.5000

Control limits of S^2-Chart		(LCL, UCL)	Width
Frequentist		(0.13100, 2.73710)	2.60610
IN prior based	$v_1 = 10$	(0.36234, 6.58564)	6.22330
	$v_1 = 15$	(0.48714, 4.89559)	4.40845
	$v_1 = 20$	(0.57338, 4.12246)	3.54908
	$v_1 = 40$	(0.76764, 3.01483)	2.24719
JNI prior based		(0.29299, 7.89839)	7.60540
UNI prior based		(0.34179, 17.54890)	17.20711

more accurate our believe) increases, the width of the control limits obtained based on informative prior distribution decreases.

4.2. OC curve comparison

In this subsection, we evaluate the proposed control limits of S^2 -Chart using OC curve as performance measures. Using the control limits, calculated in Section 4.1, OC function is calculated for different amounts of shifts (i.e. different values of α_2), $\alpha = 0.0027$, $n = 9$, and OC curves are made and provided here in Figures 1–3.

where "*F*" denotes OC function based on usual control limits, "JNI" OC function based on Jeffrey's non-informative prior distribution limits, "UNI" OC function based on uniform non-informative prior distribution limits, and "10, 15, 20, 40" based on informative prior distribution degrees of freedom limits.

From Figures 1–3, it is obvious that the proposed control limits based on informative prior distribution perform better than the frequentist control limits of S^2-Chart in the sense that the discriminatory power of detecting a shift in the parameter of interest is high for proposed control limits than the existing control limits of S^2-Chart. The higher power of detecting a shift results in low value of OC function; therefore, the curve of OC function for the proposed control limits is less than the frequentist control limits as it is clear from Figures 1–3.

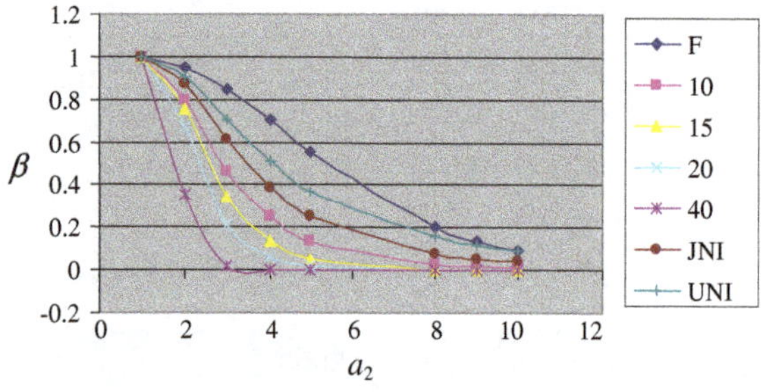

Figure 1. OC curve for S^2-Chart at $S_x^2 = S_0^2 = 0.92884$.

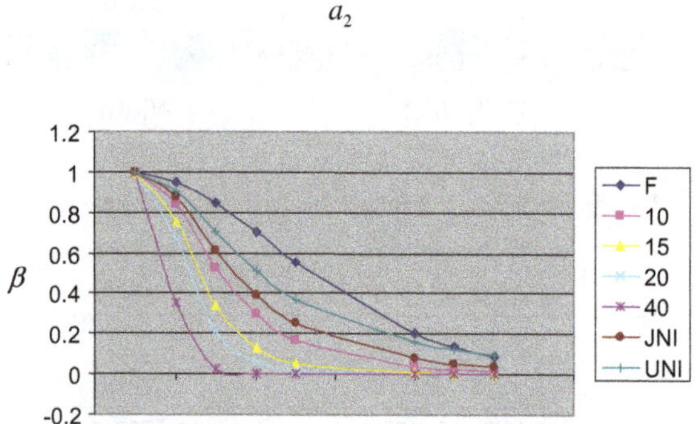

Figure 2. OC curve for S^2-Chart at $S_0^2 = 0.50000$ and $S_x^2 = 0.92884$.

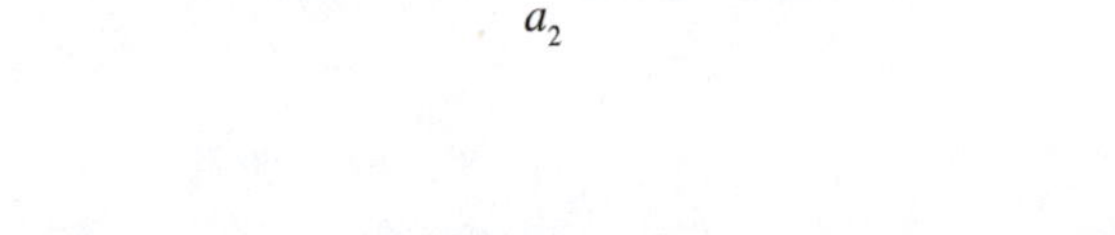

Figure 3. OC curve for S^2-Chart at $S_0^2 = 1.50000$ and $S_x^2 = 0.92884$.

The performance of non-informative priors-based control limits of S^2-Chart to detect a shift in the parameter is also better than the frequentist control limits as it is obvious from above figures but less than informative prior-based control limits' performance. Therefore, control limits of S^2-Chart based on informative prior perform better to detect a shift in a parameter of the continuous process for larger value of posterior degree of freedom v_1. The more the posterior degree of freedom v_1, more the power of detecting a shift for informative prior distribution limits, as it can be seen from Figures 1–3. A similar behavior has been observed for other choices of sample sizes and in-control false alarm rate α.

5. Control limits for $\bar{X}$-Chart when variance is unknown

In this section, we are constructing and evaluating the control limits of Bayesian $\bar{X}$-Chart based on posterior distribution for Phase-I process monitoring following Saghir (2015).

5.1. Construction of the limits

The control limits for Bayesian $\bar{X}$-Chart, based on the updated posterior distribution defined in Equation (7), when process mean as well as process variance is unknown, are:

$$\begin{aligned} LCL_B &= \frac{n_0 m_0 + n\bar{x}}{n+n_0} - Z_{\left(\frac{\alpha}{2}\right)} \frac{\sigma}{\sqrt{n+n_0}} \\ CL &= \frac{n_0 m_0 + n\bar{x}}{n+n_0} \\ UCL_B &= \frac{n_0 m_0 + n\bar{x}}{n+n_0} + Z_{\left(\frac{\alpha}{2}\right)} \frac{\sigma}{\sqrt{n+n_0}}, \end{aligned} \tag{20}$$

where $Z_{(\alpha/2)}$ is a $(\alpha/2)$th quantile point of the standard normal distribution. For the usual choice of in-control probability of false alarm rate $\alpha = 0.0027$, $Z_{(\alpha/2)} = 3.00$ and control limits defined in Equation (20) are know as 3σ-control limits. As the process standard deviation or variance is unknown, therefore, we replace unknown standard deviation by any unbiased estimator like $\hat{\sigma} = \frac{\bar{R}}{d_2}$ or $\hat{\sigma} = \frac{\bar{S}}{c_4}$ given in many textbooks including Montgomery (2004). We are updating the control limits of $\bar{X}$-Chart for monitoring the unknown process mean. So, the plotted statistic is the sample average $\bar{X}$ because it is an unbiased estimator of unknown population mean μ, which is the variable of interest to be monitored.

5.2. Evaluation of the limits

For the evaluation of the proposed control limits of $\bar{X}$-Chart, we have used OC function as performance measures under the hypothetical situation that the mean of the normal distribution does not remain at level μ following Saghir (2007, 2015). Let the sampling distribution for the next sample mean be $N(M, \frac{\sigma^2}{n})$ where M is the value of the in-control or out-of-control process mean and the process standard deviation is in-control at $\hat{\sigma} = \frac{\bar{R}}{d_2}$ or $\hat{\sigma} = \frac{\bar{S}}{c_4}$. The OC function in the corresponding situations provides a measure of the sensitivity of the control limits, i.e. their ability to not detect a shift in the mean of the process quality characteristic following Menzefricke (2002, 2007, 2010), Saghir (2007), and Saghir (2015). The OC function for $\bar{X}$-Chart based on the control limits defined in (20) is defined as:

$$\begin{aligned} &B_I N = \text{Pr [not detecting a shift in the sample mean} \\ &\text{(using informative Prior) on the first sample following the shift]} \\ &\Pr\left[LCL_B \le \bar{X} \le UCL_B | M\right] \\ &\phi\left(\frac{UCL_B - M}{\hat{\sigma}/\sqrt{n}}\right) - \phi\left(\frac{LCL_B - M}{\hat{\sigma}/\sqrt{n}}\right) \end{aligned} \tag{21}$$

Let us assume that (i) $M = m_0 + b$ where b is an amount of shift occurring in the in-control mean m_0, (ii) $\alpha = 0.0027$, (iii) $\hat{\sigma} = 1$ (without loss of generality), and the posterior mean of μ, m_1 to be equal m_0; then, Equation (21) reduces to

$$B_I N = \phi\left(-b\sqrt{n} + 3\sqrt{\frac{n}{n+n_0}}\right) - \phi\left(-b\sqrt{n} - 3\sqrt{\frac{n}{n+n_0}}\right) \quad (22)$$

For the given values of b and n, the OC function decreases with prior sample size n_0 and approaches 0 as $n_0 \to \infty$. A larger value of n_0 implies more precise knowledge about m_0 and produces thus a narrower control region.

6. Conclusions and recommendations

The proposed posterior distribution-based design structure of S^2-Chart, which incorporates the parameter uncertainty by considering a suitable prior distribution of unknown parameter, is more efficient than the frequentist design structure, which ignores this uncertainty and assumes that $\sigma^2 = S^2$, with reference to the width of the limits, lowest type-I error, and more power of detecting a shift in the parameter. Larger values of posterior degree of freedom v_1 provided more efficient control limits in terms of lowest width of control region as well as more discriminatory power of detecting a shift in the parameter when actually the shift occurs in the parameter. The control limits of S^2-Chart based on informative prior are more efficient than non-informative prior-based control limits and usual control limits. The performance of the usual control limits is least among the compared control limits. These control limits must be calculated for Phase-I data and when the process variance is statistically in-control, the control limits proposed by Menzefricke (2010) should be used. The control limits of $\bar{X}$-Chart are also constructed when mean and standard deviation of the normal process are unknown. The constructed control limits are evaluated and it has been observed that a larger value of n_0 implies more precise knowledge about m_0. When the process mean is statistically in-control for Phase-I samples based on these Bayesian limits, the Phase-II monitoring of the sample mean using Menzefricke (2002) control limits could be used.

Funding
The author received no direct funding for this research.

Author details
Aamir Saghir[1]
E-mail: aamirstat@yahoo.com
[1] Department of Mathematics, Mirpur University of Science and Technology (MUST), Mirpur, Pakistan.

References

Aitchison, J., & Dunsmore, I. R. (1975). *Statistical prediction analysis*. Cambridge: Cambridge University Press. http://dx.doi.org/10.1017/CBO9780511569647

Banerjee, A. K., & Bhattacharyya, G. K. (1979). Bayesian results for the inverse Gaussian distribution with an application. *Technometrics, 21*, 247–251. http://dx.doi.org/10.1080/00401706.1979.10489756

Bernardo, M. J., & Smith, M. F. A. (2004). *Bayesian theory. Wiley series in probability and statistics* (Ist ed.). New York: Wiley.

Box, G. E. P. (1980). Sampling and Bayes' inference in scientific modelling and robustness. *Journal of the Royal Statistical Society. Series A (General), 143*, 383–430. http://dx.doi.org/10.2307/2982063

Box, G. E. P., & Tiao, G. C. (1973). *Bayesian inference in statistical analysis*. Reading, MA: Addison-Wesley.

Chen, T., Morris, J., & Martin, E. (2005). Bayesian control limits for statistical process monitoring. In *ICCA*. Budapest.

Chhikara, R. S., & Guttman, I. (1982). Prediction limits for the inverse Gaussian distribution. *Technometrics, 24*, 319–324. http://dx.doi.org/10.2307/1267827

Gelfand, A. E., & Smith, A. F. M. (1990). Sampling-based approaches to calculating marginal densities. *Journal of the American Statistical Association, 85*, 398–409. http://dx.doi.org/10.1080/01621459.1990.10476213

Gelman, A. (2006). Prior distributions for variance parameters in hierarchical models. *Bayesian Analysis, 1*, 515–533.

Hill, B. M. (1965). Inference about variance components in the one-way model. *Journal of the American Statistical Association, 60*, 806–825. http://dx.doi.org/10.1080/01621459.1965.10480829

Menzefricke, U. (2002). On the evaluation of control chart limits based on predictive distributions. *Communications in Statistics - Theory and Methods, 31*, 1423–1440. http://dx.doi.org/10.1081/STA-120006077

Menzefricke, U. (2007). Control charts for the generalized variance based on its predictive distribution. *Communications in Statistics - Theory and Methods, 36*, 1031–1038. http://dx.doi.org/10.1080/03610920601036176

Menzefricke, U. (2010). Control charts for the variance and coefficient of variation based on their predictive distribution. *Communications in Statistics - Theory and Methods, 39*, 2930–2941. http://dx.doi.org/10.1080/03610920903168610

Montgomery, D. C. (2004). *Introduction to statistical quality control* (4th ed.). New York, NY: Wiley.

Saghir, A. (2007). Evaluation of ¯X and S charts when standards vary randomly. *Interstat, 4*.

Saghir, A. (2015). Phase-I design scheme for -chart based on posterior distribution. *Communications in Statistics - Theory and Methods, 44*, 644–655. http://dx.doi.org/10.1080/03610926.2012.752846

Sharma, K. K., & Bhutani, R. K. (1992). A comparison of classical and Bayes risks when the quality varies randomly. *Microelectronics Reliability, 32*, 493–495. http://dx.doi.org/10.1016/0026-2714(92)90479-5

Sharma, K. K., Singh, B., & Geol, J. (2007). Analysis of ¯X and R charts when standards vary randomly. In B. N. Pandey (Ed.), *Statistical techniques in life testing, reliability, sampling theory and quality control*.

Spiegelhalter, D. J., Thomas, A., Best, N. G., Gilks, W. R., & Lunn, D. (1994/2003). *BUGS: Bayesian inference using Gibbs sampling*. Cambridge: MRC Biostatistics Unit.

Tiao, G. C., & Tan, W. Y. (1965). Bayesian analysis of random-effect models in the analysis of variance. I: Posterior distribution of variance components. *Biometrika, 52*, 37–54. http://dx.doi.org/10.1093/biomet/52.1-2.37

Woodall, W. H., & Montgomery, D. C. (1999). Research issues and ideas in statistical process control. *Journal of Quality Technology, 31*, 376–386.

Appendix A

In this Appendix, we have given the derivation of the distributions used in the development and comparison of the proposed Bayesian control limits of S^2-Chart.

(1) Posterior distribution of $\sigma^2 \left| S_x^2 \right.$ based on informative prior

The sampling distribution of sufficient statistics $\bar{x}$and S^2, Menzefricke (2002, 2010) is:

$$f(\bar{x}\left|\mu,\sigma^2\right.) = N(\mu, \tfrac{\sigma^2}{n}), \qquad -\infty < \bar{x} < \infty.$$
$$f(s_x^2\left|\sigma^2\right.) = Ga(\tfrac{n-1}{2}, \tfrac{n-1}{2\sigma^2}), \qquad 0 < S^2 < \infty.$$

The prior distributions of μ and σ^2 are defined in Equation (15), so the posterior distribution of μ and σ^2 given that $\bar{x}$and S^2 will be:

$$p(\mu,\sigma^2\left|\bar{x},S_x^2\right.)\infty p(\bar{x},S_x^2\left|\mu,\sigma^2\right.)p(\mu\left|\sigma^2\right.)p(\sigma^2)$$

$$\Rightarrow p(\mu,\sigma^2\left|\bar{x},S_x^2\right.)\infty e^{-\frac{n}{2\sigma^2}(\bar{x}-\mu)^2} e^{-\frac{n_0}{2\sigma^2}(\mu-m_0)^2}\left(\frac{1}{\sigma^2}\right)^{(\frac{n-1}{2})} e^{-\left(\frac{n-1}{2\sigma^2}\right)S_x^2}\left(\frac{1}{\sigma^2}\right)^{\frac{v_0}{2}+1} e^{-\left(\frac{v_0 s_0^2}{2\sigma^2}\right)} \qquad (1.1)$$

$$\Rightarrow p(\mu,\sigma^2\left|\bar{x},S_x^2\right.)\infty e^{-\frac{n_c}{2\sigma^2}(\mu-m_1)^2}\left(\frac{1}{\sigma^2}\right)^{(\frac{n-1}{2}+\frac{v_0}{2}+1)} e^{-\left(\frac{n-1}{2\sigma^2}\right)S_x^2-\left(\frac{v_0 s_0^2}{2\sigma^2}\right)}$$

$$\Rightarrow p(\mu,\sigma^2\left|\bar{x},S_x^2\right.)\infty e^{-\frac{n_c}{2\sigma^2}(\mu-m_1)^2}\left(\frac{1}{\sigma^2}\right)^{(\frac{v_1}{2}+1)} e^{-\left(\frac{(n-1)S_x^2+v_0 s_0^2}{2\sigma^2}\right)}$$

where $v_1 = v_0 + n - 1, m_1 = \frac{(n_0 m_0 + n\bar{x})}{n_c}$ and $n_c = n + n_0$.

The proportional density of $(\mu,\sigma^2\left|\bar{x},S_x^2\right.)$ is a product of two random variables: (i) $\mu|(\bar{x},\sigma^2)$ is normally distributed random variable with mean $E(\mu\left|(\bar{x},\sigma^2)\right.) = m_1$ and variance $V(\mu\left|(\bar{x},\sigma^2)\right.) = \frac{\sigma^2}{n_1}$ and (ii) $\sigma^2\left|S_x^2\right.$ is inverse-gamma distributed random variable with parameters $\alpha = \frac{v_1}{2}, \beta = (\frac{v_0}{2}S_0^2 + \frac{n-1}{2}s_x^2 + \frac{n_0 n}{n_0+n}(m_0-\bar{x})^2)$.

(2) Posterior distribution of $\sigma^2 \left| S_x^2 \right.$ based on Jeffrey's non-informative prior

Considering the sampling distribution of sufficient statistics $\bar{x}$and S^2 defined above and the prior distributions of μ and σ^2as defined in Equation (11), then the posterior distribution of μ and σ^2 given that $\bar{x}$and S^2 is:

$$p(\mu,\sigma^2\left|\bar{x},S_x^2\right.)\infty p(\bar{x},S^2\left|\mu,\sigma^2\right.)p(\mu,\sigma^2)$$

$$\Rightarrow p(\mu,\sigma^2\left|\bar{x},S_x^2\right.)\infty e^{-\frac{n}{2\sigma^2}(\bar{x}-\mu)^2}\left(\frac{1}{\sigma^2}\right)^{(\frac{n-1}{2})} e^{-\left(\frac{n-1}{2\sigma^2}\right)S_x^2}\left(\frac{1}{\sigma^2}\right)$$

$$\Rightarrow p(\mu,\sigma^2\left|\bar{x},S_x^2\right.)\infty e^{-\frac{n}{2\sigma^2}(\bar{x}-\mu)^2}\left(\frac{1}{\sigma^2}\right)^{(\frac{n-1}{2})+1} e^{-\left(\frac{n-1}{2\sigma^2}\right)S_x^2} \qquad (2.1)$$

$$\Rightarrow p(\mu,\sigma^2\left|\bar{x},S_x^2\right.)\infty e^{-\frac{n}{2\sigma^2}(\bar{x}-\mu)^2}\left(\frac{1}{\sigma^2}\right)^{\alpha+1} e^{-\left(\frac{\beta}{\sigma^2}\right)}$$

where the proportional density is a combination of two densities: (i) $\mu|(\bar{x},\sigma^2)$ is normally distributed random variable with mean $E(\mu|(\bar{x},\sigma^2))=\bar{x}$ and variance $V(\mu|(\bar{x},\sigma^2))=\frac{\sigma^2}{n}$ and (ii) $\sigma^2|S_x^2$ is inverse-gamma distributed random variable with parameters $\alpha=\frac{n-1}{2}$, $\beta=(\frac{n-1}{2}S_x^2)$.

(3) Posterior distribution of $\sigma^2|S_x^2$ based on uniform non-informative prior

The sampling distribution of sufficient statistics $\bar{x}$ and S^2, Menzefricke (2002) is:

$f(\bar{x}|\mu,\sigma^2)=N(\mu,\frac{\sigma^2}{n}),\ -\infty<\bar{x}<\infty.$
$f(s_x^2|\sigma^2)=Ga(\frac{n-1}{2},\frac{n-1}{2\sigma^2}),\ 0<S^2<\infty.$

The posterior distribution of μ and $\bar{x}$ and S^2, by considering the prior distributions of μ and σ^2 as defined in Section 2.2.2, is:

$$p(\mu,\sigma^2|\bar{x},S_x^2)\propto p(\bar{x},S^2|\mu,\sigma^2)p(\mu,\sigma^2)$$

$$\Rightarrow p(\mu,\sigma^2|\bar{x},S_x^2)\propto e^{-\frac{n}{2\sigma^2}(\bar{x}-\mu)^2}\left(\frac{1}{\sigma^2}\right)^{(\frac{n-1}{2})}e^{-\left(\frac{n-1}{2\sigma^2}\right)S_x^2} \tag{3.1}$$

$$\Rightarrow p(\mu,\sigma^2|\bar{x},S_x^2)\propto e^{-\frac{n}{2\sigma^2}(\mu-\bar{x})^2}\left(\frac{1}{\sigma^2}\right)^{(\frac{n-3}{2})+1}e^{-\left(\frac{n-1}{2\sigma^2}\right)S_x^2}$$

$$\Rightarrow p(\mu,\sigma^2|\bar{x},S_x^2)\propto e^{-\frac{n}{2\sigma^2}(\bar{x}-\mu)^2}\left(\frac{1}{\sigma^2}\right)^{\alpha+1}e^{-\left(\frac{\beta}{\sigma^2}\right)}$$

where the proportional density is a combination of two densities: (i) $\mu|(\bar{x},\sigma^2)$ is normally distributed random variable with mean $E(\mu|(\bar{x},\sigma^2))=\bar{x}$ and variance $V(\mu|(\bar{x},\sigma^2))=\frac{\sigma^2}{n}$ and (ii) $\sigma^2|S_x^2$ is inverse-gamma distributed random variable with parameters, $\alpha=\frac{n-3}{2}$, $\beta=(\frac{n-1}{2}S_x^2)$.

(4) Relative distribution of $\sigma^2|S_x^2$ to shifted variance $\alpha_2\sigma_0^2$

The p.d.f of is:

$$p(\sigma^2|S_x^2)=\frac{\beta^\alpha}{\Gamma\alpha}\left(\frac{1}{\sigma^2}\right)^{\alpha+1}e^{-\frac{\beta}{\sigma^2}}\qquad 0<\sigma^2<\infty.$$

The distribution of a new random variable $P=\frac{(\sigma^2|s_x^2)}{a_2\sigma_0^2}$ will be:

$$P=\frac{(\sigma^2|s_x^2)}{a_2\sigma_0^2}$$

$$\Rightarrow a_2\sigma_0^2 p=\sigma^2|s_x^2$$

$$\Rightarrow \frac{d\sigma^2|s_x^2}{dp}=a_2\sigma_0^2 \tag{4.1}$$

$$f(p)=\frac{\beta^\alpha}{\Gamma\alpha}\left(\frac{1}{a_2\sigma_0^2 p}\right)^{\alpha+1}e^{-\frac{\beta}{a_2\sigma_0^2 p}}a_2\sigma_0^2$$

$$f(p)=\frac{(\frac{\beta}{a_2\sigma_0^2})^\alpha}{\Gamma\alpha}\left(\frac{1}{p}\right)^{\alpha+1}e^{-\frac{\left(\frac{\beta}{a_2\sigma_0^2}\right)}{p}}\qquad 0<p<\infty$$

which is known as inverse-gamma distribution with parameters $\{\alpha,\frac{\beta}{a_2\sigma_0^2}\}$ and the values of α and β are defined above, respectively, for informative and non-informative prior-based posterior parameters.

6

Mitigating collinearity in linear regression models using ridge, surrogate and raised estimators

Diarmuid O'Driscoll[1] and Donald E. Ramirez[2*]

*Corresponding author: Donald E. Ramirez, Department of Mathematics, University of Virginia, Charlottesville, VA, USA
E-mail: der@virginia.edu
Reviewing editor: Quanxi Shao, CSIRO, Australia

Abstract: Collinearity in the design matrix is a frequent problem in linear regression models, for example, with economic or medical data. Previous standard procedures to mitigate the effects of collinearity included ridge regression and surrogate regression. Ridge regression perturbs the moment matrix $\mathbf{X}'\mathbf{X} \to \mathbf{X}'\mathbf{X} + k\mathbf{I}_p$, while surrogate regression perturbs the design matrix $\mathbf{X} \to \mathbf{X}_S$. More recently, the raise estimators have been introduced, which allow the user to track geometrically the perturbation in the data with $\mathbf{X} \to \widetilde{\mathbf{X}}$. The raise estimators are used to reduce collinearity in linear regression models by raising a column in the experimental data matrix, which may be nearly linear with the other columns, while keeping the basic *OLS* regression model. We give a brief overview of these three ridge-type estimators and discuss practical ways of choosing the required perturbation parameters for each procedure.

Subjects: Mathematical Statistics; Mathematics & Statistics; Science; Statistical Computing; Statistics; Statistics & Probability

Keywords: collinearity; ridge estimators; surrogate estimators; raise estimators

AMS Subject Classifications: 62J05; 62J07

1. Introduction

The standard linear regression model can be written as $\mathbf{Y} = \mathbf{X}\beta + \varepsilon$ with uncorrelated, zero-mean and homoscedastic errors ε. Here $\mathbf{X}$ is a full rank $n \times p$ matrix containing the explanatory variables

ABOUT THE AUTHORS

Diarmuid O'Driscoll is the head of the Mathematics and Computer Studies Department at Mary Immaculate College, Limerick. He was awarded a Travelling Studentship for his MSc at University College Cork in 1977. He has taught at University College Cork, Cork Institute of Technology, University of Virginia, and Frostburg State University. His research interests are in mathematical education, errors in variables regression, ridge regression and design criteria. In 2014, he was awarded a Teaching Heroes Award by the National Forum for the Enhancement of Teaching and Learning (Ireland).

Donald E. Ramirez is a full professor in the Department of Mathematics at the University of Virginia in Charlottesville, Virginia. He received his PhD in Mathematics from Tulane University in New Orleans, Louisiana. His research is in harmonic analysis and mathematical statistics. His current research interests are in statistical outliers and ridge regression.

PUBLIC INTEREST STATEMENT

Collinearity is a frequent problem in statistical analysis of data, for example, with ordinary least square linear regression models of economic or medical data. Standard procedures to mitigate the effects of collinearity include ridge regression and surrogate regression. Ridge regression is based on a standard numerical technique that is used in computing an inverse of a nearly singular matrix. Surrogate regression is based on perturbing the data in a way to allow for more accurate numerical solutions. More recently, the raise estimators have been introduced. This technique also perturbs the data while allowing the researcher to track the changes in the data while retaining the basic ordinary least square regression model. We give a brief overview of these three ridge-type estimators and discuss practical ways of choosing the required perturbation parameters for each procedure. Our case study indicates an advantage for using the raise estimators.

and the response vector $\mathbf{y}$ is $n \times 1$ consisting of the observed data. The Ordinary Least Squared *OLS* estimators $\widehat{\boldsymbol{\beta}}_L$ are solutions of

$$\mathbf{X}'\mathbf{X}\widehat{\boldsymbol{\beta}} = \mathbf{X}'\mathbf{y} \tag{1}$$

given by

$$\widehat{\boldsymbol{\beta}}_L = (\mathbf{X}'\mathbf{X})^{-1}\mathbf{X}'\mathbf{y}. \tag{2}$$

The solutions $\widehat{\boldsymbol{\beta}}_L$ are unbiased with variance matrix $V(\widehat{\boldsymbol{\beta}}_L) = \sigma^2(\mathbf{X}'\mathbf{X})^{-1}$. For convenience, we take $\sigma^2 = 1$. The *OLS* solutions require that $(\mathbf{X}'\mathbf{X})^{-1}$ be accurately computed.

2. Ridge and surrogate estimators

With economic or medical data, the predictor variables in the columns of $\mathbf{X}$ may have a high level of collinearity; that is, there may be a nearly linear relationship among the predictor variables. In this case, $\mathbf{X}'\mathbf{X}$ in Equation (1) is nearly singular and thus $(\mathbf{X}'\mathbf{X})^{-1}$ will be numerically difficult to evaluate. It was observed by Riley (1955) that the perturbed matrix $\mathbf{X}'\mathbf{X} + k\mathbf{I}_p$ with $k > 0$ is better conditioned than the matrix $\mathbf{X}'\mathbf{X}$ and he suggested using the perturbed matrix in Equation (1). With $k > 0$ large enough, $(\mathbf{X}'\mathbf{X} + k\mathbf{I}_p)^{-1}$ can be accurately computed with standard numerical procedures. Using $\mathbf{X}'\mathbf{X} \to \mathbf{X}'\mathbf{X} + k\mathbf{I}_p$, Hoerl (1964) dubbed this procedure *ridge regression* with *ridge estimators*

$$\widehat{\boldsymbol{\beta}}_R(k) = (\mathbf{X}'\mathbf{X} + k\mathbf{I}_p)^{-1}\mathbf{X}'\mathbf{y}. \tag{3}$$

Near dependency among the columns of $\mathbf{X}$ causes ill-conditioning in $\mathbf{X}'\mathbf{X}$ which results in *OLS* solutions with inflated squared lengths $||\widehat{\boldsymbol{\beta}}_L||^2$, with $\widehat{\boldsymbol{\beta}}_L$ of questionable signs ($\pm$) and with $\widehat{\boldsymbol{\beta}}_L$ being "very sensitive to small changes in $\mathbf{X}$" (Belsley, 1986). With ill-conditioning in $\mathbf{X}'\mathbf{X}$, the *OLS* solutions at $k = 0$ in Equation (3) are known to be unstable with a slight movement away from $k = 0$ giving completely different estimates of the coefficients $\boldsymbol{\beta}$.

In *The International Encyclopedia of Statistical Science*, Hadi (2011) discusses two standard remedies for addressing collinearity in linear regression; namely (1) the *ridge system* $\{(\mathbf{X}'\mathbf{X} + k\mathbf{I}_p)\beta = \mathbf{X}'\mathbf{y}; k \geq 0\}$ (Hoerl & Kennard, 1970) with solutions $\{\widehat{\boldsymbol{\beta}}_R(k); k \geq 0\}$ and (2) the *surrogate system* $\{(\mathbf{X}'\mathbf{X} + k\mathbf{I}_p)\beta = (\mathbf{X}_k'\mathbf{X}_k)\boldsymbol{\beta} = \mathbf{X}_k'\mathbf{y}; k \geq 0\}$ (Jensen & Ramirez, 2008) with solutions $\{\widehat{\boldsymbol{\beta}}_S(k); k \geq 0\}$. The ridge estimators come from modifying $\mathbf{X}'\mathbf{X} \to \mathbf{X}'\mathbf{X} + k\mathbf{I}_p$ on the left side of Equation (1) while the Jensen and Ramirez surrogate estimators modify the design matrix $\mathbf{X} \to \mathbf{X}_k$ on both sides of Equation (1). In matrix notation, ridge regression comes from perturbing the eigenvalues of $\mathbf{X}'\mathbf{X}$ as $\lambda_i \to \lambda_i + k$, while surrogate regression comes from perturbing the singular values of $\mathbf{X}$ as $\xi_i \to \sqrt{\xi_i^2 + k}$. From the singular value decomposition $\mathbf{X} = PD(\xi_i)\mathbf{Q}'$, the surrogate design is $\mathbf{X}_k = \mathbf{P}D(\sqrt{\xi_i^2 + k})\mathbf{Q}'$, with $\mathbf{D}$ a diagonal matrix of dimension $n \times p$, the columns of $\mathbf{P}$ the left-singular vectors and the columns of $\mathbf{Q}$ the right singular vectors. The surrogate transformation $\mathbf{X} \to \mathbf{X}_k$ preserves the ridge moments, with $\mathbf{X}_k'\mathbf{X}_k = \mathbf{X}'\mathbf{X} + k\mathbf{I}_p$ allowing for comparison between the two methods. Ridge regression has a long history of use in the statistical literature. The earliest detailed expositions of ridge estimators are found in Marquardt (1963) and Hoerl and Kennard (1970), with Marquardt (1963) acknowledging that Levenberg (1944) had observed that a perturbation of the diagonal improved convergence in steepest descent algorithms. The history of the early use of matrix diagonal increments in statistical problems is given in the article by Piegorsch and Casella (1989).

To alleviate the problems inherent with a singular value, say ξ_p, which is indicating collinearity in $\mathbf{X}$, the surrogate transformation converts $\xi_p \to \sqrt{\xi_p^2 + k}$ moving the singular value away from zero. Principal Component Regression (PCR) does the opposite and replaces ξ_p with 0 and regresses $\mathbf{Y} = \mathbf{P}D(\xi_1, \ldots, \xi_{p-1}, 0)\boldsymbol{\alpha} + \boldsymbol{\varepsilon}$ with $\boldsymbol{\beta} = Q\alpha$. Hadi and Ling (1998) have noted "that it is possible for the PCR to fail miserably." Their example is constructed with the response variable $\mathbf{Y}$ being highly correlated with the deleted eigenvector associated with the deleted singular value. This deletion results in the remaining explanatory variables being unable to provide a good fit for the response variable.

Since ridge regression is based on a *numerical analysis* technique, the ridge estimators may lack desirable *statistical* properties. Three such desirable statistical properties follow.

(1) The *condition number* for a square $p \times p$ matrix $\mathbf{A}$ is a measure of the ill-conditioning in $\mathbf{A}$ and is defined as the ratio of the largest to smallest eigenvalues, denoted $\kappa(\mathbf{A}) = \lambda_1/\lambda_p$. Since perturbation procedures are designed to *improve* the regression model, one would expect that as $k \to \infty$ that $\kappa(V(\widehat{\boldsymbol{\beta}}_R(k)) \to 1$. However, as shown in Jensen and Ramirez (2010a), $\kappa(V(\widehat{\boldsymbol{\beta}}_R(k)) \to \kappa(V(\widehat{\boldsymbol{\beta}}_R(0))$. Initially, as k increases, the ill-conditioning in the variance matrix starts to get better but then returns to the original (bad) value. However, the surrogate system does have the desirable monotone property that $\kappa(V(\widehat{\boldsymbol{\beta}}_S(k)) \to 1$ as $k \to \infty$. This allows the user of surrogate estimators to be assured that, regardless of the chosen value for k, the variance matrix for the surrogate estimators will be more "orthogonal" than the original *OLS* variance matrix.

(2) Denote $V(\widehat{\boldsymbol{\beta}}_L) = \sigma^2(\mathbf{X}'\mathbf{X})^{-1} = \sigma^2\mathbf{V}$ so v_{jj} is the actual variance for $\widehat{\boldsymbol{\beta}}_{L\,j}$ and denote $\mathbf{X}'\mathbf{X} = \mathbf{W}$. An "ideal" predictor variable in column j would be orthogonal to the other predictor variables in $\mathbf{X}$, with $\mathbf{W}$ being zero for all off-diagonal values in the jth row and j^{th} column. In this "ideal" case, the "ideal" variance for $\widehat{\boldsymbol{\beta}}_{L\,j}$ would be $\sigma^2(\mathbf{W})^{-1}[j,j] = \sigma^2 w_{jj}^{-1}$. The *Variance Inflation Factors* (*VIFs*) of $\widehat{\beta}_L = [\widehat{\beta}_{L1}, \ldots, \widehat{\beta}_{Lp}]'$ are given by $\{VIF(\widehat{\beta}_{L\,j}) = v_{jj}/w_{jj}^{-1}; 1 \le j \le p\}$; i.e. the ratios of actual variances to "ideal" variances had the columns of $\mathbf{X}$ been orthogonal, with $VIF(\widehat{\beta}_{L\,j}) = 1$ for the ideal orthogonal case. Marquardt and Snee (1975) have identified *VIF* as "the best single measure of the conditioning of the data." Again since perturbation procedures are designed to *improve* the regression model, one would expect that as $k \to \infty$ that $VIF(V(\widehat{\boldsymbol{\beta}}_{R\,j}(k)) \to 1$. Jensen and Ramirez (2010a) also showed that $VIF(V(\widehat{\boldsymbol{\beta}}_{R\,j}(k)) \to VIF(V(\widehat{\boldsymbol{\beta}}_{R\,j}(0))$ for the ridge estimators but that $VIF(V(\widehat{\boldsymbol{\beta}}_{S\,j}(k)) \to 1$ as $k \to \infty$ for the surrogate estimators, resulting in less collinearity between the surrogate estimators than exists between the *OLS* estimators.

(3) Hoerl and Kennard (1970) established that the ridge estimators satisfy the *MSE Admissibility Condition* assuring an improvement in Mean Squared Error $MSE(\widehat{\boldsymbol{\beta}}_R(k))$ for some $k \in (0, \infty)$. With $\widehat{\mathbf{y}}_R(k)$, the predicted values for ridge regression, the statistic $MSE(\widehat{\boldsymbol{\beta}}_R(k)) = \sum_j (y_j - \hat{y}_{R\,j}(k))^2 = ||\mathbf{y} - \widehat{\mathbf{y}}_R(k)||^2$ measures how close the predicted values in the ridge regression model are to the observed values. However, Jensen and Ramirez (2010b) have shown the existence of cross-over values k_0 for which, if $k > k_0$ then $MSE(\widehat{\boldsymbol{\beta}}_R(k)) > MSE(\widehat{\boldsymbol{\beta}}_L(k))$, indicating that the ridge model should not be used. The Hoerl and Kennard (1970) result assures that for some positive value of k, the ridge model is an improved model. Jensen and Ramirez (2010a) have shown that for any $k \in (0, \infty)$ the corresponding result holds for surrogate estimators. A further improvement with surrogate estimators is given by $MSE(\widehat{\boldsymbol{\beta}}_S(k)) \le MSE(\widehat{\boldsymbol{\beta}}_R(k))$; that is, for any value of k, the surrogate estimators have predicted values closer to the original data than the ridge estimators. As the ridge and surrogate estimators are not equivariant under scaling, the common convention is to scale $\mathbf{X}'\mathbf{X}$ to *correlation form* with the explanatory variables centered and scaled to unit length.

Remark 1 Scaling $\mathbf{X}'\mathbf{X}$ to correlation form can lead to some anomalies. as noted in Jensen and Ramirez (2008). For example, the map $k \to ||\widehat{\boldsymbol{\beta}}_R(k)||^2$ is known to be monotonically decreasing with $\mathbf{X}$ centered but unscaled. Using Proc Reg in SAS with the Ridge option, this monotone property can be lost as the original $\mathbf{X}'\mathbf{X}$ moment matrix is (1) scaled into correlation form and (2) the ridge estimators are computed using the correlation form for $\mathbf{X}'\mathbf{X}$ and (3) the ridge solutions are mapped back into the original scale. This scaling-rescaling can cause $k \to ||\widehat{\boldsymbol{\beta}}_R(k)||^2$ to lose its monotonicity as in the example in Jensen and Ramirez (2008).

Remark 2 Let $\mathbf{X}$ be mean-centered. Let $\mathbf{D}^2$ be the diagonal matrix with entries $1/\mathbf{X}'\mathbf{X}_{j,j}$, $1 \leq j \leq p$, then the scaling $\mathbf{X} \to \mathbf{XD}$ has $(\mathbf{XD})'(\mathbf{XD})$ in correlation form., that is with diagonal entries all having value one. This is the scaling we have used. Sardy (2008) has suggested a covariance-based scaling using the diagonal matrix $\mathbf{D}_\Sigma^2$ with entries $(\mathbf{X}'\mathbf{X})_{j,j}^{-1}$, $1 \leq j \leq p$. We note that in this case $(\mathbf{XD}_\Sigma)'(\mathbf{XD}_\Sigma)$ has diagonal entries which are the variance inflation factors $VIF(\widehat{\beta}_j)$. The variance inflation factors are the ratios of the variances of $\widehat{\beta}_j$ to the "ideal" variances of $\widehat{\beta}_j$ assuming the explanatory variables are orthogonal; that is, $VIF(\widehat{\beta}_j) = (\mathbf{X}'\mathbf{X})_{j,j}^{-1}/(1/\mathbf{X}'\mathbf{X}_{j,j}) = (\mathbf{X}'\mathbf{X})_{j,j}^{-1}\mathbf{X}'\mathbf{X}_{j,j} = (\mathbf{X}'\mathbf{X})_{j,j}^{-1/2}\mathbf{X}'\mathbf{X}_{j,j}(\mathbf{X}'\mathbf{X})_{j,j}^{-1/2}$. In our Case Study, $p = 2$ so $VIF(\widehat{\beta}_1) = VIF(\widehat{\beta}_2)$ and $c(\mathbf{XD})'(\mathbf{XD}) = (\mathbf{XD}_\Sigma)'(\mathbf{XD}_\Sigma)$ with c the common value of the variance inflation factors.

Remark 3 When the regression model retains the parameter β_0 for the constant term with the design matrix $\mathbf{X}$ containing a unit constant column, the user needs to be careful with defining $VIF(\widehat{\beta}_j)$ when the data have not been mean-centered. In short, $VIF(\widehat{\beta}_j)$ is based on comparing the (j, j) entry of the variance matrix to the corresponding entry of an "ideal" covariance matrix. The inverse of the "ideal" covariance matrix is denoted as the "ideal" moment matrix. The "ideal" $\widehat{\beta}_j$ is uncorrelated with the other explanatory variables $\widehat{\beta}_i$, $0 < i \neq j$. Thus, the constraints on the "ideal" covariance matrix are that (1) the off-diagonal (i, j) and (j, i) entries for $cov(\widehat{\beta}_i, \widehat{\beta}_j)$ are zero where $0 < i \neq j$. Note that the "ideal" covariance matrix is not a diagonal matrix as the entries relating to $\widehat{\beta}_0$ in the first row and column are retained as the data have not been centered. Additionally, the constraints on the "ideal" moment matrix are that (2) the entries in first row and first column are the first order moments determined from the data and (3) the entries down the diagonal (j, j) with $j \geq 0$ are the second order moments determined from the data. Jensen and Ramirez (2013) have given an easy to compute algorithm for computing the "ideal" covariance matrix that satisfies constraints (1), (2) and (3).

The variance inflation factors, which are the standard measure for collinearity, have a geometric interpretation which allows them to be conveniently computed as a ratio of determinants. We assume that the variables are *centered*. Reorder $\mathbf{X} = [\mathbf{X}_{[p]}, \mathbf{X}_{(p)}]$ with $\mathbf{X}_{(p)} = \mathbf{x}_p$ the pth column and $\mathbf{X}_{[p]}$, the design matrix $\mathbf{X}$ without the pth column, dubbed the *resting columns*. Garcia, Garcia and Soto (2011) introduced the *metric number* to measure the effect of adding the last column $\mathbf{X}_{(p)}$ to the resting columns $\mathbf{X}_{[p]}$. An ideal p^{th} column would be orthogonal to the other columns with the entries in the off diagonal elements of the pth row and p^{th} column of $\mathbf{X}'\mathbf{X}$ all zeros, with idealized $\mathbf{X}'\mathbf{X}$ moment matrix

$$\mathbf{M}_p = \begin{bmatrix} \mathbf{X}'_{[p]}\mathbf{X}_{[p]} & \mathbf{0}_{p-1} \\ \mathbf{0}'_{p-1} & \mathbf{x}'_p\mathbf{x}_p \end{bmatrix}.$$

The metric number is defined by $MN(\mathbf{x}_p) = \left(\det(\mathbf{X}'\mathbf{X})/\det(\mathbf{M}_p)\right)^{1/2}$ and it measures the effect of enlarging the design matrix with the adding of the pth exploratory column. The metric number is easy to compute and is functionally equivalent to the *VIF* statistics with

$$VIF(\widehat{\beta}_p) = \frac{\det(\mathbf{M}_p)}{\det(\mathbf{X}'\mathbf{X})},$$

for example, O'Driscoll and Ramirez (2015).

In spite of the established usage of ridge regression, it is now known that the surrogate estimators have superior statistical properties over the ridge estimators. Indeed, for their statistical analysis, Woods et al. (2012) used the Jensen–Ramirez surrogate estimates for modelling of diabetes in stock rats.

A crucial question for both the ridge estimators and the surrogate estimators is: What value of k should be used? McDonald (2009, 2010) has suggested that k can be determined by controlling the correlation between the observed values and the predicted values from ridge regression. We extend this methodology to surrogate regression and will compare the two procedures.

McDonald (2009, 2010) showed that the square of the correlation coefficient $R^2(\widehat{\boldsymbol{\beta}}_R(k))$ between the observed values $\mathbf{y}$ and the ridge predicted values $\widehat{\mathbf{y}}_R(k) = \mathbf{X}\widehat{\boldsymbol{\beta}}_R(k)$ is a monotone decreasing function in the ridge parameter k. The corresponding result for the square of the correlation coefficient $R^2(\widehat{\boldsymbol{\beta}}_S(k))$ of the observed values $\mathbf{y}$ and the surrogate predicted values $\widehat{\mathbf{y}}_S(k) = \mathbf{X}\widehat{\boldsymbol{\beta}}_S(k)$ for the surrogate regression is a monotone decreasing function in the surrogate parameter k, as shown in Garcia and Ramirez (in press). This allows the user to determine a unique value for k by controlling the decrease in correlation between the observed and predicted values. The user can set a lower bound for the reduction in $R^2(\widehat{\boldsymbol{\beta}}_R(k))$ and $R^2(\widehat{\boldsymbol{\beta}}_S(k))$ and numerically compute the associated ridge and surrogate parameters, For example, to preserve 95% of the *OLS* correlation, we solve $R^2(k) = 0.95R^2(0)$. With the computed value for k, we can measure the reduction in collinearity using the *VIF* statistic or using the condition number κ of $\mathbf{X}'\mathbf{X} + k\mathbf{I}_p$. For our case study, we use the example in McDonald (2010) which is known to have severe collinearity. We report the improvements in collinearity for both methods.

3. Raise estimators

We assume that the columns of $\mathbf{X} = \left(\mathbf{x}_1, \mathbf{x}_2, \ldots, x_p\right)$ are centered and standardized, that is, $\mathbf{X}'\mathbf{X}$ is in correlation form with $||\mathbf{x}_j||^2 = 1$. For the $n \times p$ matrix $\mathbf{A} = [\mathbf{a}_1, \mathbf{a}_2, \ldots, \mathbf{a}_p]$, the column span, is denoted by $Sp(\mathbf{A})$, with $\mathbf{A}_{(j)}$ denoting the jth column vector $\mathbf{a}_j$ and $\mathbf{A}_{[j]}$ denoting the $n \times (p-1)$ matrix formed by deleting $\mathbf{A}_{(j)}$ from $\mathbf{A}$. For the linear model $\mathbf{y} = \mathbf{X}\beta + \boldsymbol{\varepsilon}$, central to a study of collinearity is the relationship between $\mathbf{X}_{(j)}$ and $Sp(\mathbf{X}_{[j]})$.

The raise estimators are based on perturbing a column $\mathbf{x}_j \rightarrow \widetilde{\mathbf{x}}_j = \mathbf{x}_j + \lambda_j\mathbf{e}_j$ by a λ_j multiple of a vector $\mathbf{e}_j$ orthogonal to the span of the remaining *resting* columns. We follow the notation from Garcia and Ramirez (in press). The regression of $\mathbf{x}_j$, viewed as the response vector using the remaining resting columns as the explanatory vectors, has an error vector $\mathbf{e}_j$ with the required properties. The raise estimators are constructed sequentially as follows.

- Step 1: We raise the vector $\mathbf{x}_1$ from the regression of $\mathbf{x}_1$ using the resting vectors $\mathbf{X}_{[1]} = \left(\mathbf{x}_2, \mathbf{x}_3, \ldots x_p\right)$. From this regression, we take the error vector $\mathbf{e}_1$ with $\mathbf{e}_1 \perp Sp(\mathbf{X}_{[1]})$ to construct $\tilde{\mathbf{x}}_1(\lambda_1) = \mathbf{x}_1 + \lambda_1\mathbf{e}_1$. The raised design matrix is denoted $\mathbf{X}_{<1>} = \left(\tilde{\mathbf{x}}_1(\lambda_1), \mathbf{x}_2, \ldots, x_p\right)$.
- Step j: we raise the vector $\mathbf{x}_j$ from the regression of $\mathbf{x}_j$ using the resting vectors from $\mathbf{X}_{<1,\ldots,j-1>}$, namely $\mathbf{X}_{<1,\ldots,j-1>[j]} = \left(\tilde{\mathbf{x}}_1(\lambda_1), \ldots, \tilde{x}_{j-1}(\lambda_{j-1}), \mathbf{x}_{j+1}, \ldots, x_p\right)$. From this regression, we take the residual vector $\mathbf{e}_j$ with $\mathbf{e}_j \perp Sp(\mathbf{X}_{<1,\ldots,j-1>[j]})$ to construct $\tilde{\mathbf{x}}_j(\lambda_j) = \mathbf{x}_j + \lambda_j\mathbf{e}_j$. The raised design matrix is denoted $\mathbf{X}_{<1,\ldots,j>} = \left(\tilde{\mathbf{x}}_1(\lambda_1), \ldots, \tilde{x}_j(\lambda_j), \mathbf{x}_{j+1}, \ldots, x_p\right)$.
- Step p: we raise the vector $\mathbf{x}_p$ from the regression of $\mathbf{x}_p$ with the resting vectors from $\mathbf{X}_{<1,\ldots,p-1>}$, namely $\mathbf{X}_{<1,\ldots,p-1>[p]} = \left(\tilde{\mathbf{x}}_1(\lambda_1), \ldots, \tilde{x}_{p-1}(\lambda_{p-1})\right)$. Then, we take the residual $\mathbf{e}_p$ with $\mathbf{e}_p \perp Sp(\mathbf{X}_{<1,\ldots,p-1>[p]})$ to construct $\tilde{\mathbf{x}}_p = \mathbf{x}_p + \lambda_p\mathbf{e}_p$. The raised design matrix is denoted $\mathbf{X}_{<1,\ldots,p>} = \left(\tilde{\mathbf{x}}_1(\lambda_1), \ldots, \tilde{x}_p(\lambda_p)\right)$ with parameters vector $\boldsymbol{\lambda} = (\lambda_1, \ldots, \lambda_p)'$ to be chosen by the user. For convenience, we denote the final raise design $\mathbf{X}_{<1,\ldots,p>}$ by $\widetilde{\mathbf{X}}$.

There is a monotone relationship between the variance inflation factors, VIF_j, and the angle between $(\mathbf{x}_j, Sp(\mathbf{X}_{[j]}))$, for example, Jensen and Ramirez (2013, Theorem 4). Let $\mathbf{P}_{[j]} = \mathbf{X}_{[j]}(\mathbf{X}'_{[j]}\mathbf{X}_{[j]})^{-1}\mathbf{X}'_{[j]}$ be the projection operator onto the subspace $Sp(\mathbf{X}_{[j]}) \subset \mathbb{R}^n$ spanned by the columns of the reduced (or relaxed) matrix $\mathbf{X}_{[j]}$. From the geometry of the right triangle formed by $(\mathbf{x}_j, \mathbf{P}_{[j]}\mathbf{X}_{(j)})$, it can be shown that the angle θ_j between $\mathbf{x}_j$ and $\mathbf{P}_{[j]}\mathbf{X}_{(j)}$ satisfies $\cos(\theta_j) = ||\mathbf{P}_{[j]}\mathbf{x}_j||/||\mathbf{x}_j||$ and similarly the angle between $\widetilde{\mathbf{x}}_j$ and $\mathbf{P}_{[j]}\mathbf{X}_{(j)}$ satisfies $\cos(\widetilde{\theta}_j) = ||\mathbf{P}_{[j]}\mathbf{x}_j||/||\widetilde{\mathbf{x}}_j||$ since $\mathbf{e}_j \perp \mathbf{P}_{[j]}\mathbf{x}_j$. Each Variance Inflation Factor, VIF_j, for $\widetilde{\mathbf{X}}$ is functionally related to the angle $\widetilde{\theta}_j$ by the rule $\widetilde{\theta}_j = \arccos(\sqrt{1 - 1/VIF_j})$, for example Jensen and Ramirez (2013). Thus as $\lambda_j \to \infty$, $\widetilde{\theta}_j \to 90°$ and the variance inflation factor VIF_j converges to one indicating that collinearity is being diminished, as in Garcia et al. (2011, Theorem 4.2).

Some desirable properties of the raised regression method are as follows.

(1) Raising a column vector in $\mathbf{X}$ does not effect the basic *OLS* regression model as the raised vector remains in the original $Sp(\mathbf{X})$, $\mathbf{e}_j = \mathbf{X}_{(j)} - \mathbf{P}_{[j]}\mathbf{X}_{(j)} \in Sp(\mathbf{X})$ so $Sp(\widetilde{\mathbf{X}}) = Sp(\mathbf{X})$, as shown in Garcia et al. (2011).

(2) Garcia et al. (2011) has shown that the raise estimators satisfy the *MSE Admissibility Condition* assuring an improvement in Mean Squared Error $MSE(\widetilde{\boldsymbol{\beta}}(\lambda))$ for some $\lambda \in (0, \infty)$ and thus the raise estimators can be said to be of ridge-type.

(3) The *VIFs* associated with the raise estimators are monotone functions, decreasing with λ_j, see Garcia, Garcia, López Martin, and Salmeron (2015).

(4) Starting with $\mathbf{X}'\mathbf{X}$ in correlation form with

$$\mathbf{X}'\mathbf{X} = \begin{pmatrix} 1 & \rho_{12} & \rho_{13} & \cdots & \rho_{1p} \\ \rho_{12} & 1 & \rho_{23} & \cdots & \rho_{2p} \\ \rho_{13} & \rho_{23} & 1 & \cdots & \rho_{3p} \\ \cdots & \cdots & \cdots & \cdots & \cdots \\ \rho_{1p} & \rho_{2p} & \rho_{3p} & \cdots & 1 \end{pmatrix}$$

results in the final raising matrix $\widetilde{\mathbf{X}}$ having moment matrix

$$\widetilde{\mathbf{X}}'\widetilde{\mathbf{X}} = \begin{pmatrix} 1+\lambda_1 & \rho_{12} & \rho_{13} & \cdots & \rho_{1p} \\ \rho_{12} & 1+\lambda_2 & \rho_{23} & \cdots & \rho_{2p} \\ \rho_{13} & \rho_{23} & 1+\lambda_3 & \cdots & \rho_{3p} \\ \cdots & \cdots & \cdots & \cdots & \cdots \\ \rho_{1p} & \rho_{2p} & \rho_{3p} & \cdots & 1+\lambda_p \end{pmatrix} \tag{4}$$

Thus, the raised regression perturbation matrix is equivalent to a *generalized ridge* regression perturbation matrix. And conversely, any generalized ridge regression matrix has a corresponding raised regression matrix as in Garcia and Ramirez (in press).

The raise estimators allow the user to specify, for each of the variables, a precision π_j that the data will retain during the raising stages by restricting the mean absolute deviation *MAD* in the jth column of $\mathbf{X} - \widetilde{\mathbf{X}}$ from

$$\lambda_j \frac{1}{n} \sum_{i=1}^{n} |\mathbf{e}_{j,i}| = \pi_j. \tag{5}$$

Thus, given a specified precision $\pi_j > 0$, the user can raise column j in $\mathbf{X}_{<1,\ldots,j>}$ to $\widetilde{\mathbf{x}}_j(\lambda_j) = \mathbf{x}_j + \lambda_j \mathbf{e}_j$, where λ_j is solved from Equation (5). The precision values should be based on the researcher's belief in the accuracy of the data. The raised parameters λ_j are thus constrained to assure that the original data have not been perturbed more than what the researcher has permitted.

Table 1. ***OLS*****, ridge, and surrogate regression with squared correlation $R^2(\mathbf{y}, \hat{\mathbf{y}}(k))$, computed parameters k, estimated coefficients $\hat{\boldsymbol{\beta}}$, squared lengths $\hat{\boldsymbol{\beta}}'\hat{\boldsymbol{\beta}}$, condition numbers κ, variance inflation factors *VIF*, and mean absolute deviation for $\mathbf{X}-\mathbf{X}_k$ for surrogate design**

	OLS	Ridge	Surrogate
$R^2(\mathbf{y}, \hat{\mathbf{y}}(k))$	0.3249	0.3086	0.3086
k	0	0.01822	0.05775
$\hat{\boldsymbol{\beta}}$	$\begin{bmatrix} 3.0255 \\ -3.1539 \end{bmatrix}$	$\begin{bmatrix} 1.3886 \\ -1.5158 \end{bmatrix}$	$\begin{bmatrix} 1.0860 \\ -1.212 \end{bmatrix}$
$\hat{\boldsymbol{\beta}}'\hat{\boldsymbol{\beta}}$	19.10	4.23	4.18
$\kappa(\mathbf{X}'\mathbf{X}+k\mathbf{I}_p)$	122.76	58.23	27.62
VIF	31.19	15.06	12.53
MAD	0	none	0.009158

Table 2. ***OLS*** **and raise regression with precision $\pi_j = 0.009158$ squared correlation $R^2(\mathbf{y}, \hat{\mathbf{y}}(k))$, computed parameters λ_j, estimated coefficients $\hat{\boldsymbol{\beta}}$, squared lengths $\hat{\boldsymbol{\beta}}'\hat{\boldsymbol{\beta}}$, condition numbers κ, variance inflation factors *VIF*, and mean absolute deviation for $\mathbf{X}-\widetilde{\mathbf{X}}$ for raise design**

	OLS	Step 1	Step 2
π_j	0	0.009158	0.009158
$R^2(\mathbf{y}, \hat{\mathbf{y}}(\lambda))$	0.3249	0.3169	0.3147
λ	0	0.5671	0.3898
$\hat{\boldsymbol{\beta}}$	$\begin{bmatrix} 3.0255 \\ -3.1539 \end{bmatrix}$	$\begin{bmatrix} 1.9307 \\ -2.0767 \end{bmatrix}$	$\begin{bmatrix} 1.3831 \\ -1.4942 \end{bmatrix}$
$\hat{\boldsymbol{\beta}}'\hat{\boldsymbol{\beta}}$	19.10	8.04	4.15
$\kappa(\widetilde{\mathbf{X}}'\widetilde{\mathbf{X}})$	122.76	51.19	27.43
VIF	31.19	13.29	7.36
θ_j	10.31°	15.92°	21.62°
MAD	0	0.009158	0.009158

Remark 4 The ridge and surrogate procedures do not require $\mathbf{X}$ to be of full rank. For example, with the surrogate transformation $\xi_i \to \sqrt{\xi_i^2 + k}$ any zero singular value will be mapped to $\sqrt{k} > 0$ with $\mathbf{X}_k$ now full rank. On the other hand, the raise procedure does require the columns of $\mathbf{X}$ to be independent as the crucial step $\mathbf{x}_1 \to \tilde{\mathbf{x}}_1(\lambda_1) = \mathbf{x}_1 + \lambda_1\mathbf{e}_1$ moves $\mathbf{x}_1$ in the direction of the orthogonal complement of $Sp(\mathbf{X}_{[1]}) \subset Sp(\mathbf{X}) \subset \mathbb{R}^n$ in $Sp(\mathbf{X})$, the span of the other columns. Thus if $\mathbf{x}_1 \in Sp(\mathbf{X}_{[1]}) = Sp(\mathbf{X})$ then $Sp(\mathbf{X}_{[1]})^{\perp} \cap Sp(\mathbf{X}) = \{0\}$ and $\mathbf{x}_1$ cannot be raised.

4. Case study

Our case study is the numerical example in McDonald (2010). Here, $n = 60$ and $p = 2$ with $\mathbf{X} = [\mathbf{x}_1, \mathbf{x}_2]$ with $\mathbf{x}_1$ the nitrogen oxide pollution potential and $\mathbf{x}_2$ the hydrocarbon pollution potential and $\mathbf{y}$ the total mortality rate in 60 US metropolitan areas. The original data-set had 15 explanatory variables. Following McDonald (2010), we concentrate on the two variables which have the highest correlation $p = 0.9838$. Since, $\mathbf{X}$ is assumed to contain only explanatory variables, the vectors $\mathbf{x}_1$, $\mathbf{x}_2$, $\mathbf{y}$ are all mean-centered and scaled to have unit length.

Table 1 reports results for *OLS*, ridge and surrogate regression for the case study. The squared correlation between the original data $\mathbf{y}$ and the predicted values $\hat{\mathbf{y}}_L(k)$ for *OLS* is 0.3249. The perturbation parameter for ridge and surrogate are chosen to retain 95% of this value for the squared correlation between $\mathbf{y}$ and $\widehat{\mathbf{y}}_R = \mathbf{X}\beta_R(k)$ for the ridge parameter and between $\mathbf{y}$ and $\widehat{\mathbf{y}}_S = \mathbf{X}\beta_S(k)$ for the surrogate parameter. Thus, both methods have the same small decrease in $R^2(\mathbf{y}, \widehat{\mathbf{y}}(k))$ down to 0.3086 shown in Row 1 of Table 1 and with the associated parameters in Row 2 of Table 1. This allows us to compare the improvement in collinearity between the two procedures. The estimated coefficients are shown in Row 3 of Table 1.

Ridge-type procedures are designed to (1) decrease the squared length of the estimated coefficient $\boldsymbol{\beta}'\boldsymbol{\beta}$ which is given in Row 4 of Table 1; (2) to decrease the condition number $\kappa(\mathbf{X}'\mathbf{X} + k\mathbf{I}_p)$ of the matrix which needs to be inverted which is given in Row 5 of Table 1; (3) to decrease the variance inflation factors *VIF* given in Row 6. Since $p = 2$, both *VIFs* have a common value so only one value appears in Row 6. For each of these three criteria, surrogate regression is shown to be a superior procedure achieving a model with smaller collinearity with comparable loss of squared correlation $R^2(\mathbf{y}, \widehat{\mathbf{y}}(k))$.

The standard method for computing *VIF* for ridge regression in correlation form follows the procedure suggested by Marquardt (1970), which is to use the values on the main diagonal values of $(\mathbf{X}'\mathbf{X} + k\mathbf{I}_\mathbf{p})^{-1}\mathbf{X}'\mathbf{X}(\mathbf{X}'\mathbf{X} + k\mathbf{I}_\mathbf{p})^{-1}$. Although this is the correct expression for $k = 0$, it has been shown to be in error for $k > 0$ by Garcia et al. (2015) as the Marquardt expression allows inadmissible values less than one. Thus, we have used Equation (22) and (25) in Garcia et al. (2015) to compute the corrected values for *VIF* for ridge regression.

From Table 1, we see that the mean absolute deviation *MAD* for $\mathbf{X} - \widetilde{\mathbf{X}}$ from the surrogate system is 0.009158. To compute a comparable raise system of estimators, we will set the precision $\pi_j = 0.009158$ in Equation (5). The *OLS* values from Table 1 are shown in Column 1 of Table 2 for comparisons. Using $\pi_1 = 0.009158$ in Step 1, we solve for $\lambda_1 = 0.5671$ to raise $\mathbf{x}_1 \to \widetilde{\mathbf{x}}_1$. With this value, the squared lengths $\widehat{\boldsymbol{\beta}}'\widehat{\boldsymbol{\beta}} = 8.04$, $\kappa = 51.19$ and $VIF = 13.29$ all showing an improvement in collinearity. The angle between the two column vectors in the design has improved from 10.31° to 15.92°. The corresponding $R^2(\mathbf{y}, \widehat{\mathbf{y}}(\lambda)) = 0.3169$ indicates that 97.5% for the squared correlation has been retained. For Step 2, we solve for $\lambda_2 = 0.3898$ to raise $\mathbf{x}_2 \to \widetilde{\mathbf{x}}_2$. This is the final raised design $\widetilde{\mathbf{X}} = [\widetilde{\mathbf{x}}_1, \widetilde{\mathbf{x}}_2]$. With these values, the squared lengths $\widehat{\boldsymbol{\beta}}'\widehat{\boldsymbol{\beta}} = 4.15$, $\kappa = 27.43$ and $VIF = 7.36$ all showing an improvement in collinearity. The angle between the two column vectors in the design has improved to 21.62°. The corresponding $R^2(\mathbf{y}, \widehat{\mathbf{y}}(\lambda)) = 0.3147$ indicates that 96.9% for the squared correlation has been retained. Row 8 records *MAD* which is 0.009158 by construction.

The values in Column 3 of Table 2 are comparable to the values for the surrogate model from Table 1. However, following Marquardt (1970, p. 610), *VIF* should be less than 10 and thus, for this example, we would favor the raise estimators as the ridge-type method to be used.

Funding

The authors received no direct funding for this research.

Author details

Diarmuid O'Driscoll[1]

E-mail: diarmuid.odriscoll@mic.ul.ie

Donald E. Ramirez[2]

E-mail: der@virginia.edu

[1] Department of Mathematics and Computer Studies, Mary Immaculate College, Limerick, Ireland.

[2] Department of Mathematics, University of Virginia, Charlottesville, VA, USA.

References

Belsley, D. A. (1986). Centering, the constant, first-differencing, and assessing conditioning. In E. Kuh & D. A. Belsley (Eds.), *Model reliability* (pp. 117–153). Cambridge: MIT Press.

Garcia, C. B., Garcia, J., López Martin, M. M., & Salmeron, R. (2015). Collinearity: Revisiting the variance inflation factor in ridge regression. *Journal of Applied Statistics, 42*, 648–661.

Garcia, C. B., Garcia, J., & Soto, J. (2011). The raise method: An alternative procedure to estimate the parameters in presence of collinearity. *Quality and Quantity, 45*, 403–423.

Garcia, J., & Ramirez, D. (in press). *The successive raising estimator and its relation with the ridge estimator* (under review).

Hadi, A. S. (2011). Ridge and surrogate ridge regressions. In M. Lovric (Ed.), *International encyclopedia of statistical science* (pp. 1232–1234). Berlin: Springer.

Hadi, A. S., & Ling, R. F. (1998). Some cautionary notes on the use of principal component regression. *The American Statistical Association, 52*, 15–19.

Hoerl, A. E. (1964). Ridge analysis. *Chemical Engineering Progress, Symposium Series, 60*, 67–77.

Hoerl, A. E., & Kennard, R. W. (1970). Ridge regression: Biased estimation for nonorthogonal problems. *Technometrics, 12*, 55–67.

Jensen, D. R., & Ramirez, D. E. (2008). Anomalies in the foundations of ridge regression. *International Statistical Review, 76*, 89–105.

Jensen, D. R., & Ramirez, D. E. (2010a). Surrogate models in ill-conditioned systems. *Journal of Statistical Planning and Inference, 140*, 2069–2077.

Jensen, D. R., & Ramirez, D. E. (2010b). Tracking MSE efficiencies in ridge regression. *Advances and Applications in Statistical Sciences, 1*, 381–398.

Jensen, D. R., & Ramirez, D. E. (2013). Revision: Variance inflation in regression. *Advances in Decision Sciences*, 1–15. 671204.

Levenberg, K. (1944). A method for the solution of certain non-linear problems in least-squares. *Quarterly of Applied Mathematics, 2*, 164–168.

Marquardt, D. W. (1963). An algorithm for least-squares estimation for nonlinear parameters. *Journal of the Society for Industrial and Applied Mathematics, 11*, 431–441.

Marquardt, D. W. (1970). Generalized inverses, ridge regression, biased linear estimation and nonlinear estimation. *Technometrics, 12*, 591–612.

Marquardt, D. W., & Snee, R. D. (1975). Ridge regression in practice. *The American Statistical Association, 29*, 3–20.

McDonald, G. C. (2009). ridge regression. *Wiley Interdisciplinary Reviews: Computational Statistics, 1*, 93–100.

McDonald, G. C. (2010). Tracing ridge regression coefficients. *Wiley Interdisciplinary Review: Computational Statistics, 2*, 695–793.

O'Driscoll, D., & Ramirez, D. (2015). Response surface design using the generalized variance inflation factors. *Cogent Mathematics, 2*, 1–11.

Piegorsch, W., & Casella, G. (1989). The early use of matrix diagonal increments in statistical problems. *SIAM Review, 31*, 428–434.

Riley, J. (1955). Solving Systems of linear equations with a positive definite, symmetric but possibly ill-conditioned matrix. *Math Tables Mathematical Tables and Other Aids to Computation, 9*, 96–101.

Sardy, S. (2008). On the practice of rescaling covariates. *International Statistical Review, 76*, 285–297.

Woods, L. C. S., Holl, K. L., Oreper, D., Xie, Y., Tsaih, S.-W., & Valdar, W. (2012). Fine-mapping diabetes-related traits including insulin resistance, in heterogeneous stock rats. *Physiological Genomics, 44*, 1013–1026.

7

Minimax-robust filtering of functionals from periodically correlated random fields

Iryna Golichenko[1], Oleksandr Masyutka[2] and Mikhail Moklyachuk[3*]

*Corresponding author: Mikhail Moklyachuk, Department of Probability Theory, Statistics and Actuarial Mathematics, Taras Shevchenko National University of Kyiv, Kyiv 01601, Ukraine

E-mail: moklyachuk@gmail.com

Reviewing editor: Zudi Lu, University of Southampton, UK

Abstract: The problem of optimal estimation of linear functionals depending on unknown values of periodically correlated random field from observations of the field with noise is considered. Formulas for calculating mean square errors and spectral characteristics of optimal linear estimates of the functionals are derived in the case where spectral densities are exactly known. Formulas that determine least favourable spectral densities and minimax (robust) spectral characteristics are proposed in the case where spectral densities are not exactly known but a class of admissible spectral densities is given.

Subjects: Mathematics & Statistics; Multivariate Statistics; Probability; Science; Statistics; Statistics & Probability; Stochastic Models & Processes

Keywords: periodically correlated random field; robust estimate; mean square error; least favourable spectral density; minimax spectral characteristic

AMS Subject classifications: 60G60; 62M40; 62M20; 93E10; 93E11

ABOUT THE AUTHORS

Iryna Golichenko is an assistant professor, Department of Mathematical Analysis and Probability Theory, National Technical University of Ukraine. She received PhD in Physics and Mathematical Sciences from Taras Shevchenko National University of Kyiv in 2014. Her research interests include estimation problems for periodically correlated stochastic processes and fields.

Oleksandr Masyutka is an assistant professor, Department of Mathematics and Theoretical Radiophysics, Taras Shevchenko National University of Kyiv. He received PhD in Physics and Mathematical Sciences from Taras Shevchenko National University of Kyiv in 2011. His research interests include estimation problems for multidimensional stochastic processes and random fields.

Mikhail Moklyachuk is a professor, Department of Probability Theory, Statistics and Actuarial Mathematics, Taras Shevchenko National University of Kyiv. He received PhD in Physics and Mathematical Sciences from the Taras Shevchenko University of Kyiv in 1977. His research interests are statistical problems for stochastic processes and random fields. He is also a member of editorial boards of several international journals.

PUBLIC INTEREST STATEMENT

Cosmological Principle (first formulated by Einstein): the Universe is, in the large, homogeneous and isotropic. Last decades indicate growing interest to the spatio-temporal data measured on the surface of a sphere. These data include cosmic microwave background (CMB) anisotropies, medical imaging, global and land-based temperature data, gravitational and geomagnetic data, and climate model. Periodically correlated processes and fields are not homogeneous but have numerous properties similar to properties of stationary processes and fields. They describe appropriate models of numerous physical and man-made processes. In this article, we considered the problem of optimal estimation of functionals depending on unknown values of periodically correlated spatial temporal isotropic random fields from observations of the field with noise in the case of spectral certainty where spectral densities are known exactly as well as in the case of spectral uncertainty where spectral densities are not known exactly but a class of admissible spectral densities is given. Formulas that determine least favourable spectral densities and minimax (robust) spectral characteristics are derived.

1. Introduction

Cosmological Principle (first formulated by Einstein): the Universe is, in the large, homogeneous and isotropic (Bartlett, 1999). Last decades indicate growing interest to the spatio-temporal data measured on the surface of a sphere. These data include cosmic microwave background anisotropies (Adshead & Hu, 2012; Bartlett, 1999; Hu & Dodelson, 2002; Kogo & Komatsu, 2006; Okamoto & Hu, 2002), medical imaging (Kakarala, 2012), global and land-based temperature data (Jones, 1994; Subba Rao & Terdik, 2006), gravitational and geomagnetic data and climate model (North & Cahalan, 1981). Some basic results and references on the theory of isotropic random fields on a sphere can be found in the books by Yadrenko (1983) and Yaglom (1987). For more recent applications and results see new books by Cressie and Wikle (2011), Gaetan and Guyon (2010), Marinucci and Peccati (2011) and several papers covering a number of problems in general for spatial-time observations (Subba Rao & Terdik, 2012; Terdik, 2013).

Periodically correlated processes and fields are not homogeneous but have numerous properties similar to properties of stationary processes and fields. They describe appropriate models of numerous physical and man-made processes. A comprehensive list of the existing references up to the year 2005 on periodically correlated processes and their applications was proposed by Serpedin, Panduru, Sari, and Giannakis (2005). See also a review by Antoni (2009). For more details see a survey paper by Gardner (1994) and book by Hurd and Miamee (2007). Note that in the literature periodically correlated processes are named in multiple different ways such as cyclostationary, periodically nonstationary or cyclic correlated processes.

Among the current trends of the theory of stochastic processes and fields important is the direction which focuses on the problem of estimation of unknown values of random processes and fields. The problem of estimation of random processes and fields includes interpolation, extrapolation and filtering problems.

The mean square optimal estimation problems for periodically correlated with respect to time isotropic on sphere random fields are natural generalization of the linear extrapolation, interpolation and filtering problems for stationary stochastic processes and homogeneous random fields. Effective methods of solution of the linear extrapolation, interpolation and filtering problems for stationary stochastic processes were developed under the condition of certainty where spectral densities of processes and fields are known exactly [see, for example, selected works of Kolmogorov (1992), survey article by Kailath (1974), books by Rozanov (1967), Wiener (1966), Yaglom (1987), Yadrenko (1983) and articles by Moklyachuk and Yadrenko (1979, 1980)].

Particularly relevant in recent years is the problem of estimation of values of processes and fields under uncertainty where spectral densities of processes and fields are not known exactly. Such problems arise when considering problems of automatic control theory, coding and signal processing in radar and sonar, pattern recognition problems of speech signals and images.

The classical approach to the problems of interpolation, extrapolation and filtering of stochastic processes and random fields is based on the assumption that the spectral densities of processes and fields are known. In practice, however, complete information about the spectral density is impossible in most cases. To overcome this complication one finds parametric or nonparametric estimates of the unknown spectral densities or select these densities by other reasoning. Then applies the classical estimation method provided that the estimated or selected density is the true one. This procedure can result in a significant increasing of the value of error as Vastola and Poor (1983) have demonstrated with the help of some examples. This is a reason to search estimates which are optimal for all densities from a certain class of admissible spectral densities. These estimates are called minimax since they minimize the maximal value of the error of estimates. A survey of results in minimax (robust) methods of data processing can be found in the paper by Kassam and Poor (1985). The paper by Grenander (1957) should be marked as the first one where the minimax approach to extrapolation problem for stationary processes was proposed. For more details see, for example, books by Moklyachuk (2008), Moklyachuk and Masyutka (2012), Golichenko and Moklyachuk (2014).

In papers by Dubovets'ka, Masyutka, and Moklyachuk (2012), Dubovets'ka and Moklyachuk (2013, 2014a), the minimax-robust estimation problems (extrapolation, interpolation and filtering) are investigated for the linear functionals which depend on unknown values of periodically correlated stochastic processes. Methods of solution of the minimax-robust estimation problems for time-homogeneous isotropic random fields on a sphere were developed by Moklyachuk (1994, 1995, 1996). In papers by Dubovets'ka, Masyutka, and Moklyachuk (2014, 2015) results of investigation of minimax-robust estimation problems for periodically correlated isotropic random fields are described.

In this article, we considered the problem of optimal linear estimation of the functional

$$A\zeta = \sum_{j=0}^{\infty} \int_{S_n} a(j,x)\zeta(-j,x)\, m_n(dx)$$

which depends on unknown values of a periodically correlated (cyclostationary with period T) with respect to time isotropic on the sphere S_n in Euclidean space $\mathbb{E}^n$ random field $\zeta(j,x)$, $j \in Z$, $x \in S_n$. Estimates are based on observations of the field $\zeta(j,x)+\theta(j,x)$ at points (j,x), $j=-1,-2,\ldots$, $x \in S_n$, where $\theta(j,x)$ is an uncorrelated with $\zeta(t,x)$ periodically correlated with respect to time isotropic on the sphere S_n random field. Formulas are derived for computing the value of the mean square error and the spectral characteristic of the optimal linear estimate of the functional $A\zeta$ in the case of spectral certainty where spectral densities of the fields are known. Formulas are proposed that determine the least favourable spectral densities and minimax-robust spectral characteristic of the optimal estimate of the functional $A\zeta$ in the case of spectral uncertainty where spectral densities are not known exactly, but classes $D = D_f \times D_g$ of admissible spectral densities are given.

We use the Kolmogorov (see, e.g. Kolmogorov, 1992) Hilbert space projection method based on properties of Fourier coefficients of the inverse to spectral density matrices. While in the paper by Dubovets'ka, Masyutka and Moklyachuk (2014a) the filtering problem is investigated with the help of the method based on factorization of spectral density matrices.

2. Spectral properties of periodically correlated random fields

Let S_n be a unit sphere in the n-dimensional Euclidean space $\mathbb{E}^n$, let $m_n(dx)$ be the Lebesgue measure on S_n, and let

$$S_m^l(x),\ l=1,\ldots,h(m,n);\ m=0,1,\ldots$$

be the orthonormal spherical harmonics of degree m (Müller, 1998).

A mean square continuous random field $\zeta(j,x)$, $j \in \mathbb{Z}$, $x \in S_n$ is called periodically correlated (cyclostationary with period T) with respect to time isotropic on the sphere S_n if

$$\mathbb{E}\zeta(j+T,x) = \mathbb{E}\zeta(j,x) = 0, \quad \mathbb{E}|\zeta(j,x)|^2 < \infty,$$
$$\mathbb{E}\left(\zeta(j+T,x)\overline{\zeta(k+T,y)}\right) = B(j,k,\cos\langle x,y\rangle),$$

where $\cos\langle x,y\rangle = (x,y)$ is the 'angular' distance between points $x,y \in S_n$. This random field can be represented in the form

$$\zeta(j,x) = \sum_{m=0}^{\infty}\sum_{l=1}^{h(m,n)} S_m^l(x)\zeta_m^l(j),$$
$$\zeta_m^l(j) = \int_{S_n} \zeta(j,x)S_m^l(x)\, m_n(dx),$$

where

$$\zeta^l_m(j),\ j\in\mathbb{Z},\ m=0,1,\dots;\ l=1,\dots,h(m,n)$$

are mutually uncorrelated periodically correlated stochastic sequences with the correlation functions $b^{\zeta}_m(j,k)$:

$$\mathbb{E}\left(\zeta^l_m(j+T)\overline{\zeta^v_u(k+T)}\right)=\delta^u_m\delta^v_l\ b^{\zeta}_m(j,k),$$
$$m,u=0,1,\dots;\quad l,v=1,\dots,h(m,n);\quad j,k\in\mathbb{Z}.$$

The correlation function of the random field $\zeta(t,x)$ can be represented as follows:

$$B(j,k,\cos\langle x,y\rangle)=\frac{1}{\omega_n}\sum_{m=0}^{\infty}h(m,n)\frac{C^{(n-2)/2}_m(\cos\langle x,y\rangle)}{C^{(n-2)/2}_m(1)}\,b^{\zeta}_m(j,k),$$

where $\omega_n=(2\pi)^{n/2}\Gamma(n/2)$ and $C^l_m(z)$ are the Gegenbauer polynomials (Müller, 1998).

It follows from the Gladyshev (1961) (see also Makagon, 2011) results that the stochastic sequence $\zeta^l_m(j)$, $j\in\mathbb{Z}$ is periodically correlated with period T if and only if there exists a T-variate stationary sequence

$$\vec{\xi}^l_m(j)=\left\{\xi^l_{mk}(j)\right\}_{k=0}^{T-1},\quad j\in\mathbb{Z},$$

such that $\zeta^l_m(j)$ can be represented in the form

$$\zeta^l_m(j)=\sum_{k=0}^{T-1}e^{2\pi ikj/T}\xi^l_{mk}(j),\quad j\in\mathbb{Z}.$$

The sequence $\vec{\xi}^l_m(j)=\{\xi^l_{mk}(j)\}_{k=0}^{T-1}$ is called generating sequence of the periodically correlated sequence $\zeta^l_m(j)$.

Denote by $\Phi^{\vec{\xi}}_m(d\lambda)$ the matrix spectral measure function of the T-variable vector stationary sequence $\vec{\xi}^l_m(j)=\{\xi^l_{mk}(j)\}_{k=0}^{T-1}$ resulting from the Gladyshev representation. Denote by $\Phi^{\vec{\zeta}}_m(d\lambda)$ the matrix spectral measure function of the T-variable vector stationary sequence

$$\vec{\zeta}^l_m(j)=\{\zeta^l_{mk}(j)\}_{k=0}^{T-1},\quad [\vec{\zeta}^l_m(j)]_k=\zeta^l_m(jT+k)$$
$$j\in\mathbb{Z},\quad k=0,1,\dots,T-1,$$

arising from the splitting into blocks of length T the univariate periodically correlated sequence $\zeta^l_m(j)$. The relation of spectral matrices $\Phi^{\vec{\xi}}_m(d\lambda)$ and $\Phi^{\vec{\zeta}}_m(d\lambda)$ is described by the formula

$$\Phi^{\vec{\zeta}}_m(d\lambda)=T\cdot V(\lambda)\Phi^{\vec{\xi}}_m(d\lambda/T)V^{-1}(\lambda),$$

where $V(\lambda)$ is an unitary $T\times T$ matrix whose (k,j)-th element is of the form

$$v_{kj}(\lambda)=\frac{1}{\sqrt{T}}e^{2\pi ijk/T+ik\lambda/T},\quad k,j=0,1,\dots,T-1.$$

This relation can also be expressed as

$$\Phi^{\vec{\xi}}_m(d\lambda)=\frac{1}{T}\cdot V^{-1}(T\lambda)\Phi^{\vec{\zeta}}_m(Td\lambda)V(T\lambda).$$

Consequently, if there exists the spectral density matrix $F_m^{\vec{\xi}}(\lambda)$ of the T-variate stationary sequence $\vec{\xi}_m^l(j)$ then there exists the spectral density matrix $F_m^{\vec{\zeta}}(\lambda)$of the T-variate stationary sequence $\vec{\zeta}_m^l(j)$ and these two density matrices satisfy the relation

$$F_m^{\vec{\zeta}}(\lambda) = T \cdot V(\lambda)F_m^{\vec{\xi}}(\lambda/T)V^{-1}(\lambda).$$

3. Hilbert space projection method of filtering

Consider the problem of mean square optimal linear estimation of the unknown value of the functional

$$A\zeta = \sum_{j=0}^{\infty} \int_{S_n} a(j,x)\zeta(-j,x)\, m_n(dx)$$

which depends on unknown values of a periodically correlated (cyclostationary with period T) with respect to time isotropic on the sphere S_n in Euclidean space E^n random field $\zeta(j,x)$, $j \in Z$, $x \in S_n$. Estimates are based on observations of the field $\zeta(j,x) + \theta(j,x)$ at points (j,x), $j \leqslant 0$, $x \in S_n$, where $\theta(j,x)$ is an uncorrelated with $\zeta(j,x)$ periodically correlated with respect to time isotropic on the sphere S_n random field which has the representation

$$\begin{aligned}\theta(j,x) &= \sum_{m=0}^{\infty}\sum_{l=1}^{h(m,n)} S_m^l(x)\theta_m^l(j)\\ &= \sum_{m=0}^{\infty}\sum_{l=1}^{h(m,n)} S_m^l(x)\sum_{k=0}^{T-1} e^{2\pi ikj/T}\eta_{mk}^l(j).\end{aligned}$$

In this representation

$$\theta_m^l(j) = \int_{S_n} \theta(j,x)S_m^l(x)\, m_n(dx), \quad j \in \mathbb{Z}, \quad m = 0,1,\dots; \quad l = 1,\dots,h(m,n),$$

are mutually uncorrelated periodically correlated stochastic sequences with the correlation functions $b_m^\theta(j,\ k)$:

$$\mathbb{E}\left(\theta_m^l(j+T)\overline{\theta_u^v(k+T)}\right) = \delta_m^u\delta_l^v\ b_m^\theta(j,k),$$
$$m,u = 0,1,\dots; \quad l,v = 1,\dots,h(m,n); \quad j,k \in \mathbb{Z},$$

and $\vec{\eta}_m^l(j) = \{\eta_{mk}^l(j)\}_{k=0}^{T-1}$ are vector-valued stationary sequences generating the periodically correlated sequences $\theta_m^l(j)$.

Assume that the function $a(j,x)$ which determines the functional

$$\begin{aligned}A\zeta &= \sum_{j=0}^{\infty} \int_{S_n} a(j,x)\zeta(-j,x)m_n(dx)\\ &= \sum_{m=0}^{\infty}\sum_{l=1}^{h(m,n)}\sum_{j=0}^{\infty} a_m^l(j)\zeta_m^l(-j) \qquad (1)\end{aligned}$$

has components

$$a_m^l(j) = \int_{S_n} a(j,x)S_m^l(x)m_n(dx)$$

which satisfy the following condition:

$$\sum_{m=0}^{\infty}\sum_{l=1}^{h(m,n)}\sum_{j=0}^{\infty}\left|a_m^l(j)\right| < \infty. \tag{2}$$

Condition (Equation 2) ensures convergence of the series representation (Equation 1) of the functional $A\zeta$ as well as finiteness of the second moment of the functional: $E|A\zeta|^2 < \infty$.

We will consider random fields which satisfy the following minimality condition

$$\int_{-\pi}^{\pi} Tr\left[\left(F_m(\lambda)+G_m(\lambda)\right)^{-1}\right]d\lambda < \infty. \tag{3}$$

Making use the Gladyshev (1961) results we can represent the functional $A\zeta$ in the form

$$A\zeta = \sum_{m=0}^{\infty}\sum_{l=1}^{h(m,n)}\sum_{j=0}^{\infty}(\vec{a}_m^l(j))^{\top}\vec{\xi}_m^l(j),$$
$$\vec{a}_m^l(j) = (a_{m0}^l(j), a_{m1}^l(j), \dots, a_{m(T-1)}^l(j))^{\top},$$
$$a_{mk}^l(j) = a_m^l(j)e^{2\pi ikj/T},\ k = 0, 1, \dots, T-1,$$

where $\vec{\xi}_m^l(j) = \{\xi_{mk}^l(j)\}_{k=0}^{T-1}$ are vector-valued stationary sequences generating the periodically correlated sequences $\zeta_m^l(j)$.

Every linear estimate $\widehat{A\zeta}$ of the functional $A\zeta$ is determined by spectral stochastic measures

$$(Z_{\vec{\xi}})_m^l(d\lambda) = \left\{(Z_{\vec{\xi}})_{mk}^l(d\lambda)\right\}_{k=0}^{T-1},\quad (Z_{\vec{\eta}})_m^l(d\lambda) = \left\{(Z_{\vec{\eta}})_{mk}^l(d\lambda)\right\}_{k=0}^{T-1},$$

of the generating sequences $\vec{\xi}_m^l(j) = \{\xi_{mk}^l(j)\}_{k=0}^{T-1}$ and $\vec{\eta}_m^l(j) = \{\eta_{mk}^l(j)\}_{k=0}^{T-1}$, and the spectral characteristic

$$h\left(e^{i\lambda}\right) = \left\{h_{ml}\left(e^{i\lambda}\right): l = 1, 2, \dots, h(m,n); m = 0, 1, \dots\right\},$$
$$h_{ml}\left(e^{i\lambda}\right) = \left\{h_{ml}^k\left(e^{i\lambda}\right)\right\}_{k=0}^{T-1},$$

which is from the space $L_2^-(F+G)$ generated by functions

$$h_{ml}(e^{i\lambda}) = \sum_{j=0}^{\infty}h_{ml}(j)e^{-ij\lambda},\ l = 1, \dots, h(m,n), m = 0, 1, \dots,$$

that satisfy condition

$$\sum_{m=0}^{\infty}\sum_{l=1}^{h(m,n)}\int_{-\pi}^{\pi}h_{ml}\left(e^{i\lambda}\right)^{\top}[F_m(\lambda)+G_m(\lambda)]\overline{h_{ml}\left(e^{i\lambda}\right)}d\lambda < \infty.$$

The estimate is of the form

$$\widehat{A\zeta} = \sum_{m=0}^{\infty} \sum_{l=1}^{h(m,n)} \int_{-\pi}^{\pi} h_{ml}^{\top}\left(e^{i\lambda}\right)\left((Z_{\vec{\xi}})_m^l(d\lambda) + (Z_{\vec{\eta}})_m^l(d\lambda)\right).$$

The mean square error $\Delta(h;F,G) = \mathbf{E}|A\zeta - \widehat{A\zeta}|^2$ of the estimate $\widehat{A\zeta}$ is determined by matrices of spectral densities

$$F(\lambda) = \{F_m(\lambda):m = 0,1\ldots\}, \quad G(\lambda) = \{G_m(\lambda):m = 0,1\ldots\}$$

of the generating sequences $\vec{\xi}_m^l(j) = \{\xi_{mk}^l(j)\}_{k=0}^{T-1}$ and $\vec{\eta}_m^l(j) = \{\eta_{mk}^l(j)\}_{k=0}^{T-1}$, and the spectral characteristic $h\left(e^{i\lambda}\right)$ of the estimate

$$\begin{aligned}\Delta(h;F,G) &= E|A\zeta - \widehat{A\zeta}|^2 \\ &= \sum_{m=0}^{\infty} \sum_{l=1}^{h(m,n)} \left\{\frac{1}{2\pi}\int_{-\pi}^{\pi} \left(A_{ml}^{\top}\left(e^{i\lambda}\right) - h_{ml}^{\top}\left(e^{i\lambda}\right)\right)F_m(\lambda)\left(A_{ml}^{\top}\left(e^{i\lambda}\right) - h_{ml}^{\top}\left(e^{i\lambda}\right)\right)^{*} d\lambda\right\} \\ &\quad + \sum_{m=0}^{\infty} \sum_{l=1}^{h(m,n)} \left\{\frac{1}{2\pi}\int_{-\pi}^{\pi} \left(h_{ml}^{\top}\left(e^{i\lambda}\right)\right)G_m(\lambda)\left(h_{ml}^{\top}\left(e^{i\lambda}\right)\right)^{*} d\lambda\right\},\end{aligned} \tag{4}$$

$$A_{ml}\left(e^{i\lambda}\right) = \sum_{j=0}^{\infty} \vec{a}_m^l(j)e^{-ij\lambda}.$$

The spectral characteristic $h(F,G)$ of the optimal linear estimate $\widehat{A\zeta}$ minimizes the value of the mean square error. We first applied the method based on factorizations of matrices of spectral densities and found the spectral characteristic $h(F,G)$ and the mean square error of the least square optimal linear estimate $\widehat{A\zeta}$ of the functional $A\zeta$. The derived results are presented in the article by Dubovets'ka et al. (2014a). Solution of the extrapolation problem is described in the article by Dubovets'ka et al. (2015). For more relative results see articles by Moklyachuk (1994, 1995, 1996) and books by Moklyachuk (2008), Moklyachuk and Masyutka (2012), Golichenko and Moklyachuk (2014).

In this article, we use the Kolmogorov (1992) Hilbert space projection method based on properties of Fourier coefficients of the inverse to spectral density matrices. With the help of the proposed method we can find formulas for calculation the mean square error $\Delta(h;F,G)$ and the spectral characteristic $h(F,G)$ of the optimal linear estimate $\widehat{A\zeta}$ of the functional $A\zeta$ under the condition that spectral densities $F_m(\lambda), G_m(\lambda)$ are known and satisfy the minimality condition (Equation 3). Following the method we found the optimal linear estimate $\widehat{A\zeta}$ as projection of $A\zeta$ on the closed linear subspace $H^{-}_{\zeta+\theta}$ generated in the space $H = L_2(\Omega, \mathcal{F}, P)$ by values $\{\zeta(s,x) + \theta(s,x):s = 0,-1,-2,\ldots;x \in S_n\}$. This projection is determined by conditions: (1) $\widehat{A\zeta} \in H^{-}_{\zeta+\theta}$; (2) $(A\zeta - \widehat{A\zeta}) \perp H^{-}_{\zeta+\theta}$. The mean square error $\Delta(h;F,G)$ and the spectral characteristic $h(F,G)$ are calculated by formulas

$$\begin{aligned}\Delta(h;F,G) &= \sum_{m=0}^{\infty} \sum_{l=1}^{h(m,n)} \frac{1}{2\pi}\int_{-\pi}^{\pi} \Big[r_G(\lambda)(H_m(\lambda))^{-1}F_m(\lambda)(H_m(\lambda))^{-1}(r_G(\lambda))^{*} \\ &\quad + r_F(\lambda)(H_m(\lambda))^{-1}G_m(\lambda)(H_m(\lambda))^{-1}(r_F(\lambda))^{*}\Big]d\lambda \\ &= \sum_{m=0}^{\infty} \sum_{l=1}^{h(m,n)} \left\langle \vec{c}_m^l, \mathbf{B}_m\vec{c}_m^l\right\rangle + \left\langle \vec{a}_m^l, \mathbf{R}_m\vec{a}_m^l\right\rangle,\end{aligned} \tag{5}$$

$$\begin{aligned}h_{ml}(F,G)^{\top} &= r_F(\lambda)(H_m(\lambda))^{-1} \\ &= A_{ml}\left(e^{i\lambda}\right)^{\top} - r_G(\lambda)(H_m(\lambda))^{-1},\end{aligned} \tag{6}$$

where

$$r_F(\lambda) = A_{ml}\left(e^{i\lambda}\right)^{\mathsf{T}} F_m(\lambda) - C_{ml}\left(e^{i\lambda}\right)^{\mathsf{T}},$$
$$r_G(\lambda) = A_{ml}\left(e^{i\lambda}\right)^{\mathsf{T}} G_m(\lambda) + C_{ml}\left(e^{i\lambda}\right)^{\mathsf{T}},$$
$$H_m(\lambda) = F_m(\lambda) + G_m(\lambda),$$
$$\vec{a}^{\,l}_m = \left\{\vec{a}^{\,l}_m(j), j = 0, 1, 2, \dots\right\},$$
$$\vec{c}^{\,l}_m = \left\{\vec{c}^{\,l}_m(j), j = 1, 2, 3, \dots\right\},$$
$$C_{ml}\left(e^{i\lambda}\right) = \sum_{j=1}^{\infty} \vec{c}^{\,l}_m(j) e^{ij\lambda}, \quad \vec{c}^{\,l}_m = \mathbf{B}_m^{-1}\mathbf{D}_m\vec{a}^{\,l}_m,$$

$\mathbf{B}_m = \{B_m(k,j)\}_{k=1}^{\infty}\{0\}_{j=1}^{\infty}$, $\mathbf{D}_m = \{D_m(k,j)\}_{k=1}^{\infty}\{0\}_{j=0}^{\infty}$, $\mathbf{R}_m = \{R_m(k,j)\}_{k=0}^{\infty}\{0\}_{j=0}^{\infty}$ are operators determined by the following block-matrices

$$B_m(k,j) = \frac{1}{2\pi}\int_{-\pi}^{\pi} \left[(H_m(\lambda))^{-1}\right]^{\mathsf{T}} e^{i(j-k)\lambda} d\lambda,$$
$$D_m(k,j) = \frac{1}{2\pi}\int_{-\pi}^{\pi} \left[F_m(\lambda)(H_m(\lambda))^{-1}\right]^{\mathsf{T}} e^{-i(k+j)\lambda} d\lambda,$$
$$R_m(k,j) = \frac{1}{2\pi}\int_{-\pi}^{\pi} \left[F_m(\lambda)(H_m(\lambda))^{-1}G_m(\lambda)\right]^{\mathsf{T}} e^{i(j-k)\lambda} d\lambda.$$

Let us summarize our results and present them in the form of a statement.

THEOREM 3.1 *Let the function $a(j,x)$ which determines the functional $A\zeta$ satisfy conditions (Equation 2). Let $\zeta(j,x), \theta(j,x)$ be uncorrelated periodically correlated with respect to time isotropic on the sphere S_n random fields which have spectral densities $F_m(\lambda)$, $G_m(\lambda)$ that satisfy the minimality condition (Equation 3). The value of the mean square error $\Delta(h;F,G)$ and the spectral characteristic $h(F,G)$ of the optimal linear estimate of the functional $A\zeta$ based on observations of the field $\zeta(j,x) + \theta(j,x)$ for $j \leqslant 0$, $x \in S_n$, can be calculated by formulas 5 and 6.*

4. Minimax-robust method of filtering

The proposed formulas may be employed under the condition that spectral densities $F_m(\lambda)$ and $G_m(\lambda)$ of the fields $\zeta(j,x), \theta(j,x)$ are exactly known. In the case where the densities are not known exactly but a set $D = D_F \times D_G$ of possible spectral densities is given the minimax (robust) approach to estimation of functionals of the unknown values of random fields is reasonable. Instead of searching an estimate that is optimal for a given spectral densities we find an estimate that minimizes the mean square error for all spectral densities from given class simultaneously.

Definition 1 Spectral densities $F^0(\lambda), G^0(\lambda)$ are called least favourable in a given class D for the optimal linear estimation of the functional $A\zeta$ if

$$\Delta(h(F^0,G^0);F^0,G^0) = \max_{(F,G)\in D} \Delta(h(F,G);F,G).$$

It follows from the relationships (Equations 1–6) that the next theorem holds true.

THEOREM 4.1 *Spectral densities $F^0_m(\lambda), G^0_m(\lambda)$ that satisfy the minimality condition (Equation 3) are least favourable in a class D for the optimal linear estimation of the functional $A\zeta$ if they determine operators $\mathbf{B}^0_m$, $\mathbf{D}^0_m$, $\mathbf{R}^0_m$ giving solution to the extremum problem*

$$\max_{(F,G)\in D} \sum_{m=0}^{\infty}\sum_{l=1}^{h(m,n)} \left[\left\langle \mathbf{B}_m^{-1}\mathbf{D}_m\vec{a}^{\,l}_m, \mathbf{D}_m\vec{a}^{\,l}_m\right\rangle + \left\langle \vec{a}^{\,l}_m, \mathbf{R}_m\vec{a}^{\,l}_m\right\rangle\right]$$
$$= \sum_{m=0}^{\infty}\sum_{l=1}^{h(m,n)} \left[\left\langle (\mathbf{B}^0_m)^{-1}\mathbf{D}^0_m\vec{a}^{\,l}_m, \mathbf{D}^0_m\vec{a}^{\,l}_m\right\rangle + \left\langle \vec{a}^{\,l}_m, \mathbf{R}^0_m\vec{a}^{\,l}_m\right\rangle\right]. \tag{7}$$

Definition 2 The spectral characteristic

$$h^0\left(e^{i\lambda}\right)=\left\{h^0_{ml}\left(e^{i\lambda}\right):l=1,\dots,h(m,n),m=0,1,\dots\right\}$$

of the optimal linear estimation of the functional $A\zeta$ is called minimax-robust if there are satisfied conditions

$$h^0\left(e^{i\lambda}\right)\in H_D=\bigcap_{(F,G)\in D}L_2^-(F+G),$$

$$\min_{h\in H_D}\sup_{(F,G)\in D}\Delta(h;F,G)=\sup_{(F,G)\in D}\Delta(h^0;F,G).$$

The least favourable spectral densities $F^0_m(\lambda),G^0_m(\lambda)$ and the minimax (robust) spectral characteristic $h^0\left(e^{i\lambda}\right)\in H_D$ form a saddle point of the function $\Delta(h;F,G)$. The saddle point inequalities

$$\Delta(h;F^0,G^0)\geqslant\Delta(h^0;F^0,G^0)\geqslant\Delta(h^0;F,G),$$
$$\forall(F,G)\in D,\quad\forall h\in H_D$$

hold true if $h^0=h(F^0,G^0)\in H_D$ and (F^0,G^0) is a solution to the conditional extremum problem

$$\Delta(h(F^0,G^0);F^0,G^0)=\sup_{(F,G)\in D}\Delta(h(F^0,G^0);\quad F,G),\tag{8}$$

where

$$\Delta(h(F^0,G^0);F,G)$$
$$=\sum_{m=0}^{\infty}\sum_{l=1}^{h(m,n)}\frac{1}{2\pi}\int_{-\pi}^{\pi}\left[r^0_G(\lambda)(H^0_m(\lambda))^{-1}F_m(\lambda)(H^0_m(\lambda))^{-1}(r^0_G(\lambda))^*\right.$$
$$\left.+r^0_F(\lambda)(H^0_m(\lambda))^{-1}G_m(\lambda)(H^0_m(\lambda))^{-1}(r^0_F(\lambda))^*\right]d\lambda.$$

The conditional extremum problem (Equation 8) is equivalent to the unconditional extremum problem

$$\Delta_D(F,G)=-\Delta(h(F^0,G^0);\quad F,G)+\delta((F,G)|D)\to\inf,\tag{9}$$

where $\delta((F,G)|D)$ is the indicator function of the set D. A solution to the problem (Equation 9) is characterized by condition $0\in\partial\Delta_D(F^0,G^0)$, where $\partial\Delta_D(F^0,G^0)$ is a subdifferential of the convex functional $\Delta_D(F,G)$ at point (F^0,G^0).

The form of the functional $\Delta(h(F^0,G^0);F,G)$ is convenient for application the Lagrange method of indefinite multipliers for finding solution to the problem (Equation 9). Making use the method of Lagrange multipliers and the form of subdifferentials of the indicator functions we describe relations that determine least favourable spectral densities in some special classes of spectral densities (see books by Moklyachuk, 2008; Moklyachuk and Masyutka, 2012; Golichenko and Moklyachuk, 2014for more details).

5. Least favourable spectral densities in the class $D_0\times D_V^U$

Consider the problem of minimax estimation of functional $A\zeta$ from periodically correlated random field $\zeta(j,x)$ based on observations of the field $\zeta(j,x)+\theta(j,x)$ at points $j=-1,-2,\dots,x\in S_n$, under the condition that matrices of spectral densities $F(\lambda)=\{F_m(\lambda):m=0,1\dots\}$ and $G(\lambda)=\{G_m(\lambda):m=0,1\dots\}$ of the field $\zeta(j,x)$ and the field $\theta(j,x)$ are not known exactly, but the following pairs of sets of spectral densities that give restrictions on the first moment and describe the "strip" model of spectral densities are given. The first pair is

$$D_0^1 = \left\{ F(\lambda) | \frac{1}{2\pi\omega_n} \sum_{m=0}^{\infty} h(m,n) \int_{-\pi}^{\pi} TrF_m(\lambda)d\lambda = p \right\},$$

$$D_V^{U^1} = \left\{ G(\lambda) | TrV_m(\lambda) \leqslant TrG_m(\lambda) \leqslant TrU_m(\lambda), \right.$$

$$\left. \frac{1}{2\pi\omega_n} \sum_{m=0}^{\infty} h(m,n) \int_{-\pi}^{\pi} TrG_m(\lambda)d\lambda = q \right\}.$$

The second pair of sets of admissible spectral densities is

$$D_0^2 = \left\{ F(\lambda) | \frac{1}{2\pi\omega_n} \sum_{m=0}^{\infty} h(m,n) \int_{-\pi}^{\pi} f_m^{kk}(\lambda)d\lambda = p_k, k = \overline{1,T} \right\},$$

$$D_V^{U^2} = \left\{ G(\lambda) | v_m^{kk}(\lambda) \leqslant g_m^{kk}(\lambda) \leqslant u_m^{kk}(\lambda), \right.$$

$$\left. \frac{1}{2\pi\omega_n} \sum_{m=0}^{\infty} h(m,n) \int_{-\pi}^{\pi} g_m^{kk}(\lambda)d\lambda = q_k, k = \overline{1,T} \right\}.$$

The third pair of sets of admissible spectral densities is

$$D_0^3 = \left\{ F(\lambda) | \frac{1}{2\pi\omega_n} \sum_{m=0}^{\infty} h(m,n) \int_{-\pi}^{\pi} \langle B, F_m(\lambda) \rangle d\lambda = p \right\},$$

$$D_V^{U^3} = \left\{ G(\lambda) | \langle B_2, V_m(\lambda) \rangle \leqslant \langle B_2, G_m(\lambda) \rangle \leqslant \langle B_2, U_m(\lambda) \rangle, \right.$$

$$\left. \frac{1}{2\pi\omega_n} \sum_{m=0}^{\infty} h(m,n) \int_{-\pi}^{\pi} \langle B_2, G_m(\lambda) \rangle d\lambda = q \right\};$$

The fourth pair of sets of admissible spectral densities is

$$D_0^4 = \left\{ F(\lambda) | \frac{1}{2\pi\omega_n} \sum_{m=0}^{\infty} h(m,n) \int_{-\pi}^{\pi} F_m(\lambda)d\lambda = P \right\},$$

$$D_V^{U^4} = \left\{ G(\lambda) | V_m(\lambda) \leqslant G_m(\lambda) \leqslant U_m(\lambda), \right.$$

$$\left. \frac{1}{2\pi\omega_n} \sum_{m=0}^{\infty} h(m,n) \int_{-\pi}^{\pi} G_m(\lambda)d\lambda = Q \right\}.$$

Here $V_m(\lambda), U_m(\lambda)$ are given matrices of spectral densities, $p, q, p_k, q_k, k = \overline{1,T}$ are given numbers, B_1, B_2, P, Q are given positive-definite Hermitian matrices.

Making use the method of Lagrange multipliers and the form of subdifferentials of the indicator functions of the sets D^0 and D_V^U (see Franke, 1985) we can conclude that the condition $0 \in \partial\Delta_D(F^0, G^0)$ is satisfied for $D = D^0 \times D_V^U$ if components of spectral densities $F^0(\lambda) = \{F_m^0(\lambda): m = 0, 1 \ldots\}$ and $G^0(\lambda) = \{G_m^0(\lambda): m = 0, 1 \ldots\}$ satisfy the following equations.

For the first pair of sets of admissible spectral densities we have equations

$$\sum_{l=1}^{h(m,n)} (r_G^0(\lambda))^* r_G^0(\lambda) = \alpha_m^2 (H_m^0(\lambda))^2, \tag{10}$$

$$\sum_{l=1}^{h(m,n)} (r_F^0(\lambda))^* r_F^0(\lambda) = (\beta_m^2 + \gamma_{m_1}(\lambda) + \gamma_{m_2}(\lambda))(H_m^0(\lambda))^2; \tag{11}$$

For the second pair of sets of admissible spectral densities we have equations

$$\sum_{l=1}^{h(m,n)} (r_G^0(\lambda))^* r_G^0(\lambda) = H_m^0(\lambda)\left\{\alpha_{mk}^2 \delta_{kn}\right\}_{k,n=1}^{T} H_m^0(\lambda), \tag{12}$$

$$\sum_{l=1}^{h(m,n)} (r_F^0(\lambda))^* r_F^0(\lambda) = H_m^0(\lambda)\left\{(\beta_{mk}^2 + \gamma_{m_1 k}(\lambda) + \gamma_{m_2 k}(\lambda))\delta_{kn}\right\}_{k,n=1}^{T} H_m^0(\lambda); \tag{13}$$

For the third pair of sets of admissible spectral densities we have equations

$$\sum_{l=1}^{h(m,n)} (r_G^0(\lambda))^* r_G^0(\lambda) = \alpha_m^2 H_m^0(\lambda) B_1^T H_m^0(\lambda), \tag{14}$$

$$\sum_{l=1}^{h(m,n)} ((r_F^0(\lambda))^* (r_F^0(\lambda) = (\beta_m^2 + \gamma_{m_1}(\lambda) + \gamma_{m_2}(\lambda)) H_m^0(\lambda) B_2^T H_m^0(\lambda); \tag{15}$$

For the fourth pair of sets of admissible spectral densities we have equations

$$\sum_{l=1}^{h(m,n)} (r_G^0(\lambda))^* r_G^0(\lambda) = H_m^0(\lambda)\vec{\alpha}_m \cdot \vec{\alpha}_m^{\,*} H_m^0(\lambda), \tag{16}$$

$$\sum_{l=1}^{h(m,n)} (r_F^0(\lambda))^* r_F^0(\lambda) = H_m^0(\lambda)(\vec{\beta} \cdot \vec{\beta}^* + \Gamma_{m_1}(\lambda) + \Gamma_{m_2}(\lambda)) H_m^0(\lambda). \tag{17}$$

Here $\Gamma_{m_1}(\lambda), \Gamma_{m_2}(\lambda)$ are Hermitian matrices,

$$\gamma_{m_1}(\lambda) \leqslant 0, \gamma_{m_1 k}(\lambda) \leqslant 0, \quad k = \overline{1,T} \text{ a.e.},$$
$$\gamma_{m_2}(\lambda) \geqslant 0, \gamma_{m_2 k}(\lambda) \geqslant 0, \quad k = \overline{1,T} \text{ a.e.},$$
$$\Gamma_{m_1}(\lambda) \leqslant 0, \Gamma_{m_2}(\lambda) \geqslant 0 \text{ a.e.},$$
$$\gamma_{m_1}(\lambda) = 0 : TrG_m^0(\lambda) > TrV_m(\lambda),$$
$$\gamma_{m_2}(\lambda) = 0 : TrG_m^0(\lambda) < TrU_m(\lambda),$$
$$\gamma_{m_1 k}(\lambda) = 0 : g_m^{0kk}(\lambda) > v_m^{kk}(\lambda),$$
$$\gamma_{m_2 k}(\lambda) = 0 : g_m^{0kk}(\lambda) < u_m^{kk}(\lambda),$$
$$\gamma_{m_1}(\lambda) = 0 : \left\langle B_2, G_m^0(\lambda)\right\rangle > \left\langle B_2, V_m(\lambda)\right\rangle,$$
$$\gamma_{m_2}(\lambda) = 0 : \left\langle B_2, G_m^0(\lambda)\right\rangle < \left\langle B_2, U_m(\lambda)\right\rangle,$$
$$\Gamma_{m_1}(\lambda) = 0 : G_m^0(\lambda) > V_m(\lambda),$$
$$\Gamma_{m_2}(\lambda) = 0 : G_m^0(\lambda) < U_m(\lambda),$$

and $\alpha_m^2, \beta_m^2, \alpha_{mk}^2, \beta_{mk}^2, \vec{\alpha}_m, \vec{\beta}_m$ are the unknown Lagrange multipliers.

The following statement holds true.

THEOREM 5.1 *Let conditions (Equation 2), (Equation 3) hold true. The least favourable spectral densities $F^0_m(\lambda)$, $G^0_m(\lambda)$ in the first pair D_0 and D^U_V of spectral densities for the optimal linear estimation of the functional $A\zeta$ are determined by relations (Equation 7), (Equation 10), (Equation 11). The least favourable spectral densities $F^0_m(\lambda)$, $G^0_m(\lambda)$ in the second pair D_0 and D^U_V of spectral densities are determined by relations (Equation 7), (Equation 12), (Equation 13). The least favourable spectral densities $F^0_m(\lambda)$, $G^0_m(\lambda)$ in the third pair D_0 and D^U_V of spectral densities are determined by relations (Equation 7), (Equation 14) and (Equation 15). The least favourable spectral densities $F^0_m(\lambda)$, $G^0_m(\lambda)$ in the fourth pair D_0 and D^U_V of spectral densities are determined by relations (Equation 7), (Equation 16) and (Equation 17). The minimax spectral characteristic of the optimal estimate of the functional $A\zeta$ is calculated by formula 6.*

6. Conclusions

In this paper, we study the filtering problem for functionals which depend on unknown values of a periodically correlated (cyclostationary with period T) with respect to time isotropic on the sphere S_n in Euclidean space E^n random field. The problem is considered in the case of spectral certainty where the matrices of spectral densities of random fields are known exactly and in the case of spectral uncertainty where matrices of spectral densities of random fields are not known exactly, but some restrictions on matrices are given which they must satisfy. We propose formulas for calculation the spectral characteristic and the mean square error of the optimal linear estimate of the functional

$$A\zeta = \sum_{j=0}^{\infty} \int_{S_n} a(j,x)\zeta(-j,x)m_n(dx)$$

which depends on unknown values of a periodically correlated (cyclostationary with period T) with respect to time isotropic on the sphere S_n random field $\zeta(j,x)$ from observations of the field $\zeta(j,x) + \theta(j,x)$ at points (j,x), $j \leqslant 0$, $x \in S_n$, where $\theta(j,x)$ is an uncorrelated with $\zeta(j,x)$ periodically correlated with respect to time isotropic on the sphere S_n random field provided that matrices of spectral densities $F_m(\lambda)$, $G_m(\lambda)$ of the vector-valued stationary sequences that generate the random fields $\zeta(j,x)$, $\theta(j,x)$ are known exactly. We propose a representation of the mean square error in the form of linear functional in $L_1 \times L_1$ with respect to spectral densities (F, G), which allows us to solve the corresponding conditional extremum problem and describe the minimax (robust) estimates of the functional. The least favourable spectral densities and the minimax (robust) spectral characteristics of the optimal estimates of the functional $A\zeta$ are determined for some special classes of spectral densities.

Acknowledgements
The authors would like to thank the referees for careful reading of the article and giving constructive suggestions.

Funding
The authors received no direct funding for this research.

Author details
Iryna Golichenko[1]
E-mail: irinka.riabinka@gmail.com
Oleksandr Masyutka[2]
E-mail: omasyutka@gmail.com
Mikhail Moklyachuk[3]
E-mail: moklyachuk@gmail.com
[1] Department of Mathematical Analysis and Probability Theory, National Technical University of Ukraine "Kyiv Polytechnic Institute", Kyiv 03056, Ukraine.
[2] Department of Mathematics and Theoretical Radiophysics, Taras Shevchenko National University of Kyiv, Kyiv 01601, Ukraine.
[3] Department of Probability Theory, Statistics and Actuarial Mathematics, Taras Shevchenko National University of Kyiv, Kyiv 01601, Ukraine.

References

Adshead, P., & Hu, W. (2012). Fast computation of first-order feature-bispectrum corrections. *Physical Review D, 85*, 103531.

Antoni, J. (2009). Cyclostationarity by examples. *Mechanical Systems and Signal Processing, 23*, 987–1036.

Bartlett, J. G. (1999). The standard cosmological model and CMB anisotropies. *New Astronomy Reviews, 43*, 83–109.

Cressie, N., & Wikle, C. K. (2011). *Statistics for spatio-temporal data* (Wiley series in probability and statistics). Hoboken, NJ: Wiley.

Dubovets'ka, I. I., Masyutka, O. Y., & Moklyachuk, M. P. (2012). Interpolation of periodically correlated stochastic sequences. *Theory of Probability and Mathematical Statistics, 84*, 43–56.

Dubovets'ka, I. I., Masyutka, O. Y., & Moklyachuk, M. P. (2014). Filtering problems for periodically correlated isotropic random fields. *Mathematics and Statistics, 2*, 162–171.

Dubovets'ka, I. I., Masyutka, O. Y., & Moklyachuk, M. P. (2015). Estimation problems for periodically correlated isotropic random fields. *Methodology and Computing in Applied Probability, 17*, 41–57.

Dubovets'ka, I. I., & Moklyachuk, M. P. (2013). Filtration of linear functionals of periodically correlated sequences. *Theory of Probability and Mathematical Statistics, 86*, 51–64.

Dubovets'ka, I. I., & Moklyachuk, M. P. (2014a). Extrapolation of periodically correlated processes from observations with noise. *Theory of Probability and Mathematical Statistics, 88*, 67–83.

Dubovets'ka, I. I., & Moklyachuk, M. P. (2014b). On minimax estimation problems for periodically correlated stochastic processes. *Contemporary Mathematics and Statistics, 2*, 123–150.

Franke, J. (1985). Minimax robust prediction of discrete time series. *Zeitschrift fur Wahrscheinlichkeitstheorie und Verwandte Gebiete, 68*, 337–364.

Gaetan, C., & Guyon, X. (2010). *Spatial statistics and modeling* (Springer series in statistics, Vol. 81). New York, NY: Springer.

Gardner, W. A. (1994). *Cyclostationarity in communications and signal processing*. New York, NY: IEEE Press.

Gladyshev, E. G. (1961). Periodically correlated random sequences. *Soviet Mathematics. Doklady, 2*, 385–388.

Golichenko, I. I., & Moklyachuk, M. P. (2014). *Estimates of functionals of periodically correlated processes*. Kyiv: NVP "Interservis".

Grenander, U. (1957). A prediction problem in game theory. *Arkiv för Matematik, 3*, 371–379.

Hu, W., & Dodelson, S. (2002). Cosmic microwave background anisotropies. *Annual Review of Astronomy and Astrophysics, 40*, 171–216.

Hurd, H. L., & Miamee, A. (2007). *Periodically correlated random sequences: Spectral theory and practice* (Wiley Series in Probability and Statistics, Wiley Interscience). Hoboken, NJ: Wiley.

Jones, P. D. (1994). Hemispheric surface air temperature variations: A reanalysis and an update to 1993. *Journal of Climate, 7*, 1794–1802.

Kailath, T. (1974). A view of three decades of linear filtering theory. *IEEE Transactions on Information Theory, 20*, 146–181.

Kakarala, R. (2012). The bispectrum as a source of phase-sensitive invariants for Fourier descriptors: A group-theoretic approach. *Journal of Mathematical Imaging and Vision, 44*, 341–353.

Kassam, S. A., & Poor, H. V. (1985). Robust techniques for signal processing: A survey. *Proceedings of the IEEE, 73*, 433–481.

Kogo, N., & Komatsu, E. (2006). Angular trispectrum of CMB temperature anisotropy from primordial non-Gaussianity with the full radiation transfer function. *Physical Review D, 73*, 083007–083012.

Kolmogorov, A. N. (1992). *Selected works of A. N. Kolmogorov. Volume II: Probability theory and mathematical statistics*. (G. Lindquist, Trans.). Dordrecht: Kluwer.

Makagon, A. (2011). Stationary sequences associated with a periodically correlated sequence. *Probability and Mathematical Statistics, 31*, 263–283.

Marinucci, D., & Peccati, G. (2011). *Random fields on the sphere. London mathematical society lecture notes series* (Vol. 389). Cambridge: Cambridge University Press.

Moklyachuk, M. P. (1994). Minimax filtering of time-homogeneous isotropic random fields on a sphere. *Theory of Probability and Mathematical Statistics, 49*, 137–146.

Moklyachuk, M. P. (1995). Extrapolation of time-homogeneous random fields that are isotropic on a sphere. I. *Theory of Probability and Mathematical Statistics, 51*, 137–146.

Moklyachuk, M. P. (1996). Extrapolation of time-homogeneous random fields that are isotropic on a sphere. II. *Theory of Probability and Mathematical Statistics, 53*, 137–148.

Moklyachuk, M. P. (2008). *Robust estimates for functionals of stochastic processes*. Kyiv: Kyiv University.

Moklyachuk, M., & Masyutka, O. (2012). *Minimax-robust estimation technique for stationary stochastic processes*. Saarbrücken: LAP Lambert .

Moklyachuk, M. P., & Yadrenko, M. I. (1979). Linear statistical problems for homogeneous isotropic random fields on a sphere. I. *Theory of Probability and Mathematical Statistics, 18*, 115–124.

Moklyachuk, M. P., & Yadrenko, M. I. (1980). Linear statistical problems for homogeneous isotropic random fields on a sphere. II. *Theory of Probability and Mathematical Statistics, 19*, 129–139.

Müller, C. (1998). *Analysis of spherical symmetries in Euclidean spaces*. New York, NY: Springer-Verlag.

North, G. R., & Cahalan, R. F. (1981). Predictability in a solvable stochastic climate model. *Journal of Atmospheric Sciences, 38*, 504–513.

Okamoto, T., & Hu, W. (2002). Angular trispectra of CMB temperature and polarization. *Physical Review D, 66*, 063008.

Rozanov, Y. A. (1967). *Stationary stochastic processes*. San Francisco, CA: Holden-Day.

Serpedin, E., Panduru, F., Sari, I., & Giannakis, G. B. (2005). Bibliography on cyclostationarity. *Signal Processing, 85*, 2233–2303.

Subba Rao, T., & Terdik, G. (2006). Multivariate non-linear regression with applications. In P. Bertail, P. Doukhan, & P. Soulier (Eds.), *Dependence in probability and statistics* (pp. 431–470). New York, NY: Springer Verlag.

Subba Rao, T., & Terdik, G. (2012). Statistical analysis of spatio-temporal models and their applications. In C. R. Rao (Ed.), *Handbook of statistics* (Vol. 30pp. 521–541). Amsterdam: Elsevier B.V. .

Terdik, G. (2013, February 17). *Angular spectra for non-Gaussian isotropic fields* (arXiv:1302.4049v1.pdf[stat.AP]).

Vastola, K. S., & Poor, H. V. (1983). An analysis of the effects of spectral uncertainty on Wiener filtering. *Automatica, 28*, 289–293.

Wiener, N. (1966). *Extrapolation, interpolation, and smoothing of stationary time series. With engineering applications*. Cambridge: MIT Press, Massachusetts Institute of Technology.

Yadrenko, M. I. (1983). *Spectral theory of random fields*. New York, NY: Optimization Software.

Yaglom, A. M. (1987). *Correlation theory of stationary and related random functions* (Vol. I: Basic results. Vol. II: Supplementary notes and references). New York, NY: Springer-Verlag.

8

Minimax prediction of random processes with stationary increments from observations with stationary noise

Maksym Luz[1] and Mikhail Moklyachuk[1*]

*Corresponding author: Mikhail Moklyachuk, Department of Probability Theory, Statistics and Actuarial Mathematics, Taras Shevchenko National University of Kyiv, Kyiv 01601, Ukraine

Email: Moklyachuk@gmail.com

Reviewing editor: Zudi Liu, University of Southampton, UK

Abstract: We deal with the problem of mean square optimal estimation of linear functionals which depend on the unknown values of a random process with stationary increments based on observations of the process with noise, where the noise process is a stationary process. Formulas for calculating values of the mean square errors and the spectral characteristics of the optimal linear estimates of the functionals are derived under the condition of spectral certainty, where the spectral densities of the processes are exactly known. In the case of spectral uncertainty, where the spectral densities of the processes are not exactly known while a class of admissible spectral densities is given, relations that determine the least favorable spectral densities and the minimax robust spectral characteristics are proposed.

Subjects: Mathematics Statistics; Operations Research; Optimization; Probability; Science; Statistical Theory Methods; Statistics; Statistics Probability; Stochastic Models Processes

Keywords: random process with stationary increments; minimax robust estimate; mean square error; least favorable spectral density; minimax spectral characteristic

AMS Subject classifications: 60G10; 60G25; 60G35; Secondary: 62M20; 93E10; 93E11

ABOUT THE AUTHORS

Maksym Luz is a Ph D student, Department of Probability Theory, Statistics and Actuarial Mathematics, Taras Shevchenko National University of Kyiv. His research interests include estimation problems for random processes and sequences with stationary increments.

Mikhail Moklyachuk is a Professor, Department of Probability Theory, Statistics and Actuarial Mathematics, Taras Shevchenko National University of Kyiv. He received Ph D degree in Physics and Mathematical Sciences from the Taras Shevchenko University of Kyiv in 1977. His research interests are statistical problems for stochastic processes and random fields. Member of editorial boards of several international journals

PUBLIC INTEREST STATEMENT

The crucial assumption of most of papers dedicated to the problem of estimating the unobserved values of random processes is that spectral densities of the involved processes are exactly known. However, the established results cannot be directly applied to practical estimation problems, because complete information of the spectral densities is impossible in most cases. This is a reason to derive the minimax estimates since they minimize the maximum of the mean-square errors for all spectral densities from a given set of admissible spectral densities simultaneously. In this article we deal with the problem of optimal estimation of functionals depending on unknown values of random processes from observations of the process with noise in the case of spectral uncertainty where spectral densities are not known exactly while a class of admissible spectral densities is given. Formulas that determine least favourable spectral densities and minimax (robust) spectra characteristics are derived.

1. Introduction

Traditional methods of finding solutions to problems of estimation of unobserved values of a random process based on a set of available observations of this process, or observations of the process with a noise process, are developed under the condition of spectral certainty, where the spectral densities of the processes are exactly known. Methods of solution of these problems, which are known as interpolation, extrapolation, and filtering of stochastic processes, were developed for stationary stochastic processes by A.N. Kolmogorov, N. Wiener, and A.M. Yaglom (see selected works by Kolmogorov (1992), books by Wiener (1966), Yaglom (1987a, 1987b), Rozanov (1967). Stationary stochastic processes and sequences admit some generalizations, which are properly described in books by Yaglom (1987a, 1987b). Random processes with stationary nth increments are among such generalizations. These processes were introduced in papers by Pinsker and Yaglom (1954), Yaglom (1955, 1957), and Pinsker (1955). In the indicated papers, the authors described the spectral representation of the stationary increment process and the canonical factorization of the spectral density, solved the extrapolation problem, and proposed some examples.

Traditional methods of finding solutions to extrapolation, interpolation, and filtering problems may be employed under the basic assumption that the spectral densities of the considered random processes are exactly known. In practice, however, the developed methods are not applicable since the complete information on the spectral structure of the processes is not available in most cases. To solve the problem, the parametric or nonparametric estimates of the unknown spectral densities are found or these densities are selected by other reasoning. Then, the classical estimation method is applied, provided that the estimated or selected densities are the true ones. However, as was shown by Vastola and Poor (1983) with the help of concrete examples, this method can result in significant increase of the value of the error of estimate. This is a reason to search estimates which are optimal for all densities from a certain class of the admissible spectral densities. The introduced estimates are called minimax robust since they minimize the maximum of the mean square errors for all spectral densities from a set of admissible spectral densities simultaneously. The paper by Grenander (1957) should be marked as the first one where the minimax approach to extrapolation problem for stationary processes was proposed. Franke and Poor (1984) and Franke (1985) investigated the minimax extrapolation and filtering problems for stationary sequences with the help of convex optimization methods. This approach makes it possible to find equations that determine the least favorable spectral densities for various classes of admissible densities. A survey of results in minimax (robust) methods of data processing can be found in the paper by Kassam and Poor (1985). A wide range of results in minimax robust extrapolation, interpolation, and filtering of random processes and sequences belong to Moklyachuk (2000, 2001, 2008a) . Later, Moklyachuk and Masyutka (2011 - 2012) developed the minimax technique of estimation for vector-valued stationary processes and sequences. Dubovets'ka, Masyutka, and Moklyachuk (2012) investigated the problem of minimax robust interpolation for another generalization of stationary processes—periodically correlated sequences. In the further papers, Dubovets'ka and Moklyachuk (2013a, 2013b, 2014a, 2014b) investigated the minimax robust extrapolation, interpolation, and filtering problems for periodically correlated processes and sequences.

The minimax robust extrapolation, interpolation, and filtering problems for stochastic sequences with nth stationary increments were investigated by Luz and Moklyachuk (2012,, 2013a, 2013b, 2014a, 2014b, 2015a, 2015b, 2015c); Moklyachuk and Luz (2013). In particular, the minimax robust extrapolation problem based on observations with and without noise for such sequences is investigated in papers by Luz and Moklyachuk (2015b), Moklyachuk and Luz (2013). Same estimation problems for random processes with stationary increments with continuous time are investigated in articles by Luz and Moklyachuk (2014a, 2015a, 2015b).

In this article, we deal with the problem of the mean square optimal estimation of the linear functionals $A\xi = \int_0^\infty a(t)\xi(t)dt$ and $A_T\xi = \int_0^T a(t)\xi(t)dt$ which depend on the unknown values of a random process $\xi(t)$ with stationary nth increments from observations of the process $\xi(t) + \eta(t)$ at points $t < 0$, where $\eta(t)$ is an uncorrelated with $\xi(t)$ stationary process. The case of spectral certainty as well as the case of spectral uncertainty are considered. Formulas for calculating values of the mean square errors and the spectral characteristics of the optimal linear estimates of the functionals are derived under the condition of spectral certainty, where the spectral densities of the processes are exactly known. In the case of spectral uncertainty, where the spectral densities of the processes are not exactly known while a class of admissible spectral densities is given, relations that determine the least favorable spectral densities and the minimax spectral characteristics are derived for some classes of spectral densities.

2. Stationary random increment process. Spectral representation

In this section, we present basic definitions and spectral properties of random processes with stationary increment. For more details, see the book by Yaglom (1987a, 1987b).

Definition 2.1 For a given random process $\xi(t), t \in \mathbb{R}$, the process

$$\xi^{(n)}(t,\tau) = (1-B_\tau)^n \xi(t) = \sum_{l=0}^{n} (-1)^l \binom{n}{l} \xi(t - l\tau), \tag{1}$$

where B_τ is a backward shift operator with a step $\tau \in \mathbb{R}$, such that $B_\tau \xi(t) = \xi(t-\tau)$ is called the random nth increment with step $\tau \in \mathbb{R}$ generated by the random process $\xi(t)$.

Definition 2.2 The random nth increment process $\xi^{(n)}(t,\tau)$ generated by a random process $\xi(t), t \in \mathbb{R}$, is in wide sense stationary, if the mathematical expectations

$$E\xi^{(n)}(t_0,\tau) = c^{(n)}(\tau),$$
$$E\xi^{(n)}(t_0+t,\tau_1)\overline{\xi^{(n)}(t_0,\tau_2)} = D^{(n)}(t,\tau_1,\tau_2)$$

exist for all $t_0, \tau, t, \tau_1, \tau_2$ and do not depend on t_0. The function $c^{(n)}(\tau)$ is called the mean value of the nth increment and the function $D^{(n)}(t,\tau_1,\tau_2)$ is called the structural function of the stationary nth increment (or the structural function of nth order of the random process $\xi(t), t \in \mathbb{R}$).

The random process $\xi(t), t \in \mathbb{R}$, which determines the stationary nth increment process $\xi^{(n)}(t,\tau)$ by formula (1) is called the process with stationary nth increments.

The following theorem describes representations of the mean value and the structural function of the random stationary nth increment process $\xi^{(n)}(t,\tau)$.

THEOREM 2.1 The mean value $c^{(n)}(\tau)$ and the structural function $D^{(n)}(t,\tau_1,\tau_2)$ of the random stationary nth increment process $\xi^{(n)}(t,\tau)$ can be represented in the following forms:

$$c^{(n)}(\tau) = c\tau^n, \tag{2}$$

$$D^{(n)}(t,\tau_1,\tau_2) = \int_{-\infty}^{\infty} e^{i\lambda t}(1-e^{-i\tau_1\lambda})^n(1-e^{i\tau_2\lambda})^n \frac{(1+\lambda^2)^n}{\lambda^{2n}} dF(\lambda), \tag{3}$$

where c is a constant and $F(\lambda)$ is a left-continuous nondecreasing bounded function, such that $F(-\infty)=0$. The constant c and the function $F(\lambda)$ are determined uniquely by the increment process $\xi^{(n)}(t,\tau)$.

The representation (3) of the structural function $D^{(n)}(t,\tau_1,\tau_2)$ and the Karhunen theorem (see Karhunen, 1947) allow us to write the following spectral representation of the stationary nth increment process $\xi^{(n)}(t,\tau)$:

$$\xi^{(n)}(t,\tau) = \int_{-\infty}^{\infty} e^{it\lambda}(1-e^{-i\lambda\tau})^n \frac{(1+i\lambda)^n}{(i\lambda)^n} dZ_{\xi^{(n)}}(\lambda), \tag{4}$$

where $Z_{\xi^{(n)}}(\lambda)$ is a random process with uncorrelated increments on $\mathbb{R}$ connected with the *spectral function* $F(\lambda)$ from representation (3) by the relation

$$\mathsf{E}|Z_{\xi^{(n)}}(t_2) - Z_{\xi^{(n)}}(t_1)|^2 = F(t_2) - F(t_1) < \infty \quad \text{for all } t_2 > t_1,\ t_1 \in \mathbb{R},\ t_2 \in \mathbb{R}. \tag{5}$$

3. The Hilbert space projection method of extrapolation

Consider a random process $\xi(t)$, $t \in \mathbb{R}$, which generates a stationary random increment process $\xi^{(n)}(t, \tau)$ with the absolutely continuous spectral function $F(\lambda)$ and the spectral density function $f(\lambda)$. Let $\eta(t)$, $t \in \mathbb{R}$, be another random process which is stationary and uncorrelated with $\xi(t)$. Suppose that the process $\eta(t)$ has absolutely continuous spectral function $G(\lambda)$ and the spectral density $g(\lambda)$. Without loss of generality, we can assume that the increment step $\tau > 0$ and both processes $\xi^{(n)}(t, \tau)$ and $\eta(t)$ have zero mean values: $\mathsf{E}\xi^{(n)}(t, \tau) = 0, \mathsf{E}\eta(t) = 0$.

The main purpose of this paper is to find optimal, in the mean square sense, linear estimates of the functionals

$$A\xi = \int_0^\infty a(t)\xi(t)dt, \quad A_T\xi = \int_0^T a(t)\xi(t)dt$$

which depend on the unknown values of the random process $\xi(t)$ at time $t \geq 0$ based on observations of the process $\zeta(t) = \xi(t) + \eta(t)$ at time $t < 0$.

For further analysis, we need to make the following assumptions. Let the function $a(t)$, $t \geq 0$, which determines the functionals $A\xi$, $A_T\xi$, and the linear transformation D^τ, being defined below, satisfy the conditions

$$\int_0^\infty |a(t)|dt < \infty, \quad \int_0^\infty t|a(t)|^2dt < \infty, \tag{6}$$

and

$$\int_0^\infty |D^\tau \mathbf{a}(t)|dt < \infty, \quad \int_0^\infty t|D^\tau \mathbf{a}(t)|^2dt < \infty. \tag{7}$$

Suppose also that the spectral densities $f(\lambda)$ and $g(\lambda)$ satisfy the minimality condition

$$\int_{-\infty}^{\infty} \frac{|\gamma(\lambda)|^2\lambda^{2n}}{|1-e^{i\lambda\tau}|^{2n}(1+\lambda^2)^n((1+\lambda^2)^n f(\lambda)+\lambda^{2n}g(\lambda))}d\lambda < \infty, \tag{8}$$

for some function $\gamma(\lambda)$ of the form $= \int_0^\infty \alpha(t)e^{i\lambda t}dt$. Assumption (8) guarantees that the mean square errors of estimates of the considered functionals are greater than 0.

Following the classical estimation theory developed for stationary processes, it is reasonable to apply the method proposed by Kolmogorov (see selected works by Kolmogorov (1992)), where the estimate is a projection of an element of the Hilbert space $H = L_2(\Omega, \mathfrak{F}, \mathsf{P})$ of the random variables γ with zero mean value, $\mathsf{E}\gamma = 0$, and finite variance, $\mathsf{E}|\gamma|^2 < \infty$ on a subspace of the space $H = L_2(\Omega, \mathfrak{F}, \mathsf{P})$. The inner product in the space $H = L_2(\Omega, \mathfrak{F}, \mathsf{P})$ is defined as $(\gamma_1; \gamma_2) = \mathsf{E}\gamma_1\overline{\gamma_2}$. Since we have no observations of the process $\xi(t)$ to take as initial values, the issue is that both functionals $A\xi$ and $A_T\xi$ have infinite variance. Thus, we need to derive other objects from the space $H = L_2(\Omega, \mathfrak{F}, \mathsf{P})$ to proceed with the Hilbert space projection method.

Consider a representation of the functional $A\xi$ in the form

$$A\xi = A\zeta - A\eta,$$

where

$$A\zeta = \int_0^\infty a(t)\zeta(t)dt, \quad A\eta = \int_0^\infty a(t)\eta(t)dt.$$

Under the condition (6), the functional $A\eta$ has finite variance and, hence, it belongs to the space $H = L_2(\Omega, \mathfrak{F}, \mathrm{P})$. A representation of the functional $A\zeta$ is described in the following lemma.

LEMMA 3.1 *The linear functional $A\zeta$ admits a representation*

$$A\zeta = B\zeta - V\zeta,$$

where

$$B\zeta = \int_0^{\infty} b_\tau(t)\zeta^{(n)}(t,\tau)dt, \quad V\zeta = \int_{-\tau n}^{0} v_\tau(t)\zeta(t)dt,$$

$$v_\tau(t) = \sum_{l=\left[-\frac{t}{\tau}\right]'}^{n} (-1)^l \binom{n}{l} b_\tau(t+l\tau), \quad t \in [-\tau n; 0), \tag{9}$$

$$b_\tau(t) = \sum_{k=0}^{\infty} a(t+\tau k)d(k) = D^\tau \mathbf{a}(t), \quad t \geq 0, \tag{10}$$

$[x]'$ denotes the least integer number among numbers that are greater than or equal to x, coefficients $\{d(k):k \geq 0\}$ are determined from the relation

$$\sum_{k=0}^{\infty} d(k)x^k = \left(\sum_{j=0}^{\infty} x^j\right)^n,$$

D^τ is a linear transformation of a function $x(t), t \geq 0$, defined by the formula

$$D^\tau \mathbf{x}(t) = \sum_{k=0}^{\infty} x(t+\tau k)d(k).$$

COROLLARY 3.1 The linear functional $A_T\zeta$ admits a representation

$$A_T\zeta = B_T\zeta - V_T\zeta,$$

where

$$B_T\zeta = \int_0^{T} b_{\tau,T}(t)\zeta^{(n)}(t,\tau)dt, \quad V_T\zeta = \int_{-\tau n}^{0} v_{\tau,T}(t)\zeta(t)dt,$$

$$v_{\tau,T}(t) = \sum_{l=\left[-\frac{t}{\tau}\right]'}^{\min\left\{\left[\frac{T-t}{\tau}\right],n\right\}} (-1)^l \binom{n}{l} b_{\tau,T}(t+l\tau), \quad t \in [-\tau n;0), \tag{11}$$

$$b_{\tau,T}(t) = \sum_{k=0}^{\left[\frac{T-t}{\tau}\right]} a(t+\tau k)d(k) = D_T^\tau \mathbf{a}(t), \; t \in [0;T], \tag{12}$$

D_T^τ is a linear transformation of an arbitrary function $x(t), t \in [0; T]$, defined by the formula

$$D_T^\tau \mathbf{x}(t) = \sum_{k=0}^{\left[\frac{T-t}{\tau}\right]} x(t+\tau k)d(k).$$

Under the condition (7), the functional $B\zeta$ from Lemma 3.1 belongs to the space $H = L_2(\Omega, \mathfrak{F}, \mathrm{P})$, while the functional $V\zeta$ is observed and can be considered as an initial value. Thus, Lemma 3.1 implies the following representation of the functional $A\xi$:

$$A\xi = A\zeta - A\eta = B\zeta - A\eta - V\zeta = H\xi - V\zeta,$$

where the functional $H\xi := B\zeta - A\eta$ belongs to the space $H = L_2(\Omega, \mathfrak{F}, \mathrm{P})$ and the Hilbert space projection method can be applied. Since the functional $V\zeta$ depends on the observations $\zeta(t)$, $-\tau n \leq t < 0$, the following relations hold true for the estimates $\widehat{A}\xi$, $\widehat{H}\xi$ and the mean square errors $\Delta(f, g; \widehat{A}\xi)$, $\Delta(f, g; \widehat{H}\xi)$:

$$\begin{aligned} &\widehat{A}\xi = \widehat{H}\xi - V\zeta, \\ &\Delta(f,g;\widehat{A}\xi) := \mathrm{E}|A\xi - \widehat{A}\xi|^2 = \mathrm{E}|H\xi - V\zeta - \widehat{H}\xi + V\zeta|^2 = \mathrm{E}|H\xi - \widehat{H}\xi|^2 =: \Delta(f,g;\widehat{H}\xi). \end{aligned} \tag{13}$$

Therefore, the problem is reduced to finding the optimal mean square estimate $\widehat{H}\xi$ of the functional $H\xi$.

The next step is to describe the spectral structure of the functional $H\xi$. The stationary random process $\eta(t)$ admits the spectral representation (see Gikhman & Skorokhod, 2004).

$$\eta(t) = \int_{-\infty}^{\infty} e^{i\lambda t} dZ_{\eta}(\lambda),$$

where $Z_{\eta}(\lambda)$ is a random process with uncorrelated increments on $\mathbb{R}$ which correspond to the spectral function $G(\lambda)$. Taking into account (4), the spectral representation of the random process $\zeta^{(n)}(t,\tau)$ can be described by the formulas

$$\begin{aligned} \zeta^{(n)}(t,\tau) &= \int_{-\infty}^{\infty} e^{i\lambda t}(1 - e^{-i\lambda\tau})^n \frac{(1+i\lambda)^n}{(i\lambda)^n} dZ_{\xi^{(n)}}(\lambda) + \int_{-\infty}^{\infty} e^{i\lambda t}(1 - e^{-i\lambda\tau})^n \frac{(1+i\lambda)^n}{(i\lambda)^n} dZ_{\eta^{(n)}}(\lambda) \\ &= \int_{-\infty}^{\infty} e^{i\lambda t}(1 - e^{-i\lambda\tau})^n \frac{(1+i\lambda)^n}{(i\lambda)^n} dZ_{\xi^{(n)}}(\lambda) + \int_{-\infty}^{\infty} e^{i\lambda t}(1 - e^{-i\lambda\tau})^n dZ_{\eta}(\lambda), \end{aligned}$$

where

$$dZ_{\eta^{(n)}}(\lambda) = (i\lambda)^n(1+i\lambda)^{-n} dZ_{\eta}(\lambda), \quad \lambda \in \mathbb{R}.$$

One can easily conclude that the spectral density $p(\lambda)$ of the random process $\zeta(t)$ is the following:

$$p(\lambda) = f(\lambda) + \frac{\lambda^{2n}}{(1+\lambda^2)^n} g(\lambda).$$

The functional $H\xi$ admits the spectral representation

$$H\xi = \int_{-\infty}^{\infty} B_{\tau}(\lambda)(1 - e^{-i\lambda\tau})^n \frac{(1+i\lambda)^n}{(i\lambda)^n} dZ_{\xi^{(n)}+\eta^{(n)}}(\lambda) - \int_{-\infty}^{\infty} A(\lambda) dZ_{\eta}(\lambda),$$

where

$$B_{\tau}(\lambda) = \int_0^{\infty} b_{\tau}(t) e^{i\lambda t} dt = \int_0^{\infty} (D^{\tau}\mathbf{a})(t) e^{i\lambda t} dt, \quad A(\lambda) = \int_0^{\infty} a(t) e^{i\lambda t} dt.$$

Denote by $H^{0-}(\xi_{\tau}^{(n)} + \eta_{\tau}^{(n)})$ the closed linear subspace of the space $H = L_2(\Omega, \mathfrak{F}, \mathrm{P})$, which is generated by observations $\{\xi^{(n)}(t,\tau) + \eta^{(n)}(t,\tau) : t < 0\}$, $\tau > 0$. Denote by $L_2^{0-}(p)$ the closed linear subspace of the Hilbert space $L_2(p)$ defined by the set of functions

$$\left\{ e^{i\lambda t}(1 - e^{-i\lambda\tau})^n (1+i\lambda)^n (i\lambda)^{-n} : t < 0 \right\}.$$

It follows from the equality

$$\xi^{(n)}(t,\tau) + \eta^{(n)}(t,\tau) = \int_{-\infty}^{\infty} e^{i\lambda t}(1 - e^{-i\lambda\tau})^n \frac{(1+i\lambda)^n}{(i\lambda)^n} dZ_{\xi^{(n)}+\eta^{(n)}}(\lambda) \tag{14}$$

that the operator which maps the vector

$$e^{i\lambda t}(1-e^{-i\lambda\tau})^n(1+i\lambda)^n(i\lambda)^{-n}$$

of the space $L_2^{0-}(p)$ to the vector $\xi^{(n)}(t,\tau)+\eta^{(n)}(t,\tau)$ of the space $H^{0-}(\xi_\tau^{(n)}+\eta_\tau^{(n)})$ may be extended to a linear isometry between the above spaces. The following relation holds true:

$$E\zeta^{(n)}(t_1,\tau_1)\overline{\zeta^{(n)}(t_2,\tau_2)}=\int_{-\infty}^{\infty}e^{i\lambda(t_1-t_2)}(1-e^{-i\lambda\tau_1})^n(1-e^{i\lambda\tau_2})^n\frac{(1+\lambda^2)^n}{\lambda^{2n}}p(\lambda)d\lambda. \tag{15}$$

Every linear estimate $\widehat{A}\xi$ of the functional $A\xi$ admits the representation

$$\widehat{A}\xi=\int_{-\infty}^{\infty}h_\tau(\lambda)dZ_{\xi^{(n)}+\eta^{(n)}}(\lambda)-\int_{-\tau n}^{0}v_\tau(t)(\xi(t)+\eta(t))dt, \tag{16}$$

where $h_\tau(\lambda)$ is the spectral characteristic of the estimate $\widehat{H}\xi$. We can find the estimate $\widehat{H}\xi$ as a projection of the element $H\xi$ of the space H on the subspace $H^{0-}(\xi_\tau^{(n)}+\eta_\tau^{(n)})$. This projection is characterized by two conditions:

(1) $\widehat{H}\xi\in H^{0-}(\xi_\tau^{(n)}+\eta_\tau^{(n)})$;

(2) $(H\xi-\widehat{H}\xi)\perp H^{0-}(\xi_\tau^{(n)}+\eta_\tau^{(n)})$. Condition (2) and property (15) imply the following relations which hold true for every $t<0$:

$$\begin{aligned}E&(H\xi-\widehat{H}\xi)\overline{(\xi^{(n)}(t,\tau)+\eta^{(n)}(t,\tau))}\\&=\frac{1}{2\pi}\int_{-\infty}^{\infty}\left[B_\tau(\lambda)(1-e^{-i\lambda\tau})^n-A(\lambda)-\frac{(i\lambda)^nh_\tau(\lambda)}{(1+i\lambda)^n}\right]e^{-i\lambda t}(1-e^{i\lambda\tau})^ng(\lambda)d\lambda\\&+\frac{1}{2\pi}\int_{-\infty}^{\infty}\left[B_\tau(\lambda)(1-e^{-i\lambda\tau})^n-\frac{(i\lambda)^nh_\tau(\lambda)}{(1+i\lambda)^n}\right]e^{-i\lambda t}(1-e^{i\lambda\tau})^n\frac{(1+\lambda^2)^n}{\lambda^{2n}}f(\lambda)d\lambda=0.\end{aligned}$$

Let us define for $\lambda\in\mathbb{R}$ the function

$$C_\tau(\lambda)=\left[\left(B_\tau(\lambda)(1-e^{-i\lambda\tau})^n-\frac{(i\lambda)^nh_\tau(\lambda)}{(1+i\lambda)^n}\right)\frac{(1+\lambda^2)^n}{\lambda^{2n}}p(\lambda)-A(e^{i\lambda})g(\lambda)\right](1-e^{i\lambda\tau})^n$$

and its Fourier transform

$$\mathbf{c}_\tau(t)=\frac{1}{2\pi}\int_{-\infty}^{\infty}C^\tau(\lambda)e^{-i\lambda t}d\lambda,\quad t\in\mathbb{R}.$$

We have $\mathbf{c}_\tau(t)=0$ for $t<0$, hence

$$C_\tau(\lambda)=\int_0^{\infty}\mathbf{c}_\tau(t)e^{i\lambda t}dt,$$

which allows us to construct the representation of the spectral characteristic

$$h_\tau(\lambda)=B_\tau(\lambda)\frac{(1-e^{-i\lambda\tau})^n(1+i\lambda)^n}{(i\lambda)^n}-A(\lambda)\frac{(-i\lambda)^ng(\lambda)}{(1-i\lambda)^np(\lambda)}-\frac{(-i\lambda)^nC_\tau(\lambda)}{(1-e^{i\lambda\tau})^n(1-i\lambda)^np(\lambda)}.$$

It follows from the condition 1) that the spectral characteristic $h_\tau(\lambda)$ admits the representation

$$h_\tau(\lambda)=h(\lambda)(1-e^{-i\lambda\tau})^n\frac{(1+i\lambda)^n}{(i\lambda)^n},$$

$$h(\lambda)=\int_{-\infty}^{0}s(t)e^{i\lambda t}dt,\quad s(t)\in L_2^-,$$

which leads to the following relations holding true for every $s \geq 0$:

$$\int_{-\infty}^{\infty}\left[B_\tau(\lambda) - \frac{A(\lambda)(1-e^{-i\lambda\tau})^{-n}\lambda^{2n}g(\lambda)}{(1+\lambda^2)^n p(\lambda)} - \frac{|1-e^{i\lambda\tau}|^{-n}\lambda^{2n}C_\tau(\lambda)}{(1+\lambda^2)^n p(\lambda)}\right]e^{-i\lambda s}d\lambda = 0. \quad (17)$$

Relation (17) can be represented in terms of linear operators in the space $L_2[0;\infty)$. Let us define the operators

$$(\mathbf{T}_\tau \mathbf{x})(s) = \frac{1}{2\pi}\int_0^{\infty}\mathbf{x}(t)\int_{-\infty}^{\infty}e^{i\lambda(t-s)}\frac{\lambda^{2n}g(\lambda)}{|1-e^{i\lambda\tau}|^{2n}(1+\lambda^2)^n p(\lambda)}d\lambda dt,\ s\in[0;\infty),$$

$$(\mathbf{P}_\tau \mathbf{y})(s) = \frac{1}{2\pi}\int_0^{\infty}\mathbf{y}(t)\int_{-\infty}^{\infty}e^{i\lambda(t-s)}\frac{\lambda^{2n}}{|1-e^{i\lambda\tau}|^{2n}(1+\lambda^2)^n p(\lambda)}d\lambda dt,\ s\in[0;\infty),$$

$$(\mathbf{Q}\mathbf{z})(s) = \frac{1}{2\pi}\int_0^{\infty}\mathbf{z}(t)\int_{-\infty}^{\infty}e^{i\lambda(t-s)}\frac{f(\lambda)g(\lambda)}{p(\lambda)}d\lambda dt,\ s\in[0;\infty),$$

where $\mathbf{x}(t), \mathbf{y}(t), \mathbf{z}(t) \in L_2[0;\infty)$. The introduced operators allow us to represent relations (17) in the form

$$b_\tau(s) - (\mathbf{T}_\tau\mathbf{a}_\tau)(s) = (\mathbf{P}_\tau\mathbf{c}_\tau)(s), \quad s \geq 0,$$

where

$$\mathbf{a}_\tau(t) = \sum_{l=0}^{\min\left\{n;\left[\frac{t}{\tau}\right]\right\}}(-1)^l\binom{n}{l}a(t-\tau l), \quad t\geq 0. \quad (18)$$

Then, under the condition that the linear operator $\mathbf{P}_\tau$ is invertible, the function $\mathbf{c}_\tau(t)$, $t \geq 0$, can be found by the formula

$$\mathbf{c}_\tau(t) = (\mathbf{P}_\tau^{-1}D^\tau\mathbf{a} - \mathbf{P}_\tau^{-1}\mathbf{T}_\tau\mathbf{a}_\tau)(t), \quad t \geq 0.$$

Consequently, the spectral characteristic $h_\tau(\lambda)$ of the optimal estimate $\widehat{H}\xi$ of the functional $H\xi$ is calculated by the formula

$$h_\tau(\lambda) = B_\tau(\lambda)\frac{(1-e^{-i\lambda\tau})^n(1+i\lambda)^n}{(i\lambda)^n} - \frac{A(\lambda)(1+i\lambda)^n(-i\lambda)^n g(\lambda)}{(1+\lambda^2)^n f(\lambda)+\lambda^{2n}g(\lambda)} - \frac{(1+i\lambda)^n(-i\lambda)^n C_\tau(\lambda)}{(1-e^{i\lambda\tau})^n\left((1+\lambda^2)^n f(\lambda)+\lambda^{2n}g(\lambda)\right)}, \quad (19)$$

where

$$C_\tau(\lambda) = \int_0^{\infty}(\mathbf{P}_\tau^{-1}D^\tau\mathbf{a} - \mathbf{P}_\tau^{-1}\mathbf{T}_\tau\mathbf{a}_\tau)(t)e^{i\lambda t}dt.$$

The value of mean square error is calculated by the formula

$$\begin{aligned}\Delta(f,g;\widehat{A}\xi) &= \Delta(f,g;\widehat{H}\xi) = E|H\xi-\widehat{H}\xi|^2\\ &= \frac{1}{2\pi}\int_{-\infty}^{\infty}\frac{\left|A(\lambda)(1-e^{i\lambda\tau})^n(1+\lambda^2)^n f(\lambda) - \lambda^{2n}C_\tau(\lambda)\right|^2}{|1-e^{i\lambda\tau}|^{2n}(1+\lambda^2)^{2n}(f(\lambda)+\frac{\lambda^{2n}}{(1+\lambda^2)^n}g(\lambda))^2}g(\lambda)d\lambda\\ &+ \frac{1}{2\pi}\int_{-\infty}^{\infty}\frac{\left|A(\lambda)(1-e^{i\lambda\tau})^n(-i\lambda)^n g(\lambda) + (-i\lambda)^n C_\tau(\lambda)\right|^2}{|1-e^{i\lambda\tau}|^{2n}(1+\lambda^2)^n(f(\lambda)+\frac{\lambda^{2n}}{(1+\lambda^2)^n}g(\lambda))^2}f(\lambda)d\lambda \quad (20)\\ &= \langle D^\tau\mathbf{a} - \mathbf{T}_\tau\mathbf{a}_\tau, \mathbf{P}_\tau^{-1}D^\tau\mathbf{a} - \mathbf{P}_\tau^{-1}\mathbf{T}_\tau\mathbf{a}_\tau\rangle + \langle\mathbf{Q}\mathbf{a},\mathbf{a}\rangle.\end{aligned}$$

The obtained results can be summarized in the following theorem.

THEOREM 3.1 *Let $\xi(t)$, $t \in \mathbb{R}$, be a random process with stationary nth increment process $\xi^{(n)}(t,\tau)$ and let $\eta(t)$, $t \in \mathbb{R}$, be an uncorrelated with $\xi(t)$ stationary random process. Suppose that the spectral densities $f(\lambda)$ and $g(\lambda)$ of the random processes $\xi(t)$ and $\eta(t)$ satisfy the minimality condition (8) and the function $a(t)$, $t \geq 0$, satisfies conditions (6) and (7). Suppose also that the linear operator $\mathbf{P}_\tau$ is invertible. The optimal estimate $\widehat{A}\xi$ of the functional $A\xi$ based on observations $\xi(t)+\eta(t)$ at time $t<0$ is calculated by formula (16). The spectral characteristic $h_\tau(\lambda)$ and the value of mean square error $\Delta(f,g;\widehat{A}\xi)$ of the estimate $\widehat{A}\xi$ can be calculated by formulas (19) and (20), respectively.*

Remark 3.1 The spectral characteristic $h_\tau(\lambda)$ determined by formula (19) can be presented in the form $h_\tau(\lambda) = h_\tau^1(\lambda) - h_\tau^2(\lambda)$, where

$$h_\tau^1(\lambda) = B_\tau(\lambda)\frac{(1-e^{-i\lambda\tau})^n(1+i\lambda)^n}{(i\lambda)^n} - \frac{(1+i\lambda)^n(-i\lambda)^n\int_0^\infty(\mathbf{P}_\tau^{-1}D^\tau\mathbf{a})(t)e^{i\lambda t}dt}{(1-e^{i\lambda\tau})^n\left((1+\lambda^2)^nf(\lambda)+\lambda^{2n}g(\lambda)\right)},$$

$$h_\tau^2(\lambda) = -\frac{A(\lambda)(1+i\lambda)^n(-i\lambda)^ng(\lambda)}{(1+\lambda^2)^nf(\lambda)+\lambda^{2n}g(\lambda)} - \frac{(1+i\lambda)^n(-i\lambda)^n\int_0^\infty(\mathbf{P}_\tau^{-1}\mathbf{T}_\tau\mathbf{a}_\tau)(t)e^{i\lambda t}dt}{(1-e^{i\lambda\tau})^n\left((1+\lambda^2)^nf(\lambda)+\lambda^{2n}g(\lambda)\right)}.$$

The functions $h_\tau^1(\lambda)$ and $h_\tau^2(\lambda)$ are the spectral characteristics of the mean square optimal estimates $\widehat{B}\zeta$ and $\widehat{A}\eta$ of the functionals $B\zeta$ and $A\eta$, respectively, based on observations $\xi(t)+\eta(t)$ at time $t<0$.

In the case of observations without noise, we have the following corollary.

COROLLARY 3.2 Let $\xi(t)$, $t \in \mathbb{R}$, be a random process with stationary nth increment process $\xi^{(n)}(t,\tau)$. Suppose that the spectral density $f(\lambda)$ of the random processes $\xi(t)$ satisfies the minimality condition (8) with $g(\lambda)=0$ and the function $a(t)$, $t \geq 0$, satisfies conditions (6) and (7). Suppose also that the linear operator $\mathbf{F}_\tau$ defined below is invertible. The optimal linear estimate $\widehat{A}\xi$ of the functional $A\xi$ which depends on the unknown values $\xi(t)$, $t \geq 0$, of the random process $\xi(t)$, based on observations of the process $\xi(t)$, $t<0$, is calculated by the formula

$$\widehat{A}\xi = \int_{-\infty}^{\infty} h_\tau^\xi(\lambda)dZ_{\xi^{(n)}}(\lambda) - \int_{-\tau n}^{0} v_\tau(t)\xi(t)dt. \tag{21}$$

The spectral characteristic $h_\tau^\xi(\lambda)$ and the mean square error $\Delta(f;\widehat{A}\xi)$ of the optimal estimate $\widehat{A}\xi$ of the functional $A\xi$ are calculated by the formulas

$$h_\tau^\xi(\lambda) = B_\tau(\lambda)\frac{(1-e^{-i\lambda\tau})^n(1+i\lambda)^n}{(i\lambda)^n} - \frac{(-i\lambda)^n\int_0^\infty(\mathbf{F}_\tau^{-1}D^\tau\mathbf{a})(t)e^{i\lambda t}dt}{(1-e^{i\lambda\tau})^n(1-i\lambda)^nf(\lambda)}, \tag{22}$$

$$\Delta(f;\widehat{A}\xi) = \frac{1}{2\pi}\int_{-\infty}^{\infty}\frac{\lambda^{2n}\left|\int_0^\infty(\mathbf{F}_\tau^{-1}D^\tau\mathbf{a})(t)e^{i\lambda t}dt\right|^2}{|1-e^{i\lambda\tau}|^{2n}(1+\lambda^2)^nf(\lambda)}d\lambda = \langle\mathbf{F}_\tau^{-1}D^\tau\mathbf{a}, D^\tau\mathbf{a}\rangle, \tag{23}$$

where $\mathbf{F}_\tau$ is the linear operator in the space $L_2[0;\infty)$ determined by the formula

$$(\mathbf{F}_\tau\mathbf{y})(s) = \frac{1}{2\pi}\int_0^\infty \mathbf{y}(t)\int_{-\infty}^{\infty} e^{i\lambda(t-s)}\frac{\lambda^{2n}}{|1-e^{i\lambda\tau}|^{2n}(1+\lambda^2)^nf(\lambda)}d\lambda dt,\ s\in[0;\infty).$$

Remark 3.2 In Corollary 3.2, we provide formulas for calculating the optimal linear estimate $\widehat{A}\xi$ of the functional $A\xi$ and the value of the mean square error $\Delta(f;\widehat{A}\xi)$ of the estimate $\widehat{A}\xi$ based on observations of the process $\xi(t)$ at time $t<0$ using the Fourier transform of the function $\dfrac{\lambda^{2n}}{|1-e^{i\lambda\tau}|^{2n}(1+\lambda^2)^nf(\lambda)}$. In the article by Luz and Moklyachuk (2014a), the same problem is considered. However, a solution is derived in terms of the function $\varphi_\tau(t)$, $t \geq 0$, which is determined by the canonical factorization of the function

$$\frac{|1-e^{-i\lambda\tau}|^{2n}(1+\lambda^2)^n}{\lambda^{2n}} f(\lambda) = \left|\int_0^\infty \varphi_\tau(t) e^{-i\lambda t} dt\right|^2.$$

Theorem (3.1) can be used to obtain the optimal estimate $\widehat{A}_T\xi$ of the functional $A_T\xi$ which depends on the unknown values $\xi(t)$, $0 \leq t \leq T$, of the random process $\xi(t)$, based on observations of the process $\xi(t)+\eta(t)$ at time $t<0$. To derive the corresponding formulas, let us put $\mathbf{a}(t)=0$ if $t>T$. We get that the spectral characteristic $h_{\tau,T}(\lambda)$ of the optimal estimate

$$\widehat{A}_T\xi = \int_{-\infty}^{\infty} h_{\tau,T}(\lambda) dZ_{\xi^{(n)}+\eta^{(n)}}(\lambda) - \int_{-\tau n}^{0} v_{\tau,T}(t)(\xi(t)+\eta(t))dt, \tag{24}$$

is calculated by the formula

$$\begin{aligned} h_{\tau,T}(\lambda) &= B_T^\tau(\lambda)\frac{(1-e^{-i\lambda\tau})^n(1+i\lambda)^n}{(i\lambda)^n} - \frac{A_T(\lambda)(1+i\lambda)^n(-i\lambda)^n g(\lambda)}{(1+\lambda^2)^n f(\lambda)+\lambda^{2n}g(\lambda)} \\ &\quad - \frac{(1+i\lambda)^n(-i\lambda)^n C_T^\tau(\lambda)}{(1-e^{i\lambda\tau})^n\left((1+\lambda^2)^n f(\lambda)+\lambda^{2n}g(\lambda)\right)}, \end{aligned} \tag{25}$$

$$B_T^\tau(\lambda) = \int_0^T b_{\tau,T}(t)e^{i\lambda t}dt = \int_0^T (D_T^\tau \mathbf{a}_T)(t)e^{i\lambda t}dt,$$

$$A_T(\lambda) = \int_0^T a(t)e^{i\lambda t}dt,$$

$$C_T^\tau(\lambda) = \int_0^\infty (\mathbf{P}_\tau^{-1}D_T^\tau\mathbf{a}_T - \mathbf{P}_\tau^{-1}\mathbf{T}_{\tau,T}\mathbf{a}_{\tau,T})(t)e^{i\lambda t}dt,$$

where the linear operator $\mathbf{T}_T^\tau$ in .the space $L_2[0;\infty)$ is determined by the formula

$$(\mathbf{T}_T^\tau \mathbf{x})(s) = \frac{1}{2\pi}\int_0^{T+\tau n} \mathbf{x}(t)\int_{-\infty}^{\infty} e^{-i\lambda(t+s)}\frac{\lambda^{2n}g(\lambda)}{|1-e^{i\lambda\tau}|^{2n}(1+\lambda^2)^n p(\lambda)} d\lambda dt,\ s\in[0;\infty),$$

the function $\mathbf{a}_{\tau,T}(t)$, $t\in[0;T+\tau n]$, is calculated by formula

$$\mathbf{a}_{\tau,T}(t) = \sum_{l=\max\left\{\left[\frac{t-T}{\tau}\right]',0\right\}}^{\min\left\{\left[\frac{t}{\tau}\right],n\right\}} (-1)^l\binom{n}{l} a(t-\tau l),\quad 0\leq t\leq T+\tau n.$$

The mean square error of the optimal estimate $\widehat{A}_T\xi$ is calculated by the formula

$$\begin{aligned} \Delta(f,g;\widehat{A}_T\xi) &= \Delta(f,g;\widehat{H}_T\xi) = E|H_T\xi-\widehat{H}_T\xi|^2 \\ &= \frac{1}{2\pi}\int_{-\infty}^{\infty}\frac{\left|A_T(\lambda)(1-e^{i\lambda\tau})^n(1+\lambda^2)^n f(\lambda)-\lambda^{2n}C_T^\tau(\lambda)\right|^2}{|1-e^{i\lambda\tau}|^{2n}(1+\lambda^2)^{2n}(f(\lambda)+\frac{\lambda^{2n}}{(1+\lambda^2)^n}g(\lambda))^2}g(\lambda)d\lambda \\ &\quad + \frac{1}{2\pi}\int_{-\infty}^{\infty}\frac{\left|A_T(\lambda)(1-e^{i\lambda\tau})^n(-i\lambda)^n g(\lambda)+(-i\lambda)^n C_T^\tau(\lambda)\right|^2}{|1-e^{i\lambda\tau}|^{2n}(1+\lambda^2)^n(f(\lambda)+\frac{\lambda^{2n}}{(1+\lambda^2)^n}g(\lambda))^2}f(\lambda)d\lambda \\ &= \langle D_T^\tau\mathbf{a}_T-\mathbf{T}_{\tau,T}\mathbf{a}_{\tau,T}, \mathbf{P}_\tau^{-1}D_T^\tau\mathbf{a}_T-\mathbf{P}_\tau^{-1}\mathbf{T}_{\tau,T}\mathbf{a}_{\tau,T}\rangle + \langle \mathbf{Q}_T\mathbf{a}_T,\mathbf{a}_T\rangle, \end{aligned} \tag{26}$$

where the linear operator $\mathbf{Q}_T$ in the space $L_2[0;\infty)$ is determined by the formula

$$(\mathbf{Q}_T\mathbf{z})(s) = \frac{1}{2\pi}\int_0^T \mathbf{z}(t)\int_{-\infty}^{\infty} e^{i\lambda(t-s)}\frac{(1+\lambda^2)^n f(\lambda)g(\lambda)}{(1+\lambda^2)^n f(\lambda)+\lambda^{2n}g(\lambda)}d\lambda dt,\ s\in[0;\infty),$$

and the function $\mathbf{a}_T(t), t \in [0;T]$, is determined as $\mathbf{a}_T(t) = a(t)$.

The described results can be summarized in the following theorem.

THEOREM 3.2 *Let $\xi(t)$, $t \in \mathbb{R}$, be a random process with stationary nth increment process $\xi^{(n)}(t,\tau)$ and let $\eta(t)$, $t \in \mathbb{R}$, be an uncorrelated with $\xi(t)$ stationary random process. Suppose that the spectral densities $f(\lambda)$ and $g(\lambda)$ of the random processes $\xi(t)$ and $\eta(t)$ satisfy the minimality condition (8) and the function a(t), $0 \le t \le T$, satisfies conditions (6) and (7). Suppose also that the linear operator $\mathbf{P}_\tau$ is invertible. The optimal linear estimate $\widehat{A}_T\xi$ of the functional $A_T\xi$ based on observations of the process $\xi(t)+\eta(t)$ at time $t<0$ is calculated by formula (24). The spectral characteristic $h_{\tau,T}(\lambda)$ and the value of mean square error $\Delta(f,g;\widehat{A}_T\xi)$ of the optimal estimate $\widehat{A}_T\xi$ are calculated by formulas (25) and (26), respectively.*

4. Minimax robust method of extrapolation

The values of the mean square errors and the spectral characteristics of the optimal estimates of the functionals $A\xi$ and $A_T\xi$ based on observations of the process $\xi(t)+\eta(t)$ or observations of the process $\xi(t)$ without noise can be calculated by formulas (20), (23), (26) and (19), (22), and (25), respectively, in the case where the spectral densities $f(\lambda)$ and $g(\lambda)$ of the random processes $\xi(t)$ and $\eta(t)$ are exactly known. In the case where the spectral densities are not exactly known while sets $\mathcal{D}=\mathcal{D}_f\times\mathcal{D}_g$ or $\mathcal{D}=\mathcal{D}_f$ of admissible spectral densities are given, the minimax robust method of estimation of the functionals which depend on the unknown values of the random process with stationary increments can be applied. The method consists in determining an estimate which minimizes the value of the mean square error for all spectral densities from the given class $\mathcal{D}=\mathcal{D}_f\times\mathcal{D}_g$ or $\mathcal{D}=\mathcal{D}_f$ simultaneously. The following definitions formalize the proposed method.

Definition 4.1 For a given class of spectral densities, $\mathcal{D}=\mathcal{D}_f\times\mathcal{D}_g$ spectral densities $f_0(\lambda)\in\mathcal{D}_f$, $g_0(\lambda)\in\mathcal{D}_g$ are called the least favorable in the class $\mathcal{D}$ for the optimal linear extrapolation of the functional $A\xi$ if the following relation holds true

$$\Delta(f^0,g^0)=\Delta(h(f^0,g^0);f^0,g^0)=\max_{(f,g)\in\mathcal{D}_f\times\mathcal{D}_g}\Delta(h(f,g);f,g).$$

Definition 4.2 For a given class of spectral densities $\mathcal{D}=\mathcal{D}_f\times\mathcal{D}_g$, the spectral characteristic $h^0(\lambda)$ of the optimal linear estimate of the functional $A\xi$ is called minimax robust if there are satisfied conditions:

$$h^0(\lambda)\in H_{\mathcal{D}}=\bigcap_{(f,g)\in\mathcal{D}_f\times\mathcal{D}_g}L_2^{0-}(p(\lambda)),$$

$$\min_{h\in H_{\mathcal{D}}}\max_{(f,g)\in\mathcal{D}_f\times\mathcal{D}_g}\Delta(h;f,g)=\max_{(f,g)\in\mathcal{D}_f\times\mathcal{D}_g}\Delta(h^0;f,g).$$

Let us now formulate lemmas which follow from the introduced definitions and formulas (20) and (23) derived in the previous section.

LEMMA 4.1 *The spectral densities $f^0(\lambda)\in\mathcal{D}_f$ and $g^0(\lambda)\in\mathcal{D}_g$ which satisfy the minimality condition (8) are the least favorable in the class $\mathcal{D}$ for the optimal linear extrapolation of the functional $A\xi$ based on observations of the random process $\xi(t)+\eta(t)$ at time $t<0$ if linear operators $\mathbf{P}^0_\tau$, $\mathbf{T}^0_\tau$, $\mathbf{Q}^0$, determined by the Fourier transform of the functions*

$$\frac{\lambda^{2n}|1-e^{i\lambda\tau}|^{-2n}}{(1+\lambda^2)^n f^0(\lambda)+\lambda^{2n}g^0(\lambda)},\ \frac{\lambda^{2n}|1-e^{i\lambda\tau}|^{-2n}g^0(\lambda)}{(1+\lambda^2)^n f^0(\lambda)+\lambda^{2n}g^0(\lambda)},\ \frac{(1+\lambda^2)^n f^0(\lambda)g^0(\lambda)}{(1+\lambda^2)^n f^0(\lambda)+\lambda^{2n}g^0(\lambda)},$$

determine a solution of the constrain optimization problem

$$\max_{(f,g)\in D_f\times D_g}(\langle D^\tau\mathbf{a}-\mathbf{T}_\tau\mathbf{a}_\tau,\mathbf{P}_\tau^{-1}D^\tau\mathbf{a}-\mathbf{P}_\tau^{-1}\mathbf{T}_\tau\mathbf{a}_\tau\rangle+\langle\mathbf{Q}\mathbf{a},\mathbf{a}\rangle)$$
$$=\langle D^\tau\mathbf{a}-\mathbf{T}_\tau^0\mathbf{a}_\tau,(\mathbf{P}_\tau^0)^{-1}D^\tau\mathbf{a}-(\mathbf{P}_\tau^0)^{-1}\mathbf{T}_\tau^0\mathbf{a}_\tau\rangle+\langle\mathbf{Q}^0\mathbf{a},\mathbf{a}\rangle. \tag{27}$$

The minimax robust spectral characteristic $h^0=h_\tau(f^0,g^0)$ can be found by formula (19) if $h_\tau(f^0,g^0)\in H_{\mathcal{D}}$.

The corresponding result holds true in the case where observations of the process $\xi(t)$ at time $t<0$ are available.

LEMMA 4.2 *The spectral density $f^0\in\mathcal{D}_f$ satisfying the minimality condition*

$$\int_{-\infty}^{\infty}\frac{|\gamma(\lambda)|^2\lambda^{2n}}{|1-e^{i\lambda\tau}|^{2n}(1+\lambda^2)^nf(\lambda)}d\lambda<\infty \tag{28}$$

is the least favorable in the class $\mathcal{D}_f$ for the optimal linear extrapolation of the functional $A\xi$ based on observations of the process $\xi(t)$ at time $t<0$ if the linear operator $\mathbf{F}_\tau^0$ defined by the Fourier transformation of the function

$$\lambda^{2n}|1-e^{i\lambda\tau}|^{-2n}(1+\lambda^2)^{-n}(f^0(\lambda))^{-1}$$

determines a solution to the constrain optimization problem

$$\max_{f\in\mathcal{D}_f}\langle\mathbf{F}_\tau^{-1}D^\tau\mathbf{a},D^\tau\mathbf{a}\rangle=\langle(\mathbf{F}_\tau^0)^{-1}D^\tau\mathbf{a},D^\tau\mathbf{a}\rangle. \tag{29}$$

The minimax robust spectral characteristic $h^0=h_\tau(f^0)$ is calculated by formula (22) under the condition $h_\tau(f^0)\in H_{\mathcal{D}}$.

The least favorable spectral densities can be found directly using the definition or applying the proposed lemmas. However, there is an approach which gives us a possibility to simplify the optimization problem using the following property of the function $\Delta(h;f,g)$. This function has a saddle point on the set $H_{\mathcal{D}}\times\mathcal{D}$, which is formed by the minimax robust spectral characteristic h^0 and a pair (f^0,g^0) of the least favorable spectral densities. The saddle point inequalities

$$\Delta(h;f^0,g^0)\ge\Delta(h^0;f^0,g^0)\ge\Delta(h^0;f,g)\quad\forall f\in\mathcal{D}_f,\forall g\in\mathcal{D}_g,\forall h\in H_{\mathcal{D}}$$

hold true if $h^0=h_\tau(f^0,g^0)$, $h_\tau(f^0,g^0)\in H_{\mathcal{D}}$ and the pair (f^0,g^0) determines a solution of the constrain optimization problem

$$\widetilde{\Delta}(f,g)=-\Delta(h_\tau(f^0,g^0);f,g)\to\inf,\quad(f,g)\in\mathcal{D},$$

where

$$\Delta(h_\tau(f^0,g^0);f,g)=\frac{1}{2\pi}\int_{-\infty}^{\infty}\frac{\left|A(\lambda)(1-e^{i\lambda\tau})^n(1+\lambda^2)^nf^0(\lambda)-\lambda^{2n}C_\tau^0(\lambda)\right|^2}{|1-e^{i\lambda\tau}|^{2n}(1+\lambda^2)^{2n}(f^0(\lambda)+\frac{\lambda^{2n}}{(1+\lambda^2)^n}g^0(\lambda))^2}g(\lambda)d\lambda$$
$$+\frac{1}{2\pi}\int_{-\infty}^{\infty}\frac{\left|A(\lambda)(1-e^{i\lambda\tau})^n(-i\lambda)^ng^0(\lambda)+(-i\lambda)^nC_\tau^0(\lambda)\right|^2}{|1-e^{i\lambda\tau}|^{2n}(1+\lambda^2)^n(f^0(\lambda)+\frac{\lambda^{2n}}{(1+\lambda^2)^n}g^0(\lambda))^2}f(\lambda)d\lambda,$$
$$C_\tau^0(e^{i\lambda})=\int_0^{\infty}((\mathbf{P}_\tau^0)^{-1}D^\tau\mathbf{a}-(\mathbf{P}_\tau^0)^{-1}\mathbf{T}_\tau^0\mathbf{a}_\tau)(t)e^{i\lambda t}dt.$$

In the case of estimating the functional $A\xi$ based on the observations $\xi(t)$, $t<0$, we have the following constrain optimization problem

$$\widetilde{\Delta}(f) = -\Delta(h_\tau(f^0);f) \to \inf, \quad f \in \mathcal{D}_f,$$

where

$$\Delta(h_\tau(f^0);f) = \frac{1}{2\pi}\int_{-\infty}^{\infty} \frac{\lambda^{2n}\left|\int_0^\infty ((\mathbf{F}_\tau^0)^{-1}D^\tau\mathbf{a})(t)e^{i\lambda t}dt\right|^2}{|1-e^{i\lambda\tau}|^{2n}(1+\lambda^2)^n(f^0(\lambda))^2} f(\lambda)d\lambda.$$

Using the indicator functions $\delta(f,g|\mathcal{D}_f \times \mathcal{D}_g)$ and $\delta(f,g|\mathcal{D}_f)$ of the sets $\mathcal{D}_f \times \mathcal{D}_g$ and $\mathcal{D}_f$, the indicated constrain optimization problems can be presented as unconditional optimization problems

$$\Delta_\mathcal{D}(f,g) = \widetilde{\Delta}(f,g) + \delta(f,g|\mathcal{D}_f \times \mathcal{D}_g) \to \inf,$$
$$\Delta_\mathcal{D}(f) = \widetilde{\Delta}(f) + \delta(f|\mathcal{D}_f) \to \inf$$

respectively. In this case, solutions (f^0,g^0) and f^0 are characterized by the conditions $0 \in \partial\Delta_\mathcal{D}(f^0,g^0)$ and $0 \in \partial\Delta_\mathcal{D}(f^0)$ which are necessary and sufficient conditions that the pair (f^0,g^0) belongs to the set of minimums of the convex functional $\Delta_\mathcal{D}(f,g)$ and the function f^0 belongs to the set of minimums of the convex functional $\Delta_\mathcal{D}(f)$. By $\partial\Delta_\mathcal{D}(f^0,g^0)$ and $\partial\Delta_\mathcal{D}(f^0)$, we denote subdifferentials of the functionals $\Delta_\mathcal{D}(f,g)$ and $\Delta_\mathcal{D}(f)$ at point $(f,g)=(f^0,g^0)$ and f^0, respectively (see books by Ioffe & Tihomirov, (1979), Moklyachuk, (2008a), Pshenichnyi, (1971), Rockafellar, (1997)).

5. Least favorable densities in the class $\mathcal{D}_f^0 \times \mathcal{D}_g^0$

In this section, we consider the problem of minimax robust extrapolation of the functional $A\xi$ based on observations of the process $\xi(t)+\eta(t)$ at time $t<0$ on the set of admissible spectral densities $\mathcal{D} = \mathcal{D}_f^0 \times \mathcal{D}_g^0$, where

$$\mathcal{D}_f^0 = \left\{ f(\lambda) | \frac{1}{2\pi}\int_{-\infty}^{\infty} f(\lambda)d\lambda \le P_1 \right\}, \quad \mathcal{D}_g^0 = \left\{ g(\lambda) | \frac{1}{2\pi}\int_{-\infty}^{\infty} g(\lambda)d\lambda \le P_2 \right\}.$$

Let us suppose that the spectral densities $f^0 \in \mathcal{D}_f^0$, $g^0 \in \mathcal{D}_g^0$ and the functions

$$h_{\tau,f}(f^0,g^0) = \frac{\left|A(\lambda)(1-e^{i\lambda\tau})^n(-i\lambda)^n g^0(\lambda) + (-i\lambda)^n C_\tau^0(\lambda)\right|}{|1-e^{i\lambda\tau}|^n(1+\lambda^2)^{n/2}(f^0(\lambda) + \frac{\lambda^{2n}}{(1+\lambda^2)^n}g^0(\lambda))}, \tag{30}$$

$$h_{\tau,g}(f^0,g^0) = \frac{\left|A(\lambda)(1-e^{i\lambda\tau})^n(1+\lambda^2)^n f^0(\lambda) - \lambda^{2n} C_\tau^0(\lambda)\right|}{|1-e^{i\lambda\tau}|^n(1+\lambda^2)^n(f^0(\lambda) + \frac{\lambda^{2n}}{(1+\lambda^2)^n}g^0(\lambda))} \tag{31}$$

are bounded. These conditions ensure the functional $\Delta(h_\tau(f^0,g^0);f,g)$ is continuous and bounded in the space $L_1 \times L_1$. Condition $0 \in \partial\Delta_\mathcal{D}(f^0,g^0)$ implies the spectral densities $f^0 \in \mathcal{D}_f^0$, $g^0 \in \mathcal{D}_g^0$ satisfy the equalities

$$\left|A(\lambda)(1-e^{i\lambda\tau})^n(1+\lambda^2)^n f^0(\lambda) - \lambda^{2n}C_\tau^0(\lambda)\right| = \alpha_1|1-e^{i\lambda\tau}|^n\left((1+\lambda^2)^n f^0(\lambda) + \lambda^{2n}g^0(\lambda)\right), \tag{32}$$

$$\left|A(\lambda)(1-e^{i\lambda\tau})^n(-i\lambda)^n g^0(\lambda) + (-i\lambda)^n C_\tau^0(\lambda)\right| = \alpha_2|1-e^{i\lambda\tau}|^n(1+\lambda^2)^{-n/2}\left((1+\lambda^2)^n f^0(\lambda) + \lambda^{2n}g^0(\lambda)\right), \tag{33}$$

where the constants $\alpha_1 \ge 0$, $\alpha_2 \ge 0$, and $\alpha_1 \ne 0$ if

$$\frac{1}{2\pi}\int_{-\infty}^{\infty} f^0(\lambda)d\lambda = P_1,$$

$\alpha_2 \ne 0$ if

$$\frac{1}{2\pi}\int_{-\infty}^{\infty} g^0(\lambda)d\lambda = P_2.$$

We can summarize the obtained results in the following theorem.

THEOREM 5.1 *Let the spectral densities $f^0(\lambda) \in \mathcal{D}_f^0$ and $g^0(\lambda) \in \mathcal{D}_g^0$ satisfy condition (8) and let the functions $h_{\tau,f}(f^0, g^0)$ and $h_{\tau,g}(f^0, g^0)$ determined by formulas (30) and (31) be bounded. The spectral densities $f^0(\lambda)$ and $g^0(\lambda)$ are the least favorable in the class $\mathcal{D} = \mathcal{D}_f^0 \times \mathcal{D}_g^0$ for the optimal linear extrapolations of the functional $A\xi$ if they satisfy equations (32) and (33) and determine a solution of the optimization problem (27). The function $h_\tau(f^0, g^0)$ calculated by formula (19) is the minimax robust spectral characteristic of the optimal estimate of the functional $A\xi$.*

THEOREM 5.2 *Let the spectral density $f(\lambda)$ be known, let the spectral density $g^0(\lambda) \in \mathcal{D}_g^0$ and let the spectral densities $f(\lambda)$, $g^0(\lambda)$ satisfy the minimality condition (8). Suppose also that the function $h_{\tau,g}(f, g^0)$ determined by formula (31) is bounded. The spectral density*

$$g^0(\lambda) = \max\left\{0, f_1(\lambda) - (1+\lambda^2)^n \lambda^{-2n} f(\lambda)\right\},$$
$$f_1(\lambda) = \frac{\left|A(\lambda)(1-e^{i\lambda\tau})^n (1+\lambda^2)^n f(\lambda) - \lambda^{2n} C_\tau^0(\lambda)\right|}{\alpha_1 |1-e^{i\lambda\tau}|^n \lambda^{2n}}, \tag{34}$$

is the least favorable in the class $\mathcal{D}_g^0$ for the optimal linear extrapolation of the functional $A\xi$ if the functions $f(\lambda) + (1+\lambda^2)^{-n}\lambda^{2n} g^0(\lambda)$, $g^0(\lambda)$ determine a solution of the optimization problem (27). The function $h_\tau(f, g^0)$ calculated by formula (19) is the minimax robust spectral characteristic of the optimal estimate of the functional $A\xi$.

THEOREM 5.3 *Let the spectral density $g(\lambda)$ be known, let the spectral density $f^0(\lambda) \in \mathcal{D}_f^0$ and let the spectral densities $f^0(\lambda)$, $g(\lambda)$ satisfy the minimality condition (8). Suppose also that the function $h_{\tau,f}(f^0, g)$ determined by formula (30) is bounded. The spectral density*

$$f^0(\lambda) = \max\left\{0, g_2(\lambda) - (1+\lambda^2)^{-n} \lambda^{2n} g(\lambda)\right\},$$
$$g_2(\lambda) = \frac{\left|A(\lambda)(1-e^{i\lambda\tau})^n (-i\lambda)^n g(\lambda) + (-i\lambda)^n C_\tau^0(\lambda)\right|}{\alpha_2 |1-e^{i\lambda\tau}|^n (1+\lambda^2)^{n/2}}, \tag{35}$$

is the least favorable in the class $\mathcal{D}_f^0$ for the optimal linear extrapolation of the functional $A\xi$ if the function $f^0(\lambda) + (1+\lambda^2)^{-n}\lambda^{2n} g(\lambda)$, determines a solution of the optimization problem (27). The function $h_\tau(f^0, g)$, calculated by formula (19), is the minimax robust spectral characteristic of the optimal estimate of the functional $A\xi$.

In the case of estimating the functional $A\xi$ based on the observations of the process $\xi(t)$ at time $t < 0$ without noise, we can formulate the following theorem.

THEOREM 5.4 *Suppose that the spectral density $f^0(\lambda) \in \mathcal{D}_f^0$ satisfies condition (28). The spectral density*

$$f^0(\lambda) = \frac{|\lambda|^n \left|\int_0^\infty ((\mathbf{F}_\tau^0)^{-1} D^\tau \mathbf{a})(t) e^{i\lambda t} dt\right|}{\alpha_1 |1-e^{i\lambda\tau}|^n (1+\lambda^2)^{n/2}}$$

is the least favorable in the class $\mathcal{D} = \mathcal{D}_f^0$ for the optimal linear extrapolations of the functional $A\xi$ based on observations of the process $\xi(t)$ at time $t < 0$ if it determines a solution of the optimization problem (29). The function $h_\tau(f^0)$ calculated by formula (22) is the minimax robust spectral characteristic of the optimal estimate of the functional $A\xi$.

6. Least favorable densities in the class $\mathcal{D} = \mathcal{D}_v^u \times \mathcal{D}_\varepsilon$

Let us consider the problem of minimax robust extrapolation of the functional $A\xi$ based on observations of the process $\xi(t) + \eta(t)$ at time $t < 0$ on the set of admissible spectral densities $\mathcal{D} = \mathcal{D}_v^u \times \mathcal{D}_\varepsilon$, where

$$\mathcal{D}_v^u = \left\{ f(\lambda) | v(\lambda) \leq f(\lambda) \leq u(\lambda), \frac{1}{2\pi}\int_{-\pi}^{\pi} f(\lambda)d\lambda = P_1 \right\},$$

$$\mathcal{D}_\varepsilon = \left\{ g(\lambda) | g(\lambda) = (1-\varepsilon)g_1(\lambda) + \varepsilon w(\lambda), \frac{1}{2\pi}\int_{-\pi}^{\pi} g(\lambda)d\lambda = P_2 \right\}.$$

Here, the spectral densities $u(\lambda)$, $v(\lambda)$, and $g_1(\lambda)$ are supposed to be known and the spectral densities $u(\lambda)$, $v(\lambda)$ are assumed to be bounded.

Using the condition $0 \in \partial\Delta_{\mathcal{D}}(f^0, g^0)$, we obtain the following equalities determining the spectral densities: $f^0 \in \mathcal{D}_v^u$, $g^0 \in \mathcal{D}_\varepsilon$:

$$\left|A(\lambda)(1-e^{i\lambda\tau})^n(1+\lambda^2)^n f^0(\lambda) - \lambda^{2n}C_\tau^0(\lambda)\right| = |1-e^{i\lambda\tau}|^n\Big((1+\lambda^2)^n f^0(\lambda) + \lambda^{2n}g^0(\lambda)\Big)(\gamma_1(\lambda)+\gamma_2(\lambda)$$
$$+\alpha_1), \left|A(\lambda)(1-e^{i\lambda\tau})^n(-i\lambda)^n g^0(\lambda) + (-i\lambda)^n C_\tau^0(\lambda)\right| = |1-e^{i\lambda\tau}|^n(1+\lambda^2)^{-n/2}\Big((1+\lambda^2)^n f^0(\lambda) \quad (36)$$
$$+\lambda^{2n}g^0(\lambda)\Big)(\beta(\lambda)+\alpha_2), \quad (37)$$

where the function $\gamma_1(\lambda) \leq 0$ and $\gamma_1(\lambda) = 0$ if $f^0(\lambda) \geq v(\lambda)$; the function $\gamma_2(\lambda) \geq 0$ and $\gamma_2(\lambda) = 0$ if $f^0(\lambda) \leq u(\lambda)$; and the function $\beta(\lambda) \leq 0$ and $\beta(\lambda) = 0$ if $g^0(\lambda) \geq (1-\varepsilon)g_1(\lambda)$.

THEOREM 6.1 *Let the spectral densities $f^0(\lambda) \in \mathcal{D}_v^u$ and $g^0(\lambda) \in \mathcal{D}_\varepsilon$ satisfy the minimality condition (8) and let the functions $h_{\tau,f}(f^0, g^0)$ and $h_{\tau,g}(f^0, g^0)$ determined by formulas (30) and (31) be bounded. The spectral densities $f^0(\lambda)$ and $g^0(\lambda)$ determined by equations (36) and (37) are the least favorable in the class $\mathcal{D} = \mathcal{D}_v^u \times \mathcal{D}_\varepsilon$ for the optimal linear extrapolations of the functional $A\xi$ if they determine a solution of the optimization problem (27). The function $h_\tau(f^0, g^0)$ calculated by formula (19) is the minimax robust spectral characteristic of the optimal estimate of the functional $A\xi$.*

THEOREM 6.2 *Let the spectral density $f(\lambda)$ be known, let the spectral density $g^0(\lambda) \in \mathcal{D}_\varepsilon$ and let the spectral densities $f(\lambda)$, $g^0(\lambda)$ satisfy the minimality condition (8). Suppose also that the function $h_{\tau,g}(f, g^0)$ determined by formula (31) is bounded. The spectral density*

$$g^0(\lambda) = \max\left\{(1-\varepsilon)g_2(\lambda), f_1(\lambda) - (1+\lambda^2)^n\lambda^{-2n}f(\lambda)\right\},$$

where the function $f_1(\lambda)$ is defined by formula (34), is the least favorable in the class $\mathcal{D}_\varepsilon$ for the optimal linear extrapolation of the functional $A\xi$ if the functions $f(\lambda) + (1+\lambda^2)^{-n}\lambda^{2n}g^0(\lambda)$, determine a solution of the optimization problem (27). The function $h_\tau(f, g^0)$ calculated by formula (19) is the minimax robust spectral characteristic of the optimal estimate of the functional $A\xi$.

THEOREM 6.3 *Let the spectral density $g(\lambda)$ be known, let the spectral density $f^0(\lambda) \in \mathcal{D}_v^u$ and let the spectral densities $f^0(\lambda)$, $g(\lambda)$ satisfy the minimality condition (8). Suppose also that the function $h_{\tau,f}(f^0, g)$ determined by formula (30) is bounded. The spectral density*

$$f^0(\lambda) = \min\left\{u(\lambda), \max\left\{v(\lambda), g_2(\lambda) - (1+\lambda^2)^{-n}\lambda^{2n}g(\lambda)\right\}\right\},$$

where the function $g_2(\lambda)$ is defined by formula (35), is the least favorable in the class $\mathcal{D}_v^u$ for the optimal linear extrapolation of the functional $A\xi$ if the function $f^0(\lambda) + (1+\lambda^2)^{-n}\lambda^{2n}g(\lambda)$, determines a solution to optimization problem (27). The function $h_\tau(f^0, g)$ calculated by formula (19) is the minimax robust spectral characteristic of the optimal estimate of the functional $A\xi$.

In the case of estimating the functional $A\xi$ based on the observations of the process $\xi(t)$ at time $t < 0$ without noise, we can formulate the following theorem.

THEOREM 6.4 *Suppose that the spectral density $f^0(\lambda) \in \mathcal{D}_v^u$ satisfies condition (28). The spectral density*

$$f^0(\lambda) = \min\left\{u(\lambda), \max\left\{v(\lambda), \frac{|\lambda|^n \left|\int_0^\infty ((\mathbf{F}_\tau^0)^{-1} D^\tau \mathbf{a})(t) e^{i\lambda t} dt\right|}{\alpha_1 |1 - e^{i\lambda\tau}|^n (1+\lambda^2)^{n/2}}\right\}\right\}$$

is the least favorable in the class $\mathcal{D} = \mathcal{D}_v^u$ for the optimal linear extrapolations of the functional $A\xi$ based on observations of the process $\xi(t)$ at time $t < 0$ if it determines a solution of the optimization problem (29). The function $h_\tau(f^0)$ calculated by formula (22) is the minimax robust spectral characteristic of the optimal estimate of the functional $A\xi$.

7. Conclusions

In this paper, we present results of investigating of the problem of optimal linear estimation of the functionals $A\xi = \int_0^\infty a(t)\xi(t)dt$ and $A_T\xi = \int_0^T a(t)\xi(t)dt$ which depend on the unknown values of a random process $\xi(t)$ with nth stationary increments based on observations of the process $\xi(t) + \eta(t)$ at time $t < 0$. In the case where the spectral densities of the processes are known, we found formulas for calculating the values of the mean square errors and the spectral characteristics of the estimates of the functionals $A\xi$ and $A_T\xi$. In the case where the spectral densities are not exactly known, but a set of admissible spectral densities was available, we applied the minimax robust method to derive relations which determine the least favorable spectral densities from the given set and the minimax robust spectral characteristics.

Acknowledgements
The authors would like to thank the referees for careful reading of the article and giving constructive suggestions.

Funding
The author received no direct funding for this research.

Author details
Maksym Luz[1]
Mikhail Moklyachuk[1]
E-mail: Moklyachuk@gmail.com
ORCID ID: http://orcid.org/0000-0002-8260-1584
[1] Department of Probability Theory, Statistics and Actuarial Mathematics, Taras Shevchenko National University of Kyiv, Kyiv, 01601, Ukraine.

References

Dubovets'ka, I. I., Masyutka, O. Yu., & Moklyachuk, M. P. (2012). Interpolation of periodically correlated stochastic sequences. *Theory of Probability and Mathematical Statistics, 84*, 43–56.

Dubovets'ka, I. I. & Moklyachuk, M. P. (2013a). Filtration of linear functionals of periodically correlated sequences. *Theory of Probability and Mathematical Statistics, 86*, 51–64.

Dubovets'ka, I. I. & Moklyachuk, M. P. (2013b). Minimax estimation problem for periodically correlated stochastic processes. *Journal of Mathematics and System Science*, 3(1), 26–30.

Dubovets'ka, I. I. & Moklyachuk, M. P. (2014a). Extrapolation of periodically correlated processes from observations with noise. *Theory of Probability and Mathematical Statistics, 88*, 67–83.

Dubovets'ka, I. I., & Moklyachuk, M. P. (2014b). On minimax estimation problems for periodically correlated stochastic processes. *Contemporary Mathematics and Statistics, 2*, 123–150.

Franke, J. (1985). Minimax robust prediction of discrete time series. *Z. Wahrsch. Verw. Gebiete, 68*, 337–364.

Franke, J., & Poor, H. V. (1984). *Minimax-robust filtering and finite-length robust predictors. Robust and nonlinear time series analysis*, Lecture notes in statistics (Vol. 26, pp. 87–126). Heidelberg, Springer-Verlag.

Gikhman, I. I., & Skorokhod, A. V. (2004). *The theory of stochastic processes. I.* Berlin: Springer.

Golichenko, I. I., & Moklyachuk, M. P. (2014). *Estimates of functionals of periodically correlated processes.* Kyiv: NVP "Interservis".

Grenander, U. (1957). A prediction problem in game theory. *Arkiv för Matematik, 3*, 371–379.

Ioffe, A. D., & Tihomirov, V. M. (1979). *Theory of extremal problems* (p. 460). Amsterdam, North-Holland Publishing Company.

Karhunen, K. (1947). Uber lineare Methoden in der Wahrscheinlichkeitsrechnung. Annales Academiae Scientiarum Fennicae. *Series A I. Mathematica, 37*, 3–79.

Kassam, S. A., & Poor, H. V. (1985). Robust techniques for signal processing: A survey. *Proceedings of the IEEE, 73*, 433–481.

Kolmogorov, A.N. (1992). Selected works of A. N. Kolmogorov. Volume II: Probability theory and mathematical statistics. Edited by A. N. Shiryayev. Dordrecht etc.: Kluwer Academic Publishers.

Luz, M. M., & Moklyachuk, M. P. (2012). Interpolation of functionals of stochastic sequences with stationary increments from observations with noise. *Prykladna Statystyka. Aktuarna ta Finansova Matematyka, 2*, 131–148.

Luz, M. M., & Moklyachuk, M. P. (2013a). Interpolation of functionals of stochastic sequences with stationary increments. *Theory of Probability and Mathematical Statistics, 87*, 117–133.

Luz, M. M., & Moklyachuk, M. P. (2013b). Minimax-robust filtering problem for stochastic sequence with stationary increments. *Theory of Probability and Mathematical Statistics, 89*, 117–131.

Luz, M., & Moklyachuk, M. (2014a). Robust extrapolation problem for stochastic processes with stationary increments. *Mathematics and Statistics, 2*, 78–88.

Luz, M., & Moklyachuk, M. (2014b). Minimax-robust filtering problem for stochastic sequences with stationary increments and cointegrated sequences. *Statistics, Optimization & Information Computing, 2*, 176–199.

Luz, M., & Moklyachuk, M. (2015a). Minimax interpolation problem for random processes with stationary increments. *Statistics, Optimization & Information Computing, 3*, 30–41.

Luz, M., & Moklyachuk, M. (2015b). Filtering problem for random processes with stationary increments. *Contemporary Mathematics and Statistics, 3*, 8–27.

Luz, M., & Moklyachuk, M. (2015c). Minimax-robust prediction problem for stochastic sequences with stationary increments and cointegrated sequences. *Statistics, Optimization & Information Computing, 3*, 160–188.

Moklyachuk, M., & Luz, M. (2013). Robust extrapolation problem for stochastic sequences with stationary increments. *Contemporary Mathematics and Statistics, 1*, 123–150.

Moklyachuk, M. P. (2000). Robust procedures in time series analysis. *Theory Stoch. Process., 6*(3–4), 127–147.

Moklyachuk, M. P. (2001). Game theory and convex optimization methods in robust estimation problems. *Theory of Stochastic Processes, 7*(1–2), 253–264.

Moklyachuk, M. P. (2008a). *Robust estimates for functionals of stochastic processes*. Kyiv: Kyiv University Publishing.

Moklyachuk, M. P. (2008b). *Nonsmooth analysis and optimization* (pp. 400), Kyiv: Kyivskyi Universitet.

Moklyachuk, M. (2015). Minimax-robust estimation problems for stationary stochastic sequences. *Statistics, Optimization & Information Computing, 3*, 348–419.

Moklyachuk, M. P., & Masyutka, O. Yu (2011). Minimax prediction problem for multidimensional stationary stochastic processes. *Communications in Statistics – Theory and Methods, 40*, 3700–3710.

Moklyachuk, M. P., & Masyutka, O. Yu. (2012). *Minimax-robust estimation technique for stationary stochastic processes* (pp. 296), Saarbrücken, LAP LAMBERT Academic Publishing.

Pinsker, M. S., & Yaglom, A. M. (1954). On linear extrapolation of random processes with nth stationary increments. *Doklady Akademii Nauk SSSR, 94*, 385–388.

Pinsker, M. S. (1955). The theory of curves with nth stationary increments in Hilbert spaces. *Izvestiya Akademii Nauk SSSR. Ser. Mat., 19*, 319–344.

Pshenichnyi, B. N. (1971). *Necessary conditions for an extremum. Pure and Applied mathematics* 4 (Vol. XVIII, p. 230), New York: Marcel Dekker, Inc.

Rockafellar, R. T. (1997). *Convex Analysis* (p. 451). Princeton, NJ: Princeton University Press.

Rozanov, Y. A. (1967). *Stationary stochastic processes*. San Francisco, CA: Holden-Day.

Vastola, K. S., & Poor, H. V. (1983). An analysis of the effects of spectral uncertainty on Wiener filtering. *Automatica, 28*, 289–293.

Wiener, N. (1966). *Extrapolation, interpolation, and smoothing of stationary time series. With engineering applications*. Massachusetts: The M. I. T. Press, Massachusetts Institute of Technology.

Yaglom, A. M. (1955). Correlation theory of stationary and related random processes with stationary nth increments. *Mat. Sbornik, 37*, 141–196.

Yaglom, A. M. (1957). Some classes of random fields in n-dimensional space related with random stationary processes. *Teor. Veroyatn. Primen., 2*, 292–338.

Yaglom, A. M. (1987a). *Correlation theory of stationary and related random functions. Basic results* (Vol. 1, p. 526). Springer series in statistics, New York (NY): Springer-Verlag.

Yaglom, A. M. (1987b). *Correlation theory of stationary and related random functions. Supplementary notes and references* (Vol. 2, p. 258). Springer Series in Statistics, New York (NY): Springer-Verlag.

9

Minimax-robust filtering problem for stochastic sequences with stationary increments and cointegrated sequences

Maksym Luz[1] and Mikhail Moklyachuk[1*]

*Corresponding author: Mikhail Moklyachuk, Department of Probability Theory, Statistics and Actuarial Mathematics, Taras Shevchenko National University of Kyiv, Kyiv 01601, Ukraine

E-mails: mmp@univ.kiev.ua, Moklyachuk@gmail.com

Reviewing editor: Nengxiang Ling, Hefei University of Technology, China

Abstract: The problem of optimal estimation of the linear functional $A\xi = \sum_{k=0}^{\infty} a(k)\xi(-k)$ depending on the unknown values of a stochastic sequence $\xi(m)$ with nth stationary increments from observations of the sequence $\xi(m) + \eta(m)$ at points $m = 0, -1, -2, \ldots$, where $\eta(m)$ is a stationary sequence uncorrelated with $\xi(m)$, is considered. Formulas for calculating the mean square error and the spectral characteristic of the optimal linear estimate of the functional are derived in the case where spectral densities of stochastic sequences are exactly known and admit the canonical factorizations. In the case of spectral uncertainty, where spectral densities are not known exactly, but sets of admissible spectral densities are specified, the minimax-robust method is applied. Formulas and relations that determine the least favourable spectral densities and the minimax-robust spectral characteristics are proposed for the given sets of admissible spectral densities. The filtering problem for a class of cointegrated sequences is investigated.

Subjects: Mathematics & Statistics; Operations Research; Optimization; Probability; Science; Statistical Theory & Methods; Statistics; Statistics & Probability; Stochastic Models & Processes

Keywords: stochastic sequence with stationary increments; cointegrated sequence; minimax-robust estimate; mean square error; least favourable spectral density; minimax-robust spectral characteristic

ABOUT THE AUTHORS

Maksym Luz is a PhD student, Department of Probability Theory, Statistics and Actuarial Mathematics, Taras Shevchenko National University of Kyiv. His research interests include estimation problems for random processes and sequences with stationary increments.

Mikhail Moklyachuk is a professor, Department of Probability Theory, Statistics and Actuarial Mathematics, Taras Shevchenko National University of Kyiv. He received PhD degree in Physics and Mathematical Sciences from the Taras Shevchenko University of Kyiv in 1977. His research interests are statistical problems for stochastic processes and random fields. He is also a member of editorial boards of several international journals.

PUBLIC INTEREST STATEMENT

The crucial assumption of application of traditional methods of finding solution to the filtering problem for random processes is that spectral densities of the processes are exactly known. However, in practical situations complete information on spectral densities is impossible and the established results cannot be directly applied to practical filtering problems. This is a reason to apply the minimax-robust method of filtering and derive the minimax estimates since they minimize the maximum value of the mean-square errors for all spectral densities from a given set of admissible densities simultaneously. In this article, we deal with the problem of optimal estimation of functionals depending on the unknown values of a random process with stationary increments based on observations of the process and a noise. In the case where spectral densities of the processes are not exactly known, relations for determining least favourable spectral densities and minimax-robust spectral characteristics are proposed.

1. Introduction

Basic results of the theory of wide sense stationary and related stochastic processes found their applications in analysis of models of economic and financial time series. The most simple examples are linear stationary models such as moving average (MA), autoregressive (AR) and autoregressive-moving average (ARMA) sequences, all of which refer to stationary sequences with rational spectral densities without unit AR-roots. Time series with trends and seasonal components are modelled by integrated ARMA (ARIMA) sequences which have unit roots in their autoregressive parts and are the examples of sequences with stationary increments. Such models are investigated during the last 30 years. The main points concerning model definition, parameter estimation, forecasting and further investigation of the models are discussed in the well-known book by Box, Jenkins, and Reinsel (1994). While analysing financial data economists noticed that in some special cases linear combinations of integrated sequences become stationary. Granger (1983) called this phenomenon cointegration. Cointegrated sequences found their application in applied and theoretical econometrics and financial time series analysis (see Engle & Granger, 1987).

The problem of estimation of unknown values of stochastic processes (extrapolation, interpolation and filtering problems) is an important part of the theory of stochastic processes. Effective methods of solution of the linear extrapolation, interpolation and filtering problems for stationary stochastic processes were developed by A.N. Kolmogorov, N. Wiener and A.M. Yaglom. See selected works by Kolmogorov (1992), books by Wiener (1966) and Yaglom (1987a, 1987b). Further results one can find in the book by Rozanov (1967).

Random processes with stationary nth increments are one of generalizations of the notion of stationary process that were introduced by Pinsker and Yaglom (1954), Yaglom (1955, 1957), Pinsker (1955). They described the spectral representation of the stationary increment process and the canonical factorization of the spectral density, solved the extrapolation problem for such processes and discussed some examples. See books by Yaglom (1987a, 1987b) for more relative results and references.

Traditional methods of finding solutions to extrapolation, interpolation and filtering problems for stationary and related stochastic processes are applied under the basic assumption that the spectral densities of the considered stochastic processes are exactly known. However, in most practical situations complete information on the spectral structure of the processes isn't available. Investigators can apply the traditional methods considering the estimated spectral densities instead of the true ones. However, as it was shown by Vastola and Poor (1983) with the help of some examples, this approach can result in significant increasing of the value of the error of estimate. Therefore, it is reasonable to derive estimates which are optimal for all densities from a certain class of spectral densities. The introduced estimates are called minimax-robust since they minimize the maximum of the mean-square errors for all spectral densities from a set of admissible spectral densities simultaneously. The minimax-robust method of estimation was proposed by Grenander (1957) and later developed by Franke and Poor (1984), Franke (1984) for investigating the extrapolation and interpolation problems. For more details we refer to the survey paper by Kassam and Poor (1985) who collected results in minimax (robust) methods of data processing till 1984. A wide range of results in minimax-robust extrapolation, interpolation and filtering of stochastic processes and sequences belongs to Moklyachuk (1990, 2000, 2001, 2008a, 2015). Later Moklyachuk and Masyutka (2006a, 2006b, 2007, 2008, 2011, 2012) developed the minimax technique of estimation for vector-valued stationary processes and sequences. Dubovets'ka, Masyutka, and Moklyachuk (2012) investigated the problem of minimax-robust interpolation for another generalization of stationary processes – periodically correlated sequences. In the further papers Dubovetska and Moklyachuk (2013a, 2013b, 2014a, 2014b) investigated the minimax-robust extrapolation, interpolation and filtering problems for periodically correlated processes and sequences. See the book by Golichenko and Moklyachuk (2014) for more relative results and references. The minimax-robust extrapolation, interpolation and filtering problems for stochastic sequences and processes with nth stationary increments were solved by Luz and Moklyachuk (2012, 2013a, 2013b, 2014a, 2014b, 2015a, 2015b,

2015c, 2016a, 2016b; Moklyachuk & Luz, 2013). The obtained results are applied to find solution of the extrapolation and filtering problems for cointegrated sequences (Luz and Moklyachuk, 2014b, 2015c). The problem of extrapolation of stochastic sequences with stationary increments from observations with non-stationary noise was investigated by Bell (1984).

In the present article, we deal with the problem of optimal linear estimation of the functional $A\xi = \sum_{k=0}^{\infty} a(k)\xi(-k)$ which depends on the unknown values of a stochastic sequence $\xi(k)$ with nth stationary increments from observations of the sequence $\xi(k)+\eta(k)$ at points $k = 0, -1, -2, \ldots$, where $\eta(k)$ is a stationary stochastic sequence uncorrelated with the sequence $\xi(k)$. Solution to this problem based on the Hilbert space projection method is described in the paper by Luz and Moklyachuk (2014b). The derived formulas for calculating the spectral characteristic and the mean square error of the optimal estimate $\widehat{A}\xi$ of the functional $A\xi$ are complicated for application and need to construct the inverse operator $(\mathbf{P}_{\mu})^{-1}$ which is also a complicated problem. On the other hand, most of spectral densities of stochastic sequences applied in time series analysis admit factorization. This is a reason to derive formulas for calculating the spectral characteristic and the mean square error of the optimal estimate of the functional which use coefficients of the canonical factorizations of spectral densities. In this article, it is shown that the derived formulas can be simplified under the condition that the spectral densities are such that the canonical factorizations of the functions hold true. The case of spectral certainty as well as the case of spectral uncertainty is considered. Formulas for calculating values of the mean-square errors and the spectral characteristics of the optimal linear estimate of the functional are derived under the condition of spectral certainty, where the spectral densities of the processes are exactly known. In the case of spectral uncertainty, where the spectral densities of the processes are not exactly known, but a class of admissible spectral densities is given, relations that determine the least favourable spectral densities and the minimax spectral characteristics are derived for some classes of spectral densities. The obtained results are applied to investigate the filtering problem for cointegrated sequences.

2. Stationary increment stochastic sequences. Spectral representation

In this section, we present a brief description of the properties of stochastic sequences with nth stationary increments. More detailed description of the properties of such sequences can be found in the books by Yaglom (1987a, 1987b).

Definition 2.1 For a given stochastic sequence $\{\xi(m), m \in \mathbb{Z}\}$ the sequence

$$\xi^{(n)}(m,\mu) = (1-B_{\mu})^n \xi(m) = \sum_{l=0}^{n} (-1)^l \binom{n}{l} \xi(m - l\mu), \tag{1}$$

where B_{μ} is a backward shift operator with step $\mu \in \mathbb{Z}$, such that $B_{\mu}\xi(m) = \xi(m-\mu)$, is called a stochastic nth increment sequence with step $\mu \in \mathbb{Z}$.

For the stochastic nth increment sequence $\xi^{(n)}(m,\mu)$ the following relations hold true:

$$\xi^{(n)}(m,-\mu) = (-1)^n \xi^{(n)}(m+n\mu,\mu), \tag{2}$$

$$\xi^{(n)}(m,k\mu) = \sum_{l=0}^{(k-1)n} A_l \xi^{(n)}(m-l\mu,\mu), \quad k \in \mathbb{N}, \tag{3}$$

where coefficients $\{A_l, l = 0, 1, 2, \ldots, (k-1)n\}$ are determined by the representation

$$(1 + x + \ldots + x^{k-1})^n = \sum_{l=0}^{(k-1)n} A_l x^l.$$

Definition 2.2 The stochastic nth increment sequence $\xi^{(n)}(m,\mu)$ generated by the stochastic sequence $\{\xi(m), m \in \mathbb{Z}\}$ is wide sense stationary if the mathematical expectations

$$\mathsf{E}\xi^{(n)}(m_0,\mu) = c^{(n)}(\mu),$$

$$\mathsf{E}\xi^{(n)}(m_0+m,\mu_1)\overline{\xi^{(n)}(m_0,\mu_2)} = D^{(n)}(m,\mu_1,\mu_2)$$

exist for all $m_0, \mu, m, \mu_1, \mu_2$ and do not depend on m_0. The function $c^{(n)}(\mu)$ is called a mean value of the nth increment sequence and the function $D^{(n)}(m, \mu_1, \mu_2)$ is called a structural function of the stationary nth increment sequence (or structural function of nth order of the stochastic sequence $\{\xi(m), m \in \mathbb{Z}\}$).

The stochastic sequence $\{\xi(m), m \in \mathbb{Z}\}$ which determines the stationary nth increment sequence $\xi^{(n)}(m, \mu)$ by formula (1) is called a sequence with stationary nth increments (or integrated sequence of order n).

THEOREM 2.1 The mean value $c^{(n)}(\mu)$ and the structural function $D^{(n)}(m, \mu_1, \mu_2)$ of the stochastic stationary nth increment sequence $\xi^{(n)}(m, \mu)$ can be represented in the following forms:

$$c^{(n)}(\mu) = c\mu^n, \tag{4}$$

$$D^{(n)}(m, \mu_1, \mu_2) = \int_{-\pi}^{\pi} e^{i\lambda m}(1 - e^{-i\mu_1\lambda})^n(1 - e^{i\mu_2\lambda})^n \frac{1}{\lambda^{2n}} dF(\lambda), \tag{5}$$

where c is a constant, $F(\lambda)$ is a left-continuous nondecreasing bounded function with $F(-\pi) = 0$. The constant c and the structural function $F(\lambda)$ are determined uniquely by the increment sequence $\xi^{(n)}(m, \mu)$.

On the other hand, a function $c^{(n)}(\mu)$ of the form (4) with a constant c and a function $D^{(n)}(m, \mu_1, \mu_2)$ of the form (5) with a function $F(\lambda)$ satisfying the indicated conditions are the mean value and the structural function of some stationary nth increment sequence $\xi^{(n)}(m, \mu)$, respectively.

Using representation (5) of the structural function of a stationary nth increment sequence $\xi^{(n)}(m, \mu)$ and the Karhunen theorem (see Gikhman & Skorokhod, 2004; Karhunen, 1947), we obtain the following spectral representation of the stationary nth increment sequence $\xi^{(n)}(m, \mu)$:

$$\xi^{(n)}(m, \mu) = \int_{-\pi}^{\pi} e^{im\lambda}(1 - e^{-i\mu\lambda})^n \frac{1}{(i\lambda)^n} dZ(\lambda), \tag{6}$$

where $Z_{\xi^{(n)}}(\lambda)$ is a random process with uncorrelated increments on $[-\pi, \pi)$ with respect to the spectral function $F(\lambda)$:

$$E|Z_{\xi^{(n)}}(t_2) - Z_{\xi^{(n)}}(t_1)|^2 = F(t_2) - F(t_1), \quad -\pi \leq t_1 < t_2 < \pi. \tag{7}$$

3. The filtering problem

Consider a stochastic sequence $\{\xi(m), m \in \mathbb{Z}\}$ which generates the stationary nth increment sequence $\xi^{(n)}(m, \mu)$ with absolutely continuous spectral function $F(\lambda)$ that has spectral density $f(\lambda)$. Let $\{\eta(m), m \in \mathbb{Z}\}$ be uncorrelated with the sequence $\xi(m)$ stationary stochastic sequence with absolutely continuous spectral function $G(\lambda)$ which has spectral density $g(\lambda)$. Without loss of generality we will assume that the mean values of the increment sequence $\xi^{(n)}(m, \mu)$ and stationary sequence $\eta(m)$ equal to 0. Let us also assume that the step $\mu > 0$.

Consider the problem of mean-square optimal linear estimation of the functional

$$A\xi = \sum_{k=0}^{\infty} a(k)\xi(-k)$$

which depends on the unknown values of the sequence $\xi(m)$ from observations of the sequence $\zeta(m) = \xi(m) + \eta(m)$ at points $m = 0, -1, -2, \ldots$.

We will suppose that coefficients $a(k)$, $k \geq 0$, which determine the functional satisfy the inequalities

$$\sum_{k=0}^{\infty} |a(k)| < \infty, \quad \sum_{k=0}^{\infty} (k+1)|a(k)|^2 < \infty, \tag{8}$$

and the spectral densities $f(\lambda)$ and $g(\lambda)$ satisfy the minimality condition

$$\int_{-\pi}^{\pi} \frac{\lambda^{2n}}{|1 - e^{i\lambda\mu}|^{2n}(f(\lambda) + \lambda^{2n}g(\lambda))} d\lambda < \infty. \tag{9}$$

This condition (9) is sufficient in order that the mean-square error of the estimate of the functional $A\xi$ is not equal to zero.

Note, that

$$p(\lambda) = f(\lambda) + \lambda^{2n}g(\lambda)$$

is the spectral density of the stochastic sequence $\zeta(m)$.

The functional $A\xi$ can be represented in the form $A\xi = A\zeta - A\eta$, where $A\zeta = \sum_{k=0}^{\infty} a(k)\zeta(-k)$ and $A\eta = \sum_{k=0}^{\infty} a(k)\eta(-k)$. In the case where conditions (8), the functional $A\eta$ has a finite second moment. To construct an estimate of the functional $A\xi$ it is sufficient to have an estimate of the functional $A\eta$. Since the functional $A\zeta$ depends on the values of the stochastic sequence $\zeta(m)$, which is observed, we have the following relations:

$$\begin{aligned} &\widehat{A}\xi = A\zeta - \widehat{A}\eta, \\ &\Delta(f, g; \widehat{A}\xi) = \mathsf{E}|A\xi - \widehat{A}\xi|^2 = \mathsf{E}|A\zeta - A\eta - A\zeta + \widehat{A}\eta|^2 = \mathsf{E}|A\eta - \widehat{A}\eta|^2 = \Delta(f, g; \widehat{A}\eta). \end{aligned} \tag{10}$$

It follows from relation (10) that any estimate $\widehat{A}\xi$ of the functional $A\xi$ can be represented in the form

$$\widehat{A}\xi = A\zeta - \int_{-\pi}^{\pi} h_{\mu}(\lambda) dZ_{\xi^{(n)}+\eta^{(n)}}(\lambda), \tag{11}$$

where $h_{\mu}(\lambda)$ is the spectral characteristic of the estimate $\widehat{A}\eta$.

Denote by $H^0(\xi_{\mu}^{(n)} + \eta_{\mu}^{(n)})$ the closed linear subspace of the Hilbert space $H = L_2(\Omega, \mathfrak{F}, \mathsf{P})$ of random variables γ that have zero first and finite second moment $\mathsf{E}\gamma = 0$, $\mathsf{E}|\gamma|^2 < \infty$, which is generated by values $\{\xi^{(n)}(k,\mu) + \eta^{(n)}(k,\mu) : k \le 0\}$, $\mu > 0$. Denote by $L_2^0(p)$ the closed linear subspace of the Hilbert space $L_2(p)$ of square integrable on $[-\pi;\pi)$ functions with respect to the measure which has the density $p(\lambda)$, generated by functions $\left\{ e^{i\lambda k}(1 - e^{-i\lambda\mu})^n (i\lambda)^{-n} : k \le 0 \right\}$. It follows from the relation

$$\xi^{(n)}(k,\mu) + \eta^{(n)}(k,\mu) = \int_{-\pi}^{\pi} e^{i\lambda k}(1 - e^{-i\lambda\mu})^n \frac{1}{(i\lambda)^n} dZ_{\xi^{(n)}+\eta^{(n)}}(\lambda)$$

that there exists a one to one correspondence between elements $e^{i\lambda k}(1 - e^{-i\lambda\mu})^n (i\lambda)^{-n}$ from the space $L_2^0(p)$ and elements $\xi^{(n)}(k,\mu) + \eta^{(n)}(k,\mu)$ from the space $H^0(\xi_{\mu}^{(n)} + \eta_{\mu}^{(n)})$ correspondingly.

Let $r(m,\mu) = \max\left\{ \left[-\frac{m}{\mu}\right]', 0 \right\}$, where by $[x]'$ we denote the least integer number among the numbers that are greater than or equal to x.

The mean square optimal estimate $\widehat{A}\eta$ is a projection of the element $A\eta$ on the subspace $H^0(\xi_{\mu}^{(n)} + \eta_{\mu}^{(n)})$. This projection is described in the paper by Luz and Moklyachuk (2014b). A solution of the filtering problem is described in the following theorem.

THEOREM 3.1 *Let $\{\xi(m), m \in \mathbb{Z}\}$ be a stochastic sequence which defines stationary nth increment sequence $\xi^{(n)}(m,\mu)$ with absolutely continuous spectral function $F(\lambda)$ which has spectral density $f(\lambda)$. Let $\{\eta(m), m \in \mathbb{Z}\}$ be uncorrelated with the sequence $\xi(m)$ stationary stochastic sequence with absolutely continuous spectral function $G(\lambda)$ which has spectral density $g(\lambda)$. Let the coefficients $\{a(k) : k \geq 0\}$ satisfy conditions (8). Let the spectral densities $f(\lambda)$ and $g(\lambda)$ of stochastic sequences $\xi(m)$ and $\eta(m)$ satisfy the minimality condition (9). The mean square optimal linear estimate $\widehat{A}\xi$ of the functional $A\xi$ based on observations of values $\xi(m)+\eta(m)$ based on observations of the sequence $\xi(m)+\eta(m)$ at points $m = 0, -1, -2, \ldots$ can be calculated by formula (11). The spectral characteristic $h_{\mu}(\lambda)$ and the mean square error $\Delta(f, g; \widehat{A}\eta)$ of the optimal estimate $\widehat{A}\xi$ are calculated by the formulas*

$$h_{\mu}(\lambda) = A(e^{-i\lambda})\frac{(-i\lambda)^n g(\lambda)}{f(\lambda)+\lambda^{2n}g(\lambda)} - \frac{(-i\lambda)^n C_{\mu}(e^{i\lambda})}{(1-e^{i\lambda\mu})^n(f(\lambda)+\lambda^{2n}g(\lambda))},$$
$$C_{\mu}(e^{i\lambda}) = \sum_{k=0}^{\infty}(\mathbf{P}_{\mu}^{-1}\mathbf{S}_{\mu}\widetilde{\mathbf{a}}_{\mu})_k e^{i\lambda(k+1)}, \tag{12}$$

and

$$\begin{aligned}\Delta(f,g;\widehat{A}\xi) &= \Delta(f,g;\widehat{A}\eta) = \mathsf{E}|A\eta - \widehat{A}\eta|^2\\ &= \frac{1}{2\pi}\int_{-\pi}^{\pi}\frac{\left|A(e^{-i\lambda})(1-e^{i\lambda\mu})^n f(\lambda)+\lambda^{2n}C_{\mu}(e^{i\lambda})\right|^2}{|1-e^{i\lambda\mu}|^{2n}(f(\lambda)+\lambda^{2n}g(\lambda))^2}g(\lambda)d\lambda\\ &+\frac{1}{2\pi}\int_{-\pi}^{\pi}\frac{\left|A(e^{-i\lambda})(1-e^{i\lambda\mu})^n\lambda^{2n}g(\lambda)-\lambda^{2n}C_{\mu}(e^{i\lambda})\right|^2}{\lambda^{2n}|1-e^{i\lambda\mu}|^{2n}(f(\lambda)+\lambda^{2n}g(\lambda))^2}f(\lambda)d\lambda\\ &= \langle\mathbf{S}_{\mu}\widetilde{\mathbf{a}}_{\mu}, \mathbf{P}_{\mu}^{-1}\mathbf{S}_{\mu}\widetilde{\mathbf{a}}_{\mu}\rangle + \langle\mathbf{Qa},\mathbf{a}\rangle\end{aligned} \tag{13}$$

respectively, where $\mathbf{a} = (a(0), a(1), a(2), \ldots)'$, $\widetilde{\mathbf{a}}_{\mu} = (\widetilde{a}_{\mu}(0), \widetilde{a}_{\mu}(1), \widetilde{a}_{\mu}(2), \ldots)'$, coefficients $\widetilde{a}_{\mu}(k) = a_{-\mu}(k-\mu n), k \geq 0$, are calculated by the formula

$$a_{-\mu}(m) = \sum_{l=r(m,\mu)}^{n}(-1)^l\binom{n}{l}a(m+\mu l), \quad m \geq -\mu n. \tag{14}$$

Here $\mathbf{S}_{\mu}, \mathbf{P}_{\mu}, \mathbf{Q}$ are the linear operators in the space ℓ_2 determined with the help of matrices with the elements $(\mathbf{S}_{\mu})_{kj} = S^{\mu}_{k+1,j-\mu n}$, $(\mathbf{P}_{\mu})_{kj} = P^{\mu}_{kj}$, $(\mathbf{Q})_{kj} = Q_{kj}$ for $k,j \geq 0$,

$$S^{\mu}_{kj} = \frac{1}{2\pi}\int_{-\pi}^{\pi}e^{-i\lambda(k+j)}\frac{\lambda^{2n}g(\lambda)}{|1-e^{i\lambda\mu}|^{2n}(f(\lambda)+\lambda^{2n}g(\lambda))}d\lambda, \quad k\geq 0, j\geq -\mu n,$$
$$P^{\mu}_{kj} = \frac{1}{2\pi}\int_{-\pi}^{\pi}e^{i\lambda(j-k)}\frac{\lambda^{2n}}{|1-e^{i\lambda\mu}|^{2n}(f(\lambda)+\lambda^{2n}g(\lambda))}d\lambda, \quad k,j\geq 0,$$
$$Q_{kj} = \frac{1}{2\pi}\int_{-\pi}^{\pi}e^{i\lambda(j-k)}\frac{f(\lambda)g(\lambda)}{f(\lambda)+\lambda^{2n}g(\lambda)}d\lambda, \quad k,j\geq 0.$$

This theorem gives us a possibility to find a solution to the filtering problem with the help of the Fourier coefficients of the functions

$$\frac{\lambda^{2n}}{|1-e^{i\lambda\mu}|^{2n}(f(\lambda)+\lambda^{2n}g(\lambda))}, \quad \frac{\lambda^{2n}g(\lambda)}{|1-e^{i\lambda\mu}|^{2n}(f(\lambda)+\lambda^{2n}g(\lambda))},$$
$$\frac{f(\lambda)g(\lambda)}{f(\lambda)+\lambda^{2n}g(\lambda)}.$$

The derived formulas are complicated for application and need to construct the inverse operator $(\mathbf{P}_{\mu})^{-1}$ which is also a complicated problem. On the other hand, most of spectral densities of stochastic sequences applied in time series analysis admit the factorization. This is a reason to derive formulas for calculating the spectral characteristic and the mean square error of the optimal estimate $\widehat{A}\xi$

of the functional $A\xi$ which use the canonical factorizations of the spectral densities. In particular, the proposed formulas (12) and (13) can be simplified under the condition that the spectral densities $f(\lambda)$ and $g(\lambda)$ are such that the following canonical factorizations of the functions hold true:

$$\frac{\lambda^{2n}}{|1-e^{i\lambda\mu}|^{2n}(f(\lambda)+\lambda^{2n}g(\lambda))} = \left|\sum_{k=0}^{\infty} \psi_\mu(k)e^{-i\lambda k}\right|^2 = \left|\sum_{j=0}^{\infty} \theta_\mu(j)e^{-i\lambda j}\right|^{-2}, \tag{15}$$

$$g(\lambda) = \sum_{k=-\infty}^{\infty} g(k)e^{i\lambda k} = \left|\sum_{j=0}^{\infty} \phi(j)e^{-i\lambda j}\right|^2. \tag{16}$$

Denote by $\mathbf{G}$ the linear operator in the space ℓ_2 which is determined by the matrix with elements $(\mathbf{G})_{l,k} = g(l-k), l, k \geq 0$. The following lemmas from the paper by Luz and Moklyachuk (2015b) give formulas for calculating operators $\mathbf{P}_\mu$ and $\mathbf{G}$ with the help of the coefficients of factorizations ()-(16).

Lemma 3.1 *Let the spectral densities $f(\lambda)$ and $g(\lambda)$ be such that the canonical factorizations (15) - (16) hold true. Define linear operators Ψ_μ and Φ in the space ℓ_2 with the help of matrices with elements $(\Psi_\mu)_{k,j} = \psi_\mu(k-j)$ and $(\Phi)_{k,j} = \phi(k-j)$ for $0 \leq j \leq k$, $(\Psi_\mu)_{k,j} = 0$ and $(\Phi)_{k,j} = 0$ for $0 \leq k < j$. Then*

(a) the following factorization holds true

$$\frac{g(\lambda)\lambda^{2n}}{|1-e^{i\lambda\mu}|^{2n}(f(\lambda)+\lambda^{2n}g(\lambda))} = \sum_{k=-\infty}^{\infty} s_\mu(k)e^{i\lambda k} = \left|\sum_{k=0}^{\infty} \upsilon_\mu(k)e^{-i\lambda k}\right|^2,$$
$$\upsilon_\mu(k) = \sum_{j=0}^{k} \psi_\mu(j)\phi(k-j) = \sum_{j=0}^{k} \phi(j)\psi_\mu(k-j); \tag{17}$$

(b) linear operator Υ_μ in the space ℓ_2 determined by a matrix with elements $(\Upsilon_\mu)_{k,j} = \upsilon_\mu(k-j)$ for $0 \leq j \leq k$, $(\Upsilon_\mu)_{k,j} = 0$ for $0 \leq k < j$, admits the representation $\Upsilon_\mu = \Psi_\mu\Phi = \Phi\Psi_\mu$.

Lemma 3.2 *Let canonical factorizations (15-16) hold true. Let the linear operators Ψ_μ and Υ_μ in the space ℓ_2 determined in the same way as in the lemma 3.1 and let the linear operator Θ_μ in the space ℓ_2 determined by the matrix with elements $(\Theta_\mu)_{k,j} = \theta_\mu(k-j)$ for $0 \leq j \leq k$, $(\Theta_\mu)_{k,j} = 0$ for $0 \leq k < j$. Define also a linear operator $\mathbf{T}_\mu$ in the space ℓ_2 with the help of matrices with elements $(\mathbf{T}_\mu)_{l,k} = s_\mu(l-k)$, $l, k \geq 0$, where the coefficients $s_\mu(k)$, $k \geq 0$, are determined in (17). Then*

(a) the linear operators $\mathbf{P}_\mu$, $\mathbf{T}_\mu$ and $\mathbf{G}$ in the space ℓ_2 admit the factorizations $\mathbf{P}_\mu = \Psi'_\mu\overline{\Psi}_\mu$, $\mathbf{T}_\mu = \Upsilon'_\mu\overline{\Upsilon}_\mu$ and $\mathbf{G} = \Phi'\overline{\Phi}$;

(b) the inverse operator $(\mathbf{P}_\mu)^{-1}$ admits the factorization $(\mathbf{P}_\mu)^{-1} = \overline{\Theta}_\mu\Theta'_\mu$.

Lemma 3.3 *Let the function $g(\lambda)$ admit the factorization (16) and let the linear operators $\mathbf{S}$ and $\mathbf{K}$ in the space ℓ_2 are determined by matrix with elements $(\mathbf{S})_{k,j} = g(k+j)$ and $(\mathbf{K})_{k,j} = \phi(k+j)$, $k,j \geq 0$. Then the operators $\mathbf{S}$ and $\mathbf{K}$ satisfy the relation $\mathbf{S} = \overline{\mathbf{K}}\Phi = \Phi'\overline{\mathbf{K}}$, where the linear operator Φ is determined in the lemma 3.1.*

With the help of the introduced results we show that formulas (12) and (13) can be simplified in the case where the spectral densities $f(\lambda)$ and $g(\lambda)$ are such that the canonical factorizations (15) - (16) hold true. Denote $\mathbf{e}_\mu = \Theta'_\mu\mathbf{S}_\mu\widetilde{\mathbf{a}}_\mu$. With the help of factorization (15) we have the next transformations:

$$\frac{\lambda^{2n}C_{\mu}(e^{i\lambda})}{|1-e^{i\lambda\mu}|^{2n}p(\lambda)} = \left(\sum_{k=0}^{\infty}\psi_{\mu}(k)e^{-i\lambda k}\right)\sum_{j=0}^{\infty}\sum_{k=0}^{\infty}\overline{\psi}_{\mu}(j)(\overline{\Theta}_{\mu}\mathbf{e}_{\mu})_{k}e^{i\lambda(k+j+1)}$$
$$= \left(\sum_{k=0}^{\infty}\psi_{\mu}(k)e^{-i\lambda k}\right)\sum_{m=0}^{\infty}\sum_{p=0}^{m}\sum_{k=p}^{m}\overline{\psi}_{\mu}(m-k)\overline{\theta}_{\mu}(k-p)e_{\mu}(p)e^{i\lambda(m+1)}$$
$$= \left(\sum_{k=0}^{\infty}\psi_{\mu}(k)e^{-i\lambda k}\right)\sum_{m=0}^{\infty}e_{\mu}(m)e^{i\lambda(m+1)},$$

where $e_{\mu}(m) = (\Theta'_{\mu}\mathbf{S}_{\mu}\widetilde{\mathbf{a}}_{\mu})_m$, $m \geq 0$, is the m-th elements of the vector $\mathbf{e}_{\mu} = \Theta'_{\mu}\mathbf{S}_{\mu}\widetilde{\mathbf{a}}_{\mu}$. Since

$$(\Theta'_{\mu}\mathbf{S}_{\mu}\widetilde{\mathbf{a}}_{\mu})_m = \sum_{j=-\mu n}^{\infty}\sum_{p=m}^{\infty}\theta_{\mu}(p-m)s_{\mu}(p+j+1)a_{-\mu}(j)$$
$$= \sum_{j=-\mu n}^{\infty}\sum_{l=0}^{\infty}\theta_{\mu}(l)s_{\mu}(m+j+l+1)a_{-\mu}(j),$$

the following equality holds true

$$\frac{\lambda^{2n}C_{\mu}(e^{i\lambda})}{|1-e^{i\lambda\mu}|^{2n}p(\lambda)} = \left(\sum_{k=0}^{\infty}\psi_{\mu}(k)e^{-i\lambda k}\right)\sum_{m=1}^{\infty}\sum_{j=-\mu n}^{\infty}\sum_{l=0}^{\infty}\theta_{\mu}(l)s_{\mu}(m+j+l)a_{-\mu}(j)e^{i\lambda m}. \tag{18}$$

With the help of factorization (17) and the relation

$$1 = \left(\sum_{k=0}^{\infty}\psi_{\mu}(k)e^{-i\lambda k}\right)\left(\sum_{j=0}^{\infty}\theta_{\mu}(j)e^{-i\lambda j}\right),$$

we have the next transformations

$$\frac{A(e^{-i\lambda})(1-e^{i\lambda\mu})^{n}\lambda^{2n}g(\lambda)}{|1-e^{i\lambda\mu}|^{2n}p(\lambda)} = \left(\sum_{k=0}^{\infty}\psi_{\mu}(k)e^{-i\lambda k}\right)\left(\sum_{l=0}^{\infty}\theta_{\mu}(l)e^{-i\lambda l}\right)\sum_{j=-\mu n}^{\infty}\sum_{m=-\infty}^{\infty}s_{\mu}(m+j)a_{-\mu}(j)e^{i\lambda m}$$
$$= \left(\sum_{k=0}^{\infty}\psi_{\mu}(k)e^{-i\lambda k}\right)\sum_{m=-\infty}^{\infty}\sum_{j=-\mu n}^{\infty}\sum_{l=0}^{\infty}s_{\mu}(m+j+l)\theta_{\mu}(l)a_{-\mu}(j)e^{i\lambda m}. \tag{19}$$

Define coefficients $\{\widetilde{b}_{\mu}(k) : k \geq 0\}$ in the following way: $\widetilde{b}_{\mu}(0) = 0$, $\widetilde{b}_{\mu}(k) = a_{-\mu}(-k)$ for $1 \leq k \leq \mu n$, $\widetilde{b}_{\mu}(k) = 0$ for $k > \mu n$, where coefficients $a_{-\mu}(k)$ are determined by formula (14). Define the vectors $\mathbf{a}_{-\mu} = (a_{-\mu}(0), a_{-\mu}(1), a_{-\mu}(2), \ldots)'$ and $\widetilde{\mathbf{b}}_{\mu} = (\widetilde{b}_{\mu}(0), \widetilde{b}_{\mu}(1), \widetilde{b}_{\mu}(2), \ldots)'$. Denote by $\widetilde{\mathbf{B}}_{\mu}$ the linear operator which is determined by the matrix with elements $(\widetilde{\mathbf{B}}_{\mu})_{k,j} = \widetilde{b}_{\mu}(k-j)$ for $0 \leq j \leq k$, $(\widetilde{\mathbf{B}}_{\mu})_{k,j} = 0$ for $0 \leq k < j$.

Making use of relations (18), (19) and the relation

$$\frac{A(e^{-i\lambda})(1-e^{i\lambda\mu})^n\lambda^{2n}g(\lambda)}{|1-e^{i\lambda\mu}|^{2n}p(\lambda)}-\frac{\lambda^{2n}C_\mu(e^{i\lambda})}{|1-e^{i\lambda\mu}|^{2n}p(\lambda)}$$
$$=\left(\sum_{k=0}^{\infty}\psi_\mu(k)e^{-i\lambda k}\right)\left(\sum_{m=0}^{\infty}\sum_{j=-\mu n}^{\infty}\sum_{l=0}^{\infty}s_\mu(j+l-m)\theta_\mu(l)a_{-\mu}(j)e^{-i\lambda m}\right)$$
$$=\left(\sum_{k=0}^{\infty}\psi_\mu(k)e^{-i\lambda k}\right)\left(\sum_{m=0}^{\infty}\sum_{j=0}^{\infty}\sum_{l=0}^{\infty}\overline{s}_\mu(m-j-l)\theta_\mu(l)a_{-\mu}(j)e^{-i\lambda m}\right)$$
$$+\left(\sum_{k=0}^{\infty}\psi_\mu(k)e^{-i\lambda k}\right)\left(\sum_{m=0}^{\infty}\sum_{j=1}^{\mu n}\sum_{l=0}^{\infty}\overline{s}_\mu(m+j-l)\theta_\mu(l)a_{-\mu}(-j)e^{-i\lambda m}\right)$$
$$=\left(\sum_{k=0}^{\infty}\psi_\mu(k)e^{-i\lambda k}\right)\left(\sum_{m=0}^{\infty}(\overline{\mathbf{T}}_\mu\Theta_\mu\mathbf{a}_{-\mu})_m e^{-i\lambda m}+\sum_{m=0}^{\infty}(\widetilde{\mathbf{B}}'_\mu\overline{\mathbf{T}}_\mu\theta_\mu)_m e^{-i\lambda m}\right),$$

we derive the following formula for calculating the spectral characteristic:

$$h_\mu(\lambda)=\frac{(1-e^{-i\lambda\mu})^n}{(i\lambda)^n}\left(\sum_{k=0}^{\infty}\psi_\mu(k)e^{-i\lambda k}\right)\sum_{m=0}^{\infty}\left((\mathbf{C}_{-\mu}+\mathbf{C}_\mu)\overline{\psi}_\mu\right)_m e^{-i\lambda m}. \tag{20}$$

In the last relation $(\mathbf{C}_\mu\overline{\psi}_\mu)_m$, $m\geq 0$, is the mth element of the vector

$$\mathbf{C}_\mu\overline{\psi}_\mu=\overline{\Psi}'_\mu\widetilde{\mathbf{S}\mathbf{b}}_\mu,$$

$\psi_\mu=(\psi_\mu(0),\psi_\mu(1),\psi_\mu(2),\ldots)'$, $\mathbf{C}_\mu$ is the linear operator which is determined by the matrix with elements $(\mathbf{C}_\mu)_{k,j}=\mathbf{c}_\mu(k+j)$, $k,j\geq 0$, $\mathbf{c}_\mu=\widetilde{\mathbf{S}\mathbf{b}}_\mu$, $\mathbf{S}$ is the linear operator which is determined by the matrix with elements $(\mathbf{S})_{k,j}=g(k+j)$, $k,j\geq 0$.

It is stated in Lemma 3.3 that the operator $\mathbf{S}$ admits the representation $\mathbf{S}=\overline{\mathbf{K}}\Phi=\Phi'\overline{\mathbf{K}}$, where $\mathbf{K}$ is the linear operator which is determined by the matrix with elements $(\mathbf{K})_{k,j}=\phi(k+j)$, $k,j\geq 0$. $(\mathbf{C}_{-\mu}\overline{\psi}_\mu)_m$, $m\geq 0$, is the m-th element of the vector $\mathbf{C}_{-\mu}\overline{\psi}_\mu=\overline{\Psi}'_\mu\mathbf{G}\mathbf{a}_{-\mu}$, $\psi_\mu=(\psi_\mu(0),\psi_\mu(1),\psi_\mu(2),\ldots)'$, $\mathbf{C}_{-\mu}$ is the linear operator which is determined by the matrix with elements $(\mathbf{C}_{-\mu})_{k,j}=\mathbf{c}_{-\mu}(k+j)$, $k,j\geq 0$, $\mathbf{c}_{-\mu}=\mathbf{G}\mathbf{a}_{-\mu}$, $\mathbf{G}$ is the linear operator which is determined by the matrix with elements $(\mathbf{G})_{k,j}=g(k-j)$, $k,j\geq 0$.

It follows from the Lemma 3.2 that the operator $\mathbf{G}$ admits the representation $\mathbf{G}=\Phi'\overline{\Phi}$, where Φ is the linear operator which is determined by the matrix with elements $(\Phi)_{k,j}=\phi(k-j)$, $k,j\geq 0$.

The mean square error is calculated by the formula

$$\begin{aligned}\Delta(f,g;\widehat{A}\xi)&=\Delta(f,g;\widehat{A}\eta)=\mathsf{E}|A\eta-\widehat{A}\eta|^2\\&=\frac{1}{2\pi}\int_{-\pi}^{\pi}|A(e^{-i\lambda})|^2g(\lambda)d\lambda+\frac{1}{2\pi}\int_{-\pi}^{\pi}|h_\mu(e^{i\lambda})|^2(f(\lambda)+\lambda^{2n}g(\lambda))d\lambda\\&\quad-\frac{1}{2\pi}\int_{-\pi}^{\pi}h_\mu(e^{i\lambda})\overline{A(e^{-i\lambda})(i\lambda)^n}g(\lambda)d\lambda-\frac{1}{2\pi}\int_{-\pi}^{\pi}\overline{h_\mu(e^{i\lambda})}A(e^{-i\lambda})(-i\lambda)^n g(\lambda)d\lambda\\&=\langle\mathbf{G}\mathbf{a},\mathbf{a}\rangle-\langle(\mathbf{C}_\mu+\mathbf{C}_{-\mu})\overline{\psi}_\mu,(\mathbf{C}_\mu+\mathbf{C}_{-\mu})\overline{\psi}_\mu\rangle.\end{aligned} \tag{21}$$

These observations can be summarized in the form of the theorem.

THEOREM 3.2 *Let $\{\xi(m),m\in\mathbb{Z}\}$ be a stochastic sequence which defines the stationary nth increment sequence $\xi^{(n)}(m,\mu)$ with an absolutely continuous spectral function $F(\lambda)$ which has spectral density $f(\lambda)$. Let $\{\eta(m),m\in\mathbb{Z}\}$ be an uncorrelated with the sequence $\xi(m)$ stationary stochastic sequence with an absolutely continuous spectral function $G(\lambda)$ which has spectral density $g(\lambda)$. Let the coefficients $\{a(k):k\geq 0\}$ satisfy condition (8), and let the spectral densities $f(\lambda)$ and $g(\lambda)$ of the sequences $\xi(m)$ and*

$\eta(m)$ admit canonical factorizations (15–16). The spectral characteristic $h_\mu(\lambda)$ and the mean square error $\Delta(f,g;\widehat{A}\xi)$ of the optimal estimate $\widehat{A}\xi$ of the functional $A\xi$ based on observations of the sequence $\xi(m)+\eta(m)$ at points $m=0,-1,-2,\ldots$ can be calculated by formulas (20) and (21).

Remark 3.1 Results described in theorem can be used for finding the optimal estimate $\widehat{A}_N\xi$ of the functional $A_N\xi=\sum_{k=0}^{N}a(k)\xi(-k)$ based on observations of the sequence $\xi(m)+\eta(m)$ at points $m=0,-1,-2,\ldots$. For this purpose it is sufficient to take $a(k)=0$ for $k>N$ in the formulas (11), (20), (21). In the case where $N=0$ we have the smoothing problem. Solution of this problem is described in the following corollary.

Corollary 3.1 The optimal estimate $\widehat{\xi}(0)$ of the unknown value $\xi(0)$ based on observations of the sequence $\zeta(m)=\xi(m)+\eta(m)$ at points $m=0,-1,-2,\ldots$ is calculated by formula

$$\widehat{\xi}(0)=\zeta(0)-\int_{-\pi}^{\pi}h_{\mu,0}(\lambda)dZ_{\xi^{(n)}+\eta^{(n)}}(\lambda).$$

The spectral characteristic $h_{\mu,0}(\lambda)$ and the mean square error $\Delta(f,g;\widehat{\xi}(0))$ of the optimal estimate $\widehat{\xi}(0)$ are calculated by the formulas

$$h_{\mu,0}(\lambda)=\frac{(1-e^{-i\lambda\mu})^n}{(i\lambda)^n}\left(\sum_{k=0}^{\infty}\psi_\mu(k)e^{-i\lambda k}\right)\sum_{m=0}^{\infty}\left(\overline{\Psi}'_\mu\overline{\Phi}'\mathbf{K}\mathbf{a}_{\mu,0}\right)_m e^{-i\lambda m}$$

and

$$\Delta(f,g;\widehat{\xi}(0))=||\phi||^2-||\overline{\Psi}'_\mu\overline{\Phi}'\mathbf{K}\mathbf{a}_{\mu,0}||^2$$

correspondingly, where $\phi=(\phi(0),\phi(1),\phi(2),\ldots)'$, $\mathbf{a}_{\mu,0}=(a_{\mu,0}(0),a_{\mu,0}(1),a_{\mu,0}(2),\ldots)'$ is an infinite dimension vector with elements $a_{\mu,0}(\mu l)=(-1)^l\binom{n}{l}$ for $l=0,1,2,\ldots,n$ and $a_{\mu,0}(k)=0$ for $k\geq 0$, $k\neq \mu l$, $l=0,1,2,\ldots,n$.

Remark 3.2 Since for all $n\geq 1$ and $\mu\geq 1$ the condition

$$\int_{-\pi}^{\pi}\left|\ln\frac{|1-e^{i\lambda\mu}|^{2n}}{\lambda^{2n}}\right|d\lambda<\infty$$

holds true, then there is a function $w_\mu(z)=\sum_{k=0}^{\infty}w_\mu(k)z^k$ such that $\sum_{k=0}^{\infty}|w_\mu(k)|^2<\infty$ and $\frac{|1-e^{i\lambda\mu}|^{2n}}{\lambda^{2n}}=|w_\mu(e^{-i\lambda})|^2$ (see Gikhman & Skorokhod, 2004). In the case where factorization (15) holds true, the function $f(\lambda)+\lambda^{2n}g(\lambda)$ admits the factorization

$$f(\lambda)+\lambda^{2n}g(\lambda)=\left|\sum_{k=0}^{\infty}\theta(k)e^{-i\lambda k}\right|^2=\left|\sum_{j=0}^{\infty}\psi(j)e^{-i\lambda j}\right|^{-2}. \tag{22}$$

The spectral density $f(\lambda)$ admits the canonical factorization

$$f(\lambda)=|\Phi(e^{-i\lambda})|^2,\quad \Phi(z)=\sum_{k=0}^{\infty}\varphi(k)z^k, \tag{23}$$

where the function $\Phi(z)$ has the radius of convergence $r>1$ and does not have zeros in the region $|z|\leq 1$.

Introduce the linear operators Θ, Ψ and $\mathbf{W}_\mu$ in the space ℓ_2 with the help of the matrices with elements $(\Theta)_{kj}=\theta(k-j)$, $(\Psi)_{kj}=\psi(k-j)$ and $(\mathbf{W}_\mu)_{kj}=w_\mu(k-j)$ for $0\leq j\leq k$, $(\Theta)_{kj}=0$, $(\Psi)_{kj}=0$ and $(\mathbf{W}_\mu)_{kj}=0$ for $0\leq k<j$. Denote $\mathbf{U}_\mu=\mathbf{W}_\mu^{-1}$. The following relations hold true:

$$\Theta_\mu=\Theta\mathbf{W}_\mu=\mathbf{W}_\mu\Theta,\quad \Psi_\mu=\Psi\mathbf{U}_\mu=\mathbf{U}_\mu\Psi,$$

$$\theta_\mu=\mathbf{W}_\mu\theta,\quad \psi_\mu=\mathbf{U}_\mu\psi, \tag{24}$$

where $\theta = (\theta(0), \theta(1), \theta(2), \ldots)'$, $\psi = (\psi(0), \psi(1), \psi(2), \ldots)'$.

4. Filtering of cointegrated stochastic sequences

Consider two integrated stochastic sequences $\xi^{(n)}(m,\mu)$ and $\zeta^{(n)}(m,\mu)$ with absolutely continuous spectral functions $F(\lambda)$ and $P(\lambda)$ which have spectral densities $f(\lambda)$ and $p(\lambda)$ correspondingly.

Definition 4.1 Two integrated stochastic sequences $\{\xi(m), m \in \mathbb{Z}\}$ and $\{\zeta(m), m \in \mathbb{Z}\}$ are called cointegrated (of order 0) if there exists a constant $\beta \neq 0$ such that the sequence $\{\zeta(m) - \beta\xi(m) : m \in \mathbb{Z}\}$ is stationary.

The filtering problem for cointegrated stochastic sequences consists in finding the mean-square optimal linear estimate of the functional

$$A\xi = \sum_{k=0}^{\infty} a(k)\xi(-k)$$

which depends on the unknown values of the sequence $\xi(m)$ from observations of the sequence $\zeta(m)$ at points $m = 0, -1, -2, \ldots$. This problem can be solved by using results presented in the preceding section under the condition that sequences $\xi(m)$ and $\zeta(m) - \beta\xi(m)$ are uncorrelated.

Suppose that the spectral densities $f(\lambda)$ and $p(\lambda)$ are such that the following canonical factorizations hold true

$$\frac{|1 - e^{i\lambda\mu}|^{2n} p(\lambda)}{\lambda^{2n}} = \left|\sum_{k=0}^{\infty} \psi_\mu^\beta(k) e^{-i\lambda k}\right|^{-2}, \quad p(\lambda) = \left|\sum_{k=0}^{\infty} \psi^\beta(k) e^{-i\lambda k}\right|^{-2}, \tag{25}$$

$$p(\lambda) - \beta^2 f(\lambda) = \lambda^{2n}\left|\sum_{k=0}^{\infty} \phi^\beta(k) e^{-i\lambda k}\right|^{2}. \tag{26}$$

Detemine the linear operators $\mathbf{K}^\beta$, Ψ^β and Φ^β with the help of the canonical factorizations (25-26) in the same way as operators $\mathbf{K}, \Psi$ and Φ were defined. By using theorem 3.2, we derive that the spectral characteristic $h_\mu^\beta(\lambda)$ of the optimal estimate

$$\widehat{A}\xi = A\zeta - \int_{-\pi}^{\pi} h_\mu^\beta(\lambda) dZ_{\zeta^{(n)}}(\lambda) \tag{27}$$

of the functional $A\xi$ is calculated by the formula

$$h_\mu^\beta(\lambda) = \frac{(1 - e^{-i\lambda\mu})^n}{(i\lambda)^n} \left(\sum_{m=0}^{\infty} \left((\mathbf{C}_\mu^\beta + \mathbf{C}_{-\mu}^\beta)\overline{\psi}_\mu^\beta \right)_m e^{-i\lambda m} \right) \sum_{k=0}^{\infty} \psi_\mu^\beta(k) e^{-i\lambda k}, \tag{28}$$

where

$$\mathbf{C}_\mu^\beta \overline{\psi}_\mu^\beta = \overline{\mathbf{U}}_\mu' \overline{\Psi^\beta}' \overline{\Phi^\beta}' \mathbf{K}^\beta \widetilde{\mathbf{b}}_\mu, \quad \mathbf{C}_{-\mu}^\beta \overline{\psi}_\mu^\beta = \overline{\mathbf{U}}_\mu' \overline{\Psi^\beta}' \overline{\Phi^\beta}' \Phi^\beta \mathbf{a}_{-\mu},$$

the operator $\mathbf{U}_\mu$ is determined in remark 3.2. The value of the mean-square error is calculated by the formula

$$\Delta(f, g; \widehat{A}\xi) = \langle \mathbf{G}^\beta \mathbf{a}, \mathbf{a} \rangle - \left\langle (\mathbf{C}_\mu^\beta + \mathbf{C}_{-\mu}^\beta)\overline{\psi}_\mu^\beta, (\mathbf{C}_\mu^\beta + \mathbf{C}_{-\mu}^\beta)\overline{\psi}_\mu^\beta \right\rangle. \tag{29}$$

THEOREM 4.2 *Let $\{\xi(m), m \in \mathbb{Z}\}$ and $\{\zeta(m), m \in \mathbb{Z}\}$ be two cointegrated stochastic sequences which have absolutely continuous spectral functions $F(\lambda)$ and $G(\lambda)$ with the spectral densities $f(\lambda)$ and $p(\lambda)$, respectively. Let coefficients $\{a(k) : k \geq 0\}$ satisfy conditions (8). If the spectral densities $f(\lambda)$ and $p(\lambda)$ admit canonical factorizations (25-26), and the sequences $\xi(m)$ and $\zeta(m) - \beta\xi(m)$ are uncorrelated,* *then the spectral characteristic $h^{\beta}_{\mu}(\lambda)$ and the mean-square error $\Delta(f,g;\widehat{A}\xi)$ of the optimal linear estimate $\widehat{A}\xi$ of the functional $A\xi$ of unknown elements $\xi(m)$, $m\leq 0$, from observations of the sequence $\zeta(m)$ at points $m=0,-1,-2,\dots$ is calculated by formulas (28) and (29).*

Example 4.1 Consider two random sequences $\{(\xi(m),\zeta(m)), m\in\mathbb{Z}\}$ which are determined by the equations

$$\xi(m)=\xi(m-1)+\varepsilon_1(m)+\varphi\varepsilon_1(m-1),$$
$$\zeta(m)=\xi(m)+\varepsilon_2(m),$$

where $\{\varepsilon_1(m),\varepsilon_2(m):m\in\mathbb{Z}\}$ are two uncorrelated sequences of independent identically distributed random variables with $\mathsf{E}\varepsilon_i(m)=0$, $\mathsf{E}\varepsilon_i^2(m)=1$, $i=1,2$. Denote

$$x=\frac{1}{2}\left(3+\varphi^2\mp\sqrt{(\varphi^2-1)^2+(\varphi+1)^2}\right),$$
$$y=\frac{-1}{2-2\varphi}\left(3+\varphi^2\pm\sqrt{(\varphi^2-1)^2+(\varphi+1)^2}\right),$$

and suppose that $|\varphi|<1$, $|y|<1$. In this case the random sequences $\xi(m)$ and $\zeta(m)$ are ARIMA(0, 1, 1) sequences with the spectral densities

$$f(\lambda)=\frac{\lambda^2|1+\varphi e^{-i\lambda}|^2}{|1-e^{-i\lambda}|^2},\quad p(\lambda)=\frac{x\lambda^2|1+ye^{-i\lambda}|^2}{|1-e^{-i\lambda}|^2}.$$

The difference $\zeta(m)-\xi(m)=\varepsilon_2(m)$ is a stationary sequence. That is why the integrated random sequences $\xi(m)$ and $\zeta(m)$ are cointegrated with the parameter of cointegration $\beta=1$. Since the random sequences $\varepsilon_1(m)$ and $\varepsilon_2(m)$ are uncorrelated, then the sequences $\xi(m)$ and $\zeta(m)-\xi(m)$ are uncorrelated also.

Consider the problem of filtering of the functional $A_1\xi=\xi(0)+a\xi(-1)$ from observations of the sequence $\zeta(m)$ at points $m=0,-1,-2,\dots$. Making use of Theorem 4.1 we will have

$$\Phi^{\beta}=\begin{pmatrix}1&0&0&\cdots\\0&1&0&\cdots\\0&0&1&\cdots\\\vdots&\vdots&\vdots&\ddots\end{pmatrix},\quad \mathbf{K}^{\beta}=\begin{pmatrix}1&0&0&\cdots\\0&0&0&\cdots\\0&0&0&\cdots\\\vdots&\vdots&\vdots&\ddots\end{pmatrix},\quad \mathbf{U}_{\mu}\Psi^{\beta}=\frac{1}{\sqrt{x}}\begin{pmatrix}1&0&0&\cdots\\y&1&0&\cdots\\y^2&y&1&\cdots\\\vdots&\vdots&\vdots&\ddots\end{pmatrix},$$

$\mathbf{a}_{-\mu}=(1-a,a,0\dots)'$. Since the first coordinate of the vector $\widetilde{\mathbf{b}}_{\mu}$ is equal to 0, then

$$\mathbf{K}^{\beta}\widetilde{\mathbf{b}}_{\mu}=(0,0,0,\dots)',\quad \mathbf{C}^{\beta}_{\mu}\overline{\psi}^{\beta}_{\mu}=(0,0,0,\dots)',$$

and

$$\mathbf{C}^{\beta}_{-\mu}\overline{\psi}^{\beta}_{\mu}=\overline{\mathbf{U}}'_{\mu}\overline{\Psi^{\beta}}'\overline{\Phi^{\beta}}'\Phi^{\beta}\mathbf{a}_{-\mu}=\frac{1}{\sqrt{x}}(1+a(y-1),a,0,\dots)'.$$

That is why the spectral characteristic $h^{\beta}_{\mu}(\lambda)$ of the optimal estimate $\widehat{A}_1\xi$ of the functional $A_1\xi$ is calculated by the formula

$$h^{\beta}_{\mu}(\lambda)=\frac{(1-e^{-i\lambda\mu})^n}{(i\lambda)^n}\frac{1}{x}\left(1+a(y-1)+(ay^2+y(1-a)-a)\sum_{k=1}^{\infty}y^{k-1}e^{-i\lambda k}\right),$$

Denote by $s(0)=x^{-1}(1+a(y-1))$ and $s(k)=x^{-1}(ay^2+y(1-a)-a)y^{k-1}$, $k\geq 1$. The optimal estimate $\widehat{A}_1\xi$ of the functional $A_1\xi$ is calculated by the formula

$$\widehat{A}_1\xi = \zeta(0) + a\zeta(-1) - \sum_{k=0}^{\infty} s(k)\zeta^{(1)}(-k,1)$$

$$= (1 - s(0))\zeta(0) + (a + s(0) - s(1))\zeta(-1) - \sum_{k=2}^{\infty}(s(k) - s(k-1))\zeta(-k)$$

$$= x^{-1}(x - 1 - a(y-1))\zeta(0) + x^{-1}(1 - y - a(y^2 - 2y - x + 2))\zeta(-1)$$

$$- x^{-1}(y-1)(y + a(y^2 - y + 1)) \sum_{k=2}^{\infty} y^{k-2}\zeta(-k).$$

The value of the mean-square error $\Delta(f, g; \widehat{A}_1\xi)$ of the optimal estimate $\widehat{A}_1\xi$ of the functional $A_1\xi$ is calculated by the formula

$$\Delta(f, g; \widehat{A}_1\xi) = 1 + a^2 - x^{-1}((1 + a(y-1))^2 + a^2).$$

5. Minimax-robust method of filtering

Formulas for calculation of values of the mean-square errors and spectral characteristics of the optimal linear estimates of the functional $A\xi$ based on observations of the stochastic sequence $\xi(k) + \eta(k)$ are derived under the condition that the spectral densities $f(\lambda)$ and $g(\lambda)$ of the stochastic sequences $\xi(m)$ and $\eta(m)$ are known. In the case where the spectral densities are not exactly known, but a set $\mathcal{D} = \mathcal{D}_f \times \mathcal{D}_g$ of admissible spectral densities is given, the minimax (robust) approach to estimation of functionals which depend on the unknown values of stochastic sequence with stationary increments is reasonable. In other words, we are interested in finding an estimate that minimizes the maximum of mean-square errors for all spectral densities from a given class $\mathcal{D} = \mathcal{D}_f \times \mathcal{D}_g$ of admissible spectral densities simultaneously.

Definition 5.1 For a given class of spectral densities $\mathcal{D} = \mathcal{D}_f \times \mathcal{D}_g$ the spectral densities $f^0(\lambda) \in \mathcal{D}_f$, $g^0(\lambda) \in \mathcal{D}_g$ are called the least favourable densities in the class $\mathcal{D}$ for the optimal linear filtering of the functional $A\xi$ if the following relation holds true

$$\Delta(f^0, g^0) = \Delta(h(f^0, g^0); f^0, g^0) = \max_{(f,g)\in\mathcal{D}_f\times\mathcal{D}_g} \Delta(h(f, g); f, g).$$

Definition 5.2 For a given class of spectral densities $\mathcal{D} = \mathcal{D}_f \times \mathcal{D}_g$ the spectral characteristic $h^0(\lambda)$ of the optimal linear estimate of the functional $A\xi$ is called minimax-robust if there are satisfied conditions

$$h^0(\lambda) \in H_{\mathcal{D}} = \bigcap_{(f,g)\in\mathcal{D}_f\times\mathcal{D}_g} L_2^0(p),$$

$$\min_{h\in H_{\mathcal{D}}} \max_{(f,g)\in\mathcal{D}_f\times\mathcal{D}_g} \Delta(h; f, g) = \max_{(f,g)\in\mathcal{D}_f\times\mathcal{D}_g} \Delta(h^0; f, g).$$

The following statements are consequences of the introduced definitions of least favourable spectral densities, minimax-robust spectral characteristic and 3.2.

LEMMA 5.1 *Spectral densities $f^0 \in \mathcal{D}_f$, $g^0 \in \mathcal{D}_g$ which admit canonical factorizations (15) and (16) are least favourable in the class $\mathcal{D} = \mathcal{D}_f \times \mathcal{D}_g$ for the optimal linear filtering of the functional $A\xi$ based on observations of the sequence $\xi(m) + \eta(m)$ at points $m \leq 0$ if coefficients $\{\psi^0(k), \phi^0(k): k \geq 0\}$ of the canonical factorizations*

$$f^0(\lambda) + \lambda^{2n} g^0(\lambda) = \left|\sum_{k=0}^{\infty} \psi^0(k) e^{-i\lambda k}\right|^{-2}, \quad g^0(\lambda) = \left|\sum_{k=0}^{\infty} \phi^0(k) e^{-i\lambda k}\right|^2. \tag{30}$$

determine a solution of the constrained optimization problem

$$\langle \mathbf{Ga}, \mathbf{a}\rangle - \left\langle (\mathbf{C}_\mu + \mathbf{C}_{-\mu})\overline{\psi}_\mu, (\mathbf{C}_\mu + \mathbf{C}_{-\mu})\overline{\psi}_\mu \right\rangle \to \sup,$$
$$f(\lambda) = \left|\sum_{k=0}^{\infty} \psi(k)e^{-i\lambda k}\right|^{-2} - \lambda^{2n}\left|\sum_{k=0}^{\infty} \phi(k)e^{-i\lambda k}\right|^{2} \in \mathcal{D}_f, \tag{31}$$
$$g(\lambda) = \left|\sum_{k=0}^{\infty} \phi(k)e^{-i\lambda k}\right|^{2} \in \mathcal{D}_g.$$

The minimax spectral characteristic $h^0 = h_\mu(f^0, g^0)$ is calculated by formula (20) if $h_\mu(f^0, g^0) \in H_{\mathcal{D}}$.

Lemma 5.2 *The spectral density $g^0 \in \mathcal{D}_g$ which admits canonical factorizations (15) - (16) with the known spectral density $f(\lambda)$ is least favourable in the class $\mathcal{D}_g$ for the optimal linear filtering of the functional $A\xi$ based on observations of the sequence $\xi(m) + \eta(m)$ at points $m \leq 0$ if coefficients $\{\psi^0(k), \phi^0(k) : k \geq 0\}$ of the canonical factorizations*

$$f(\lambda) + \lambda^{2n} g^0(\lambda) = \left|\sum_{k=0}^{\infty} \psi^0(k)e^{-i\lambda k}\right|^{-2}, \quad g^0(\lambda) = \left|\sum_{k=0}^{\infty} \phi^0(k)e^{-i\lambda k}\right|^{2}. \tag{32}$$

determine a solution of the constrained optimization problem

$$\langle \mathbf{Ga}, \mathbf{a}\rangle - \left\langle (\mathbf{C}_\mu + \mathbf{C}_{-\mu})\overline{\psi}_\mu, (\mathbf{C}_\mu + \mathbf{C}_{-\mu})\overline{\psi}_\mu \right\rangle \to \sup,$$
$$g(\lambda) = \left|\sum_{k=0}^{\infty} \phi(k)e^{-i\lambda k}\right|^{2} \in \mathcal{D}_g. \tag{33}$$

The minimax spectral characteristic $h^0 = h_\mu(f, g^0)$ is calculated by formula (20) if $h_\mu(f, g^0) \in H_{\mathcal{D}}$.

Lemma 5.3 *The spectral density $f^0 \in \mathcal{D}_f$ which admits the canonical factorizations (15) with the known spectral density $g(\lambda)$ is least favourable in the class $\mathcal{D}_f$ for the optimal linear filtering of the functional $A\xi$ based on observations of the sequence $\xi(m) + \eta(m)$ at points $m \leq 0$, if coefficients $\{\psi^0(k), \phi^0(k) : k \geq 0\}$ of the canonical factorizations*

$$f^0(\lambda) + \lambda^{2n} g(\lambda) = \left|\sum_{k=0}^{\infty} \psi^0(k)e^{-i\lambda k}\right|^{-2}, \tag{34}$$

determine a solution of the constrained optimization problem

$$\left\langle (\mathbf{C}_\mu + \mathbf{C}_{-\mu})\overline{\psi}_\mu, (\mathbf{C}_\mu + \mathbf{C}_{-\mu})\overline{\psi}_\mu \right\rangle \to \inf,$$
$$f(\lambda) = \left|\sum_{k=0}^{\infty} \psi(k)e^{-i\lambda k}\right|^{-2} - \lambda^{2n}\left|\sum_{k=0}^{\infty} \phi(k)e^{-i\lambda k}\right|^{2} \in \mathcal{D}_f \tag{35}$$

with the fixed coefficients $\{\phi(k) : k \geq 0\}$. The minimax spectral characteristic $h^0 = h_\mu(f^0, g)$ is calculated by formula (20) if $h_\mu(f^0, g) \in H_{\mathcal{D}}$.

The minimax spectral characteristic h^0 and the pair (f^0, g^0) of least favourable spectral densities form a saddle point of the function $\Delta(h; f, g)$ on the set $H_{\mathcal{D}} \times \mathcal{D}$. The saddle point inequalities

$$\Delta(h; f^0, g^0) \geq \Delta(h^0; f^0, g^0) \geq \Delta(h^0; f, g) \quad \forall f \in \mathcal{D}_f, \forall g \in \mathcal{D}_g, \forall h \in H_{\mathcal{D}}$$

hold true if $h^0 = h_\mu(f^0, g^0)$ and $h_\mu(f^0, g^0) \in H_{\mathcal{D}}$, where (f^0, g^0) is a solution to the constrained optimization problem

$$
\begin{aligned}
&\widetilde{\Delta}(f,g) = -\Delta(h_\mu(f^0,g^0); f,g) \to \inf, \quad (f,g) \in \mathcal{D},\\
&\Delta(h_\mu(f^0,g^0); f,g) = \frac{1}{2\pi}\int_{-\pi}^{\pi} \frac{\left|r^0_{\mu,g}(e^{-i\lambda})\right|^2}{f^0(\lambda)+\lambda^{2n}g^0(\lambda)} f(\lambda)d\lambda + \frac{1}{2\pi}\int_{-\pi}^{\pi} \frac{\lambda^{2n}\left|r^0_{\mu,f}(e^{-i\lambda})\right|^2}{f^0(\lambda)+\lambda^{2n}g^0(\lambda)} g(\lambda)d\lambda,\\
&r^0_{\mu,g}(e^{-i\lambda}) = \sum_{k=0}^{\infty}\left((\mathbf{C}^0_\mu + \mathbf{C}^0_{-\mu})\overline{\Psi}^0_\mu\right)_k e^{-i\lambda k},\\
&r^0_{\mu,f}(e^{-i\lambda}) = \frac{A(e^{-i\lambda})}{(1-e^{-i\lambda\mu})^n}\left(\sum_{k=0}^{\infty}\psi^0_\mu(k)e^{-i\lambda k}\right)^{-1} - \sum_{k=0}^{\infty}\left((\mathbf{C}^0_\mu + \mathbf{C}^0_{-\mu})\overline{\Psi}^0_\mu\right)_k e^{-i\lambda k}.
\end{aligned}
\tag{36}
$$

This constrained optimization problem is equivalent to the unconstrained optimization problem

$$
\Delta_{\mathcal{D}}(f,g) = \widetilde{\Delta}(f,g) + \delta(f,g|\mathcal{D}_f \times \mathcal{D}_g) \to \inf, \tag{37}
$$

where $\delta(f,g|\mathcal{D}_f \times \mathcal{D}_g)$ is the indicator function of the set $\mathcal{D}_f \times \mathcal{D}_g$. A solution (f^0,g^0) to this unconstrained optimization problem is characterized by a condition $0 \in \partial\Delta_{\mathcal{D}}(f^0,g^0)$, which is the necessary and sufficient condition that the pair (f^0,g^0) belongs to the set of minimums of the convex functional $\Delta_{\mathcal{D}}(f,g)$ (see Moklyachuk, 2008b; Pshenichnyi, 1971; Rockafellar, 1997). Here the notion $\partial\Delta_{\mathcal{D}}(f^0,g^0)$ determines a subdifferential of the functional $\Delta_{\mathcal{D}}(f,g)$ at the point $(f,g)=(f^0,g^0)$, which is a set of all linear bounded functionals Λ on $\mathcal{L}_1 \times \mathcal{L}_1$ satisfying the inequality

$$
\Delta_{\mathcal{D}}(f,g) - \Delta_{\mathcal{D}}(f^0,g^0) \geq \Lambda\left((f,g)-(f^0,g^0)\right), \quad (f,g) \in \mathcal{D}.
$$

In the case of investigation the cointegrated sequences we get the following optimization problem for determination of the least favourable spectral densities

$$
\begin{aligned}
&\Delta_{\mathcal{D}}(f,p) = \widetilde{\Delta}(f,p) + \delta(f,p|\mathcal{D}_f \times \mathcal{D}_p) \to \inf,\\
&\widetilde{\Delta}(f,p) = \Delta(h^\beta_\mu(f^0,p^0); f,p)\\
&\quad = \frac{1}{2\pi}\int_{-\pi}^{\pi} \frac{|r^{\beta,0}_{\mu,p}(e^{-i\lambda})|^2 - \beta^2|r^{\beta,0}_{\mu,f}(e^{-i\lambda})|^2}{p^0(\lambda)} f(\lambda)d\lambda + \frac{1}{2\pi}\int_{-\pi}^{\pi} \frac{|r^{\beta,0}_{\mu,f}(e^{-i\lambda})|^2}{p^0(\lambda)} p(\lambda)d\lambda,\\
&r^{\beta,0}_{\mu,p}(e^{-i\lambda}) = \sum_{k=0}^{\infty}\left(\left((\mathbf{C}^\beta_\mu)^0 + (\mathbf{C}^\beta_{-\mu})^0\right)(\overline{\Psi}^\beta_\mu)^0\right)_k e^{-i\lambda k},\\
&r^{\beta,0}_{\mu,f}(e^{-i\lambda}) = \frac{A(e^{-i\lambda})}{(1-e^{-i\lambda\mu})^n}\left(\sum_{k=0}^{\infty}\left(\psi^\beta_\mu(k)\right)^0 e^{-i\lambda k}\right)^{-1} - \sum_{k=0}^{\infty}\left(\left((\mathbf{C}^\beta_\mu)^0 + (\mathbf{C}^\beta_{-\mu})^0\right)(\overline{\Psi}^\beta_\mu)^0\right)_k e^{-i\lambda k}.
\end{aligned}
\tag{38}
$$

A solution (f^0,g^0) to this unconstrained optimization problem is characterized by the condition $0 \in \partial\Delta_{\mathcal{D}}(f^0,p^0)$.

The form of the functionals $\Delta(h_\mu(f^0,g^0); f,g)$ and $\Delta(h^\beta_\mu(f^0,p^0); f,p)$ allows us to find derivatives and differentials of these functionals in the space $L_1 \times L_1$. Hence, the complexity of the optimization problems (37) and (38) is characterized by the complexity of finding subdifferentials of the indicator functions $\delta(f,g|\mathcal{D}_f \times \mathcal{D}_g)$ of the sets $\mathcal{D}_f \times \mathcal{D}_g$.

6. Least favourable spectral densities in the class $\mathcal{D}^0_f \times \mathcal{D}^0_g$

Consider the problem of minimax-robust estimation of the functional $A\xi$ based on observations of the sequence $\xi(k)+\eta(k)$ at points of time $k=0,-1,-2,\ldots$ provided the spectral densities $f(\lambda)$ and $g(\lambda)$ admit canonical factorizations (15 and 16) and belong to the set of admissible spectral densities $\mathcal{D}=\mathcal{D}_f \times \mathcal{D}_g$, where

$$\mathcal{D}_f^0 = \left\{ f(\lambda) \left| \frac{1}{2\pi} \int_{-\pi}^{\pi} f(\lambda) d\lambda \le P_1 \right. \right\}, \quad \mathcal{D}_g^0 = \left\{ g(\lambda) \left| \frac{1}{2\pi} \int_{-\pi}^{\pi} g(\lambda) d\lambda \le P_2 \right. \right\}.$$

We use the Lagrange method of indefinite multiplies to find a solution to the constrained optimization problem (36), we get the following relations for determination the least favourable spectral densities $f^0 \in \mathcal{D}_f^0, g^0 \in \mathcal{D}_g^0$:

$$f^0(\lambda) + \lambda^{2n} g^0(\lambda) = \alpha_1 \left| r^0_{\mu,g}(e^{-i\lambda}) \right|^2, \tag{39}$$

$$f^0(\lambda) + \lambda^{2n} g^0(\lambda) = \alpha_2 \lambda^{2n} \left| r^0_{\mu,f}(e^{-i\lambda}) \right|^2,$$

where the multiplies $\alpha_1, \alpha_2 \ge 0$, matrices $\mathbf{C}^0_{\mu}, \mathbf{C}^0_{-\mu}$, vector $\psi^0_{\mu} = (\psi^0_{\mu}(0), \psi^0_{\mu}(1), \psi^0_{\mu}(2), \dots)'$ are determined with the help of factorizations (16) and (22) of the functions $g^0(\lambda)$ and $f^0(\lambda) + \lambda^{2n} g^0(\lambda)$, relation (24) and condition

(40)

$$\frac{1}{2\pi} \int_{-\pi}^{\pi} f^0(\lambda) d\lambda = P_1, \quad \frac{1}{2\pi} \int_{-\pi}^{\pi} g^0(\lambda) d\lambda = P_2. \tag{41}$$

Making use the derived reasonings, we can formulate the following statements.

Proposition 6.1 The spectral densities $f^0(\lambda) \in \mathcal{D}_f^0$ and $g^0(\lambda) \in \mathcal{D}_g^0$ which admit canonical factorizations (16) and (22) are least favourable in the class $\mathcal{D} = \mathcal{D}_f^0 \times \mathcal{D}_g^0$ for the optimal linear estimation of the functional $A\xi$ based on observations of the sequence $\xi(m) + \eta(m)$ at points $m \le 0$, if they satisfy equations (39 and 40), relations (24), the problem (31) and conditions (41). The function $h_{\mu}(f^0, g^0)$ determined by formula (20), is minimax spectral characteristic of the optimal estimate of the functional $A\xi$.

Proposition 6.2 Suppose that the spectral density $f(\lambda)$ is known and admits canonical factorization (23). The spectral density

$$g^0(\lambda) = \frac{1}{\lambda^{2n}} \left[\alpha_2 \lambda^{2n} \left| r^0_{\mu,f}(e^{-i\lambda}) \right|^2 - f(\lambda) \right]_+$$

from the class $\mathcal{D}_g^0$ is least favourable for the optimal linear estimation of the functional $A\xi$ based on observations of the sequence $\xi(m) + \eta(m)$ at points $m \le 0$, if the coefficient $\alpha_2 \ge 0$, matrices $\mathbf{C}^0_{\mu}, \mathbf{C}^0_{-\mu}$, vector $\psi^0_{\mu} = (\psi^0_{\mu}(0), \psi^0_{\mu}(1), \psi^0_{\mu}(2), \dots)'$ are determined from canonical factorizations (16), (22) of the functions $g^0(\lambda)$ and $f(\lambda) + \lambda^{2n} g^0(\lambda)$, relations (24), problem (33) and condition $\int_{-\pi}^{\pi} g^0(\lambda) d\lambda = 2\pi P_2$. The function $h_{\mu}(f, g^0)$, determined by formula (20), is minimax spectral characteristic of the optimal estimate of the functional $A\xi$.

Proposition 6.3 Suppose that the spectral density $g(\lambda)$ is known and admits canonical factorization (16). The spectral density

$$f^0(\lambda) = \left[\alpha_1 \left| r^0_{\mu,g}(e^{-i\lambda}) \right|^2 - \lambda^{2n} g(\lambda) \right]_+$$

from the class $\mathcal{D}_f^0$ is least favourable for the optimal linear estimation of the functional $A\xi$ based on observations of the sequence $\xi(m) + \eta(m)$ at points $m \le 0$, if the coefficient $\alpha_1 \ge 0$, vector $\psi^0_{\mu} = (\psi^0_{\mu}(0), \psi^0_{\mu}(1), \psi^0_{\mu}(2), \dots)'$ are determined from canonical factorization (22) of the function $f^0(\lambda) + \lambda^{2n} g(\lambda)$, relation (24), problem (35) and condition $\int_{-\pi}^{\pi} f^0(\lambda) d\lambda = 2\pi P_1$. In this case, the matrices $\mathbf{C}_{\mu}, \mathbf{C}_{-\mu}$ are determined from the canonical factorization (16) of the given spectral density $g(\lambda)$. The function $h_{\mu}(f^0, g)$, determined by formula (20), is minimax spectral characteristic of the optimal estimate of the functional $A\xi$.

Consider the problem of minimax-robust estimation of the functional $A\xi$ based on observations of the cointegrated sequence $\zeta(m)$ at points of time $m = 0, -1, -2, \ldots$ provided the spectral densities $f(\lambda)$ and $p(\lambda)$ admit canonical factorizations (25 and 26) and stochastic sequences $\xi(m)$ and $\zeta(m) - \beta\xi(m)$ are uncorrelated. The least favourable spectral densities in the set of admissible spectral densities $\mathcal{D}_f^0 \times \mathcal{D}_p^0$, where

$$\mathcal{D}_f^0 = \left\{ f(\lambda) \middle| \frac{1}{2\pi} \int_{-\pi}^{\pi} f(\lambda) d\lambda \leq P_1 \right\}, \quad \mathcal{D}_p^0 = \left\{ p(\lambda) \middle| \frac{1}{2\pi} \int_{-\pi}^{\pi} p(\lambda) d\lambda \leq P_2 \right\},$$

are determined by the condition $0 \in \partial\Delta_{\mathcal{D}}(f^0, p^0)$. It follows from this condition that the least favourable spectral densities $f^0 \in \mathcal{D}_f^0, p^0 \in \mathcal{D}_p^0$ are determined by the relations

$$p^0(\lambda) = \alpha_1 \left(\left| r_{\mu,p}^{\beta,0}(e^{-i\lambda}) \right|^2 - \beta^2 |r_{\mu,f}^{\beta,0}(e^{-i\lambda})|^2 \right), \tag{42}$$

$$p^0(\lambda) = \alpha_2 |r_{\mu,f}^{\beta,0}(e^{-i\lambda})|^2, \tag{43}$$

where coefficients $\alpha_1, \alpha_2 \geq 0$, vector $\left(\psi_\mu^\beta\right)^0 = (\psi_\mu^{\beta,0}(0), \psi_\mu^{\beta,0}(1), \psi_\mu^{\beta,0}(2), \ldots)'$, matrices $\left(\mathbf{C}_\mu^\beta\right)^0$, $\left(\mathbf{C}_{-\mu}^\beta\right)^0$ are determined by factorization Equations (25) and (26) of the function $p^0(\lambda)$ and $p^0(\lambda) - \beta^2 f^0(\lambda)$, relation (24) and conditions

$$\frac{1}{2\pi} \int_{-\pi}^{\pi} f^0(\lambda) d\lambda = P_1, \quad \frac{1}{2\pi} \int_{-\pi}^{\pi} p^0(\lambda) d\lambda = P_2. \tag{44}$$

Thus, we have the following statements.

Proposition 6.4 The spectral densities $f^0(\lambda)$ and $p^0(\lambda)$, that admit canonical factorizations (25) and (26), are least favourable in the class $D_f^0 \times \mathcal{D}_p^0$ for the optimal linear estimation of the functional $A\xi$ based on observations of the cointegrated with $\xi(m)$ sequence $\zeta(m)$ at points $m \leq 0$, if these densities satisfy equations (42 and 43) and are determined by relations (24), the problem (31) with $g(\lambda) := \lambda^{-2n}(p(\lambda) - \beta^2 f(\lambda))$ and conditions (44). The function $h_\mu^\beta(f^0, p^0)$, determined by formula (28), is minimax spectral characteristic of the optimal estimate of the functional $A\xi$.

7. Least favourable densities in the class $\mathcal{D} = \mathcal{D}_u^v \times \mathcal{D}_\varepsilon$

Consider the problem of minimax-robust estimation of the functional $A\xi$ based on observations of the sequence $\xi(k) + \eta(k)$ at points of time $k = 0, -1, -2, \ldots$ provided the spectral densities $f(\lambda)$ and $g(\lambda)$ admit canonical factorizations (15 and 16) and belong to the set of admissible spectral densities $\mathcal{D} = \mathcal{D}_v^u \times \mathcal{D}_\varepsilon$, where

$$\mathcal{D}_v^u = \left\{ f(\lambda) \middle| v(\lambda) \leq f(\lambda) \leq u(\lambda), \frac{1}{2\pi} \int_{-\pi}^{\pi} f(\lambda) d\lambda = P_1 \right\},$$

$$\mathcal{D}_\varepsilon = \left\{ g(\lambda) \middle| g(\lambda) = (1-\varepsilon) g_1(\lambda) + \varepsilon w(\lambda), \frac{1}{2\pi} \int_{-\pi}^{\pi} g(\lambda) d\lambda = P_2 \right\}.$$

The spectral densities $u(\lambda), v(\lambda), g_1(\lambda)$ are known and fixed and the spectral densities $u(\lambda), v(\lambda)$ are bounded. It follows from the condition $0 \in \partial\Delta_{\mathcal{D}}(f^0, g^0)$ that the least favourable spectral densities $f^0 \in \mathcal{D}_u^u, g^0 \in \mathcal{D}_\varepsilon$ satisfy the relations

$$f^0(\lambda) + \lambda^{2n} g^0(\lambda) = \alpha_1 \left| r_{\mu,g}^0(e^{-i\lambda}) \right|^2 (\gamma_1(\lambda) + \gamma_2(\lambda) + 1)^{-1}, \tag{45}$$

$$f^0(\lambda) + \lambda^{2n} g^0(\lambda) = \alpha_2 \lambda^{2n} \left| r_{\mu,f}^0(e^{-i\lambda}) \right|^2 (\beta(\lambda) + 1)^{-1}, \tag{46}$$

where $\gamma_1(\lambda) \leq 0$ and $\gamma_1(\lambda) = 0$ if $f^0(\lambda) \geq v(\lambda)$; $\gamma_2(\lambda) \geq 0$ and $\gamma_2(\lambda) = 0$, if $f^0(\lambda) \leq u(\lambda)$; $\beta(\lambda) \leq 0$ and $\beta(\lambda) = 0$ if $g^0(\lambda) \geq (1-\varepsilon)g_1(\lambda)$. The coefficients $\alpha_1 \geq 0$, $\alpha_2 \geq 0$, matrices $\mathbf{C}^0_\mu$, $\mathbf{C}^0_{-\mu}$, vector $\psi^0_\mu = (\psi^0_\mu(0), \psi^0_\mu(1), \psi^0_\mu(2), \dots)'$ are determined with the help of factorizations (16) and (22) of functions $g^0(\lambda)$ and $f^0(\lambda) + \lambda^{2n} g^0(\lambda)$, relations (24) and conditions (41).

The following theorems hold true.

Proposition 7.1 The spectral densities $f^0(\lambda) \in \mathcal{D}^0_f$ and $g^0(\lambda) \in \mathcal{D}^0_g$ which admit the canonical factorizations (16) and (22) are least favourable in the class $\mathcal{D}^u_v \times \mathcal{D}_\varepsilon$ for the optimal linear estimation of the functional $A\xi$ based on observations of the sequence $\xi(m) + \eta(m)$ at points $m \leq 0$, if they satisfy equations (45 and 46), relations (24), problem (31) and conditions (41). The function $h_\mu(f^0, g^0)$ determined by formula (20), is minimax spectral characteristic of the optimal estimate of the functional $A\xi$.

Proposition 7.2 Suppose that the spectral density $f(\lambda)$ is known and admits canonical factorization (23). The spectral density

$$g^0(\lambda) = \frac{1}{\lambda^{2n}} \max\left\{ \alpha_2 \lambda^{2n} \left| r^0_{\mu,f}(e^{-i\lambda}) \right|^2 - f(\lambda), (1-\varepsilon) g_1(\lambda) \right\}$$

from the class $\mathcal{D}_\varepsilon$ is least favourable for the optimal linear estimation of the functional $A\xi$ based on observations of the sequence $\xi(m) + \eta(m)$ at points $m \leq 0$, if the coefficient $\alpha_2 \geq 0$, matrices $\mathbf{C}^0_\mu$, $\mathbf{C}^0_{-\mu}$, vector $\psi^0_\mu = (\psi^0_\mu(0), \psi^0_\mu(1), \psi^0_\mu(2), \dots)'$ are determined from the canonical factorizations (16), (22) of the functions $g^0(\lambda)$ and $f(\lambda) + \lambda^{2n} g^0(\lambda)$, relations (24), problem (33) and condition $\int_{-\pi}^{\pi} g^0(\lambda) d\lambda = 2\pi P_2$. The function $h_\mu(f, g^0)$, determined by formula (20), is minimax spectral characteristic of the optimal estimate of the functional $A\xi$.

Proposition 7.3 Suppose that the spectral density $g(\lambda)$ is known and admits the canonical factorization (16). The spectral density

$$f^0(\lambda) = \min\left\{ \max\left\{ \alpha_1 \left| r^0_{\mu,g}(e^{-i\lambda}) \right|^2 - \lambda^{2n} g(\lambda), v(\lambda) \right\} u(\lambda) \right\}$$

from the class $\mathcal{D}^u_v$ is least favourable for the optimal linear estimation of the functional $A\xi$ based on observations of the sequence $\xi(m) + \eta(m)$ at points $m \leq 0$, if the coefficient $\alpha_1 \geq 0$, vector $\psi^0_\mu = (\psi^0_\mu(0), \psi^0_\mu(1), \psi^0_\mu(2), \dots)'$ are determined from the canonical factorization (22) of the function $f^0(\lambda) + \lambda^{2n} g(\lambda)$, relation (24), problem (35) and condition $\int_{-\pi}^{\pi} f^0(\lambda) d\lambda = 2\pi P_1$. In this case, the matrices $\mathbf{C}_\mu$, $\mathbf{C}_{-\mu}$ are determined from canonical factorization (16) of the given spectral density $g(\lambda)$. The function $h_\mu(f^0, g)$, determined by formula (20), is minimax spectral characteristic of the optimal estimate of the functional $A\xi$.

Consider the problem of minimax-robust estimation of the functional $A\xi$ based on observations of the cointegrated sequence $\zeta(m)$ at points of time $m = 0, -1, -2, \dots$ provided the spectral densities $f(\lambda)$ and $p(\lambda)$ admit canonical factorizations (25) - (26) and stochastic sequences $\xi(m)$ and $\zeta(m) - \beta\xi(m)$ are uncorrelated. The least favourable spectral densities in the set of admissible spectral densities $\mathcal{D} = \mathcal{D}^u_v \times \mathcal{D}_\varepsilon$, where

$$\mathcal{D}^u_v = \left\{ f(\lambda) \middle| v(\lambda) \leq f(\lambda) \leq u(\lambda), \frac{1}{2\pi} \int_{-\pi}^{\pi} f(\lambda) d\lambda = P_1 \right\},$$

$$\mathcal{D}_\varepsilon = \left\{ p(\lambda) \middle| p(\lambda) = (1-\varepsilon) p_1(\lambda) + \varepsilon w(\lambda), \frac{1}{2\pi} \int_{-\pi}^{\pi} p(\lambda) d\lambda = P_2 \right\},$$

under the condition that the spectral densities $f(\lambda)$ and $p(\lambda)$ admit the canonical factorizations (25 and 26). From the condition $0 \in \partial\Delta_{\mathcal{D}}(f^0, p^0)$, we get the following relations that determine the least favourable spectral densities

$$p^0(\lambda) = \alpha_1\left(\left|r^{\beta,0}_{\mu,p}(e^{-i\lambda})\right|^2 - \beta^2\left|r^{\beta,0}_{\mu,f}(e^{-i\lambda})\right|^2\right)(\gamma_1(\lambda) + \gamma_2(\lambda) + 1)^{-1}, \tag{47}$$

$$p^0(\lambda) = \alpha_2\left|r^{\beta,0}_{\mu,f}(e^{-i\lambda})\right|^2(\beta(\lambda) + 1)^{-1}, \tag{48}$$

where $\gamma_1(\lambda) \leq 0$ and $\gamma_1(\lambda) = 0$ if $f^0(\lambda) \geq v(\lambda)$; $\gamma_2(\lambda) \geq 0$ and $\gamma_2(\lambda) = 0$ if $f^0(\lambda) \leq u(\lambda)$; $\beta(\lambda) \leq 0$ and $\beta(\lambda) = 0$ if $p^0(\lambda) \geq (1-\varepsilon)p_1(\lambda)$. The coefficients $\alpha_1 \geq 0$, $\alpha_2 \geq 0$, matrices $\left(\mathbf{C}^{\beta}_{\mu}\right)^0$, $\left(\mathbf{C}^{\beta}_{-\mu}\right)^0$, vector $\left(\psi^{\beta}_{\mu}\right)^0 = (\psi^{\beta,0}_{\mu}(0), \psi^{\beta,0}_{\mu}(1), \psi^{\beta,0}_{\mu}(2), \ldots)'$ are determined by the canonical factorizations (25) and (26) of functions $p^0(\lambda) - \beta^2 f^0(\lambda)$ and $p^0(\lambda)$, relations (24) and condition (44).

Thus, we have the following statements.

Proposition 7.4 The spectral densities $f^0(\lambda)$ and $p^0(\lambda)$, that admit canonical factorizations (25 and 26), are least favourable in the class $\mathcal{D}^u_v \times \mathcal{D}_\varepsilon$ for the optimal linear estimation of the functional $A\xi$ based on observations of the cointegrated with $\xi(m)$ sequence $\zeta(m)$ at points $m \leq 0$, if these densities satisfy equations (47 and 48) and are determined by relations (24), problem (31) with $g(\lambda) := \lambda^{-2n}(p(\lambda) - \beta^2 f(\lambda))$ and conditions 44). The function $h_\mu(f^0, p^0)$, determined by formula (28), is minimax spectral characteristic of the optimal estimate of the functional $A\xi$.

8. Conclusions

In this article, we propose a solution of the filtering problem for the functional $A\xi = \sum_{k=0}^{\infty} a(k)\xi(-k)$ which depends on unobserved values of a stochastic sequence $\xi(k)$ with stationary nth increments. Estimates are based on observations of the sequence $\xi(m) + \eta(m)$ at points of time $m = -1, -2, \ldots$, where $\eta(m)$ is a stationary sequence uncorrelated with $\xi(k)$. We derive formulas for calculating the values of the mean-square errors and the spectral characteristics of the optimal linear estimate of the functional in the case where spectral densities $f(\lambda)$ and $g(\lambda)$ of the sequences $\xi(m)$ and $\eta(m)$ are exactly known. The obtained formulas are simpler than those obtained with the help of the Fourier coefficients of some functions determined by the spectral densities. In the case of spectral uncertainty, where spectral densities are not known exactly, but a set of admissible spectral densities is specified, the minimax-robust method is applied. Formulas that determine the least favourable spectral densities and the minimax (robust) spectral characteristics are derived for some special sets of admissible spectral densities. The obtained results are applied to find a solution of the filtering problem for a class of cointegrated sequences.

Acknowledgements
The authors would like to thank the referees for careful reading of the article and giving constructive suggestions.

Funding
The authors received no direct funding for this research.

Author details
Maksym Luz[1]
E-mail: maksim_luz@ukr.net
ORCID ID: http://orcid.org/0000-0002-8260-1584
Mikhail Moklyachuk[1]
E-mails: mmp@univ.kiev.ua, moklyachuk@gmail.com
[1] Department of Probability Theory, Statistics and Actuarial Mathematics, Taras Shevchenko National University of Kyiv, Kyiv 01601, Ukraine.

References

Bell, W. (1984). Signal extraction for nonstationary time series. *The Annals of Statistics, 12*, 646–664.

Box, G. E. P., Jenkins, G. M., & Reinsel, G. C. (1994). *Time series analysis, Forecasting and control* (3rd ed.). Englewood Cliffs, NJ: Prentice Hall.

Dubovets'ka, I. I., Masyutka, O. Y., & Moklyachuk, M. P. (2012). Interpolation of periodically correlated stochastic sequences. *Theory of Probability and Mathematical Statistics, 84*, 43–56.

Dubovets'ka, I. I., & Moklyachuk, M. P. (2013a). Filtration of linear functionals of periodically correlated sequences. *Theory of Probability and Mathematical Statistics, 86*, 51–64.

Dubovets'ka, I. I., & Moklyachuk, M. P. (2013b). Minimax estimation problem for periodically correlated stochastic processes. *Journal of Mathematics and System Science, 3*, 26–30.

Dubovets'ka, I. I., & Moklyachuk, M. P. (2014a). Extrapolation of periodically correlated processes from observations with noise. *Theory of Probability and Mathematical Statistics, 88*, 67–83.

Dubovets'ka, I. I., & Moklyachuk, M. P. (2014b). On minimax estimation problems for periodically correlated stochastic processes. *Contemporary Mathematics and Statistics, 2,* 123–150.

Engle, R. F., & Granger, C. W. J. (1987). Co-integration and error correction: Representation, estimation and testing. *Econometrica, 55,* 251–276.

Franke, J. (1985). Minimax robust prediction of discrete time series. *Zeitschrift far. Wahrscheinlichkeitstheorie und verwandte Gebiete, 68,* 337–364.

Franke, J., & Poor, H. V. (1984). *Minimax-robust filtering and finite-length robust predictors* (Robust and Nonlinear Time Series Analysis. Lecture Notes in Statistics, Vol. 26, 87–126. Heidelberg: Springer-Verlag.

Gikhman, I. I., & Skorokhod, A. V. (2004). *The theory of stochastic processes. I.*. Berlin: Springer.

Golichenko, I. I., & Moklyachuk, M. P. (2014). *Estimates of functionals of periodically correlated processes*. Kyiv: NVP "Interservis".

Granger, C. W. J. (1983). *Cointegrated variables and error correction models* (UCSD Discussion paper. 83–13a).

Grenander, U. (1957). A prediction problem in game theory. *Arkiv för Matematik, 3,* 371–379.

Ioffe, A. D., & Tihomirov, V. M. (1979). *Theory of extremal problems* (p. 460). Amsterdam: North-Holland.

Karhunen, K. (1947). Uber lineare Methoden in der Wahrscheinlichkeitsrechnung. Annales Academiae Scientiarum Fennicae. *Series A I. Mathematica, 37,* 3–79.

Kassam, S. A., & Poor, H. V. (1985). Robust techniques for signal processing: A survey. *Proceedings of the IEEE, 73,* 433–481.

Kolmogorov, A. N. (1992). Selected works of A. N. Kolmogorov. In A. N. Shiryayev (Ed.), *Probability theory and mathematical statistics* (Vol. II). Dordrecht: Kluwer Academic.

Luz, M. M., & Moklyachuk, M. P. (2012). Interpolation of functionals of stochastic sequences with stationary increments from observations with noise. *Prykladna Statystyka. Aktuarna ta Finansova Matematyka, 2,* 131–148.

Luz, M. M., & Moklyachuk, M. P. (2013a). Interpolation of functionals of stochastic sequences with stationary increments. *Theory of Probability and Mathematical Statistics, 87,* 117–133.

Luz, M. M., & Moklyachuk, M. P. (2013b). Minimax-robust filtering problem for stochastic sequence with stationary increments. *Theory of Probability and Mathematical Statistics, 89,* 117–131.

Luz, M., & Moklyachuk, M. (2014a). Robust extrapolation problem for stochastic processes with stationary increments. *Mathematics and Statistics, 2,* 78–88.

Luz, M., & Moklyachuk, M. (2014b). Minimax-robust filtering problem for stochastic sequences with stationary increments and cointegrated sequences. *Statistics, Optimization & Information Computing, 2,* 176–199.

Luz, M., & Moklyachuk, M. (2015a). Minimax interpolation problem for random processes with stationary increments. *Statistics, Optimization & Information Computing, 3,* 30–41.

Luz, M., & Moklyachuk, M. (2015b). Filtering problem for random processes with stationary increments. *Contemporary Mathematics and Statistics, 3,* 8–27.

Luz, M., & Moklyachuk, M. (2015c). Minimax-robust prediction problem for stochastic sequences with stationary increments and cointegrated sequences. *Statistics, Optimization & Information Computing, 3,* 160–188.

Luz, M., & Moklyachuk, M. (2016a). Filtering problem for functionals of stationary sequences. *Statistics, Optimization & Information Computing, 4,* 68–83.

Luz, M., & Moklyachuk, M. (2016b). Minimax prediction of random processes with stationary increments from observations with stationary noise. *Cogent Mathematics, 3,* 1–17, 1133219.

Moklyachuk, M. P. (1990). Minimax extrapolation and autoregressive-moving average processes. *Theory of Probability and Mathematical Statistics, 41,* 77–84.

Moklyachuk, M. P. (2000). Robust procedures in time series analysis. *Theory of Stochastic Processes, 6,* 127–147.

Moklyachuk, M. P. (2001). Game theory and convex optimization methods in robust estimation problems. *Theory of Stochastic Processes, 7,* 253–264.

Moklyachuk, M. P. (2008a). *Robust estimates for functionals of stochastic processes*. Kyiv: Kyiv University.

Moklyachuk, M. P. (2008b). *Nonsmooth analysis and optimization* (p. 400). Kyiv: Kyivskyi Universitet.

Moklyachuk, M. P. (2015). Minimax-robust estimation problems for stationary stochastic sequences. *Statistics, Optimization & Information Computing, 3,* 348–419.

Moklyachuk, M. P., & Masyutka, O Yu (2006a). Extrapolation of multidimensional stationary processes. *Random Operators and Stochastic Equations, 14,* 233–244.

Moklyachuk, M., & Luz, M. (2013). Robust extrapolation problem for stochastic sequences with stationary increments. *Contemporary Mathematics and Statistics, 1,* 123–150.

Moklyachuk, M. P., & Masyutka, O. Y. (2006b). Robust estimation problems for stochastic processes. *Theory of Stochastic Processes, 12,* 88–113.

Moklyachuk, M. P., & Masyutka, O. Y. (2007). Robust filtering of stochastic processes. *Theory of Stochastic Processes, 13,* 166–181.

Moklyachuk, M. P., & Masyutka, O. Y. (2008). Minimax prediction problem for multidimensional stationary stochastic sequences. *Theory of Stochastic Processes, 14,* 89–103.

Moklyachuk, M. P., & Masyutka, O. Y. (2011). Minimax prediction problem for multidimensional stationary stochastic processes. *Communications in Statistics - Theory and Methods, 40,* 3700–3710.

Moklyachuk, M. P., & Masyutka, O. Y. (2012). *Minimax-robust estimation technique for stationary stochastic processes* (p. 296). Lap Lambert Academic.

Pinsker, M. S. (1955). The theory of curves with nth stationary increments in Hilbert spaces. *Izvestiya Akademii Nauk SSSR. Seriya Matematicheskaya, 19,* 319–344.

Pinsker, M. S., & Yaglom, A. M. (1954). On linear extrapolation of random processes with nth stationary increments. *Doklady Akademii Nauk SSSR, 94,* 385–388.

Pshenichnyi, B. N. (1971). Necessary conditions for an extremum. *Pure and applied mathematics*. 4 (Vol. XVIII, 230 p). New York, NY: Marcel Dekker.

Rockafellar, R. T. (1997). *Convex analysis* (p. 451). Princeton, NJ: Princeton University Press.

Rozanov, Y. A. (1967). *Stationary stochastic processes*. San Francisco, CA: Holden-Day.

Vastola, K. S., & Poor, H. V. (1983). An analysis of the effects of spectral uncertainty on Wiener filtering. *Automatica, 28,* 289–293.

Wiener, N. (1966). *Extrapolation, interpolation, and smoothing of stationary time series. With engineering applications*. Cambridge: MIT Press, Massachusetts Institute of Technology.

Yaglom, A. M. (1955). Correlation theory of stationary and related random processes with stationary nth increments. *Matematicheskii Sbornik, 37,* 141–196.

Yaglom, A. M. (1957). Some classes of random fields in n-dimensional space related with random stationary processes. *Teoriya Veroyatnostej i Ee Primeneniya, 2,* 292–338.

Yaglom, A. M. (1987a). Correlation theory of stationary and related random functions. *Basic results* (Springer Series in Statistics, Vol. 1, p. 526). New York, NY: Springer-Verlag.

Yaglom, A. M. (1987b). Correlation theory of stationary and related random functions. In *Supplementary notes and references* (Springer Series in Statistics, Vol. 2, p. 258). New York, NY: Springer-Verlag.

10

SPC methods for time-dependent processes of counts—A literature review

Christian H. Weiß[1]*

*Corresponding author: Christian H. Weiß, Department of Mathematics and Statistics, Helmut Schmidt University, Postfach 700822, 22008 Hamburg, Germany

E-mail: weissc@hsu-hh.de

Reviewing editor: Zudi Lu, University of Southampton, UK

Abstract: During the last few years, there was increasing interest in SPC methods for time-dependent processes of counts. We survey recent developments in this field: feasible models for autocorrelated counts processes are presented, approaches for corresponding control charts are considered, and also the topic of process capability indices is briefly discussed. The article is accompanied by a comprehensive list of relevant references, and it concludes by outlining promising directions for future research.

Subjects: Mathematics & Statistics; Science; Statistical Computing; Statistical Theory & Methods; Statistics & Computing; Statistics & Probability; Statistics for Business; Finance & Economics

Keywords: statistical process control; control charts; process capability; autocorrelated attributes data; count data time series; run length performance

1. Introduction

Methods of *statistical process control* (SPC) help to monitor and improve processes in manufacturing and service industries. For such a process, certain quality characteristics are measured at discrete times $t \in \mathbb{N} := \{1, 2, \ldots\}$, thus leading to a (possibly multivariate) stochastic process $(X_t)_{\mathbb{N}}$ of continuous-valued or discrete-valued random variables (*variables data* or *attributes data*, respectively). One of the most important SPC tools is the *control chart*, which requires the relevant quality characteristics to be measured online. Control charts are applied to a process operating in a stable state (*in control*), i.e. $(X_t)_{\mathbb{N}}$ is assumed to be stationary according to a specified model (in-control model). As a new measurement arrives, it is used to compute a statistic (possibly also incorporating past values of the quality characteristic) which is then plotted on the control chart with its control limits. If the statistic violates the limits, an alarm is triggered, signaling that the process may not be stable

ABOUT THE AUTHOR

Christian H. Weiß is a Professor at the Department of Mathematics and Statistics at the Helmut Schmidt University in Hamburg, Germany (since 2013). He earned his doctorate (Mathematical Statistics) in 2009 at the University of Würzburg, and from 2009 to 2013 he had a permanent post as an "Akademischer Rat" at the Department of Mathematics at Darmstadt University of Technology. His research areas include time series analysis, statistical quality control, and computational statistics.

PUBLIC INTEREST STATEMENT

In many fields of application, we are concerned with count data processes. Typical examples are counts of defects per produced item in manufacturing industry, counts of new cases of an infection per time unit in health care monitoring, or counts of complaints by customers per time unit in service industry. Often, it is important to detect changes in the process as soon as possible to be able to start preventive actions or to avoid further damages. Methods of statistical process control are a suitable tool for this purpose. The article provides a detailed survey of such methods together with a comprehensive list of relevant references, and it concludes by outlining promising directions for future research.

anymore (*out of control*) and requires corrective actions. Besides such an online monitoring to detect changes in the process, it is also important to analyze to what extent the given target values and specification limits[1] are met by the process in its in-control state. A widely used SPC solution for this purpose is *process capability indices*, which are also briefly discussed in the present article. Furthermore, also another type of application of control charts is considered. The use of control charts for online monitoring, as described before, is commonly referred to as the *Phase-II* application. But control charts may also be applied in a retrospective manner to already available in-control data, with the aim of characterizing the in-control properties of $(X_t)_{\mathbb{N}}$; this is called the *Phase-I* application of a control chart. More details about all these terms and concepts can be found in the textbook by Montgomery (2009) and in the survey papers by Woodall (2000), Woodall and Montgomery (2014).

In this article, we shall concentrate on a type of attributes data processes: count data processes, where each X_t has a range contained in the set of non-negative integers, $\mathbb{N}_0 := \{0, 1, \dots\}$. Typical examples are counts of defects per produced item in manufacturing industry, counts of new cases of an infection per time unit in health care monitoring, or counts of complaints by customers per time unit in service industry. A lot of work has been done regarding such attributes data processes, see the survey by Woodall (1997), but with one important restriction: the large majority of papers about SPC methods for attributes data assumes the underlying process to be *serially independent* in its in-control state, so the counts $X_1, X_2, \dots$ have to be independent and identically distributed (i.i.d.). Only during the last few years, increasing research activity can be observed concerning attributes data processes with serial dependence. The aim of the present paper is to present a survey of these research activities, and to outline relevant issues for future research in this area.

At this point, it is important to stress that this lack of interest in *autocorrelated* attributes data is in sharp contrast to the variables case. After few scattered works concerning the effects of autocorrelation on variables control charts' performance during the 1960s to 1980s, a lot of research activity in this direction can be observed since the 1990s, initiated, among others, by the works by Alwan and Roberts (1988), Alwan (1992, 1995). Surveys on control charts for autocorrelated variables data processes are provided by Knoth and Schmid (2004), Psarakis and Papaleonida (2007). Although not being a topic of research until a few years ago, Alwan (1995) had already shown that autocorrelation is indeed a common phenomenon if being concerned with attributes data processes. Typical reasons for counts data processes to be autocorrelated are high sampling frequency due to automated production environments in manufacturing industry, or varying service times (extending over more than one time unit) in service industry, or varying incubation times and infectivities of diseases in health care monitoring.

The delay in working on SPC methods for autocorrelated attributes data processes might have been caused by the problem that simple stochastic models for such processes, i.e. which are of comparable simplicity to the well-known autoregressive moving average (ARMA) models for autocorrelated variables data processes, were not known to a broader audience for a long time. Therefore, we start in Section 2 with a brief review of the basic approaches for modeling autocorrelated processes of counts. Section 3 then provides information about the most popular SPC tools, control charts and process capability indices. While Section 3.1 only presents a basic Shewhart chart and puts more emphasize on topics like performance evaluation and the effect of estimated parameters, details on advanced control charts like CUSUM and EWMA methods are presented in Section 4. Finally, we outline possible directions for future research in Section 5.

2. Basic models for autocorrelated counts processes

In the sequel, several common count data distributions shall be mentioned without presenting further details about them; a reader being interested in more background information is referred to the book by Johnson, Kemp, and Kotz (2005).

One of the oldest approaches toward stationary count data processes is the *INAR(1) model* by McKenzie (1985), Al-Osh and Alzaid (1987), the *in*teger-valued counterpart to the usual *a*uto*r*egressive model of order 1. This model can be understood as a special type of branching process with immigration, and it uses the binomial thinning operator by Steutel and van Harn (1979): If X is a count data random variable and if $\alpha \in (0;1)$, then the random variable $\alpha \circ X := \sum_{i=1}^{X} Z_i$ is said to arise from X by *binomial thinning*, where Z_i are i.i.d. binary random variables with $P(Z_i = 1) = \alpha$ (we abbreviate $Z \sim Bin(1, \alpha)$), which are also independent of X. So $\alpha \circ X$ is conditionally binomially distributed, $\alpha \circ X \sim Bin(X, \alpha)$.

Let $(\epsilon_t)_{\mathbb{N}}$ be an (unobservable) i.i.d. count data process with $E[\epsilon_t] = \mu_\epsilon$ and $V[\epsilon_t] = \sigma_\epsilon^2$, the *innovations* to the process. The *INAR(1) model* by McKenzie (1985), Al-Osh and Alzaid (1987) now assumes the observations $(X_t)_{\mathbb{N}_0}$ to satisfy the recursion

$$X_t = \alpha \circ X_{t-1} + \epsilon_t, \tag{2.1}$$

where all thinning operations are performed independently of each other and of $(\epsilon_t)_{\mathbb{N}}$, and where the thinning operations at each time t as well as ϵ_t are independent of $(X_s)_{s<t}$.

Except using the thinning operator "$\circ$" instead of the usual multiplication "$\cdot$", recursion (2.1) looks like the usual AR(1) recursion. In fact, it also constitutes a Markov chain with an exponentially decaying autocorrelation function (ACF), $\rho(k) := Corr[X_t, X_{t-k}] = \alpha^k$, and marginal mean and variance-mean ratio are obtained as

$$\mu_X = \frac{\mu_\epsilon}{1-\alpha}, \qquad \frac{\sigma_X^2}{\mu_X} = \frac{\frac{\sigma_\epsilon^2}{\mu_\epsilon} + \alpha}{1+\alpha}. \tag{2.2}$$

Beyond mimicing the typical AR(1)-like autocorrelation structure, the INAR(1) model is particularly relevant for typical tasks of statistical quality control due to its intuitive interpretation (see Weiß, 2007). The thinning operation $\alpha \circ X$ itself is interpreted as expressing the number of survivors from a population of size X, where each individual, independent of the other individuals, has survival probability α. So recursion (2.1) is interpreted as

$$\underbrace{X_t}_{\text{Population at time } t} = \underbrace{\alpha \circ X_{t-1}}_{\text{Survivors of generation } t-1} + \underbrace{\epsilon_t}_{\text{Immigration}}. \tag{2.3}$$

Adapted to the application scenarios sketched in Section 1, the "population" at time t might consist of faults in a system or network, of persons being infected by a certain disease, or of unanswered complaints by customers. These might be faults or infected persons or complaints that were already available at the previous time $t-1$ ("survivors"), or which newly occured at time t ("immigration").

The most popular case of the INAR(1) family is the *Poisson INAR(1) model*. Here, it is assumed that the innovations ϵ_t are Poisson-distributed according to $Poi(\lambda)$ such that $\mu_\epsilon = \sigma_\epsilon^2 = \lambda$. Then the stationary marginal distribution is also a Poisson distribution, $Poi\left(\frac{\lambda}{1-\alpha}\right)$ (see Al-Osh & Alzaid, 1987), such that also the observations have a variance being equal to the mean (the latter property is referred to as *equidispersion*). In applications, however, one often observes the counts having a variance being larger than the mean, i.e. having *overdispersion* (Weiß & Testik, 2011). According to (2.2), such a feature is easily implemented into the INAR(1) model by simply using an overdispersed distribution for the innovations, like a compound Poisson distribution (Schweer & Weiß, 2014) or the Poisson log-normal distribution (Weiß & Testik, 2015a). By the same approach, also other non-standard features like, e.g. *zero inflation* (excess of zeros) can be implemented into the INAR(1) model (see Jazi, Jones, & Lai, 2012). Finally, it should be mentioned that also higher order INARMA models have been discussed in the literature, for instance, in Du and Li (1991), Weiß (2008b).

Often motivated by the aim of defining an AR(1)-like model for counts with overdispersion, a number of modifications to the basic INAR(1) model (2.1) have been proposed where the binomial thinning operator is replaced by another type of thinning (see Weiß, 2008a for a survey). As an example, Ristić, Bakouch, and Nastić (2009) introduced the *negative binomial thinning operator* $\alpha * X := \sum_{i=1}^{X} Z_i$, where the Z_j's are geometrically distributed with "success probability" $1/(1+\alpha)$ (such that $E[Z_i] = \alpha > 0$). Then the innovations' distribution can be chosen in such a way that the *new geometric integer-valued autoregressive* (NGINAR) process of order 1, defined by

$$X_t = \alpha * X_{t-1} + \epsilon_t, \tag{2.4}$$

is stationary with geometrically distributed marginals having an arbitrary mean $\mu > 0$, provided that $\alpha \le \mu/(1+\mu)$, and with ACF $\rho(k) = \alpha^k$.

Another popular approach for modeling stationary processes of counts are the INGARCH models, which are particularly attractive for overdispersed counts. The *INGARCH model*, the *in*teger-valued counterpart to the conventional *g*eneralized *a*uto*r*egressive *c*onditional *h*eteroskedasticity model, was introduced by Heinen (2003), Ferland, Latour, and Oraichi (2006). Given the past observations, a conditional Poisson distribution with an ARMA-like recursion for the conditional means is assumed. For the special case of the INARCH(1) model, which constitutes a counterpart to the INAR(1) model discussed before, let us denote the model parameters by $\beta > 0$ and $0 < \alpha < 1$. Then the process $(X_t)_{\mathbb{Z}}$ is said to follow the *INARCH(1) model* if X_t is conditionally Poisson distributed in the following way:

$$X_t \mid X_{t-1}, X_{t-2}, \ldots \sim Poi(\beta + \alpha \cdot X_{t-1}). \tag{2.5}$$

The ACF equals $\rho(k) = \alpha^k$ like in the standard AR(1) case, and marginal mean and variance-mean ratio of the INARCH(1) process are given by

$$\mu_X = \frac{\beta}{1-\alpha}, \qquad \frac{\sigma_X^2}{\mu_X} = \frac{1}{1-\alpha^2} > 1. \tag{2.6}$$

There are certainly many alternative approaches for modeling time series of counts, e.g. regression models (Kedem & Fokianos, 2002) or hidden Markov models (Zucchini & MacDonald, 2009), but these shall not be considered further in this text, since they have not been used yet in an SPC context (to the knowledge of the author).

3. Common SPC methods

3.1. Control charts

The most common application scenario for control charts is the so-called Phase-II application (also see Section 1 before), i.e. the prospective online monitoring to detect a possible change in the process. The (unknown) time where such a process change first happens is called a *change point*. To be more precise, we consider the following (unconditional) *change point model* (Knoth, 2006):

For $\tau \in \mathbb{N}$, we assume that $(X_t)_{t<\tau}$ and $(X_t)_{t\ge\tau}$ are stationary processes with distributions abbreviated as F_0 and F_1, respectively. The time index τ is the *change point*, which is not known in practice. For $t < \tau$, the process is said to be *in control*, while it is *out of control* for $t \ge \tau$ if $F_1 \ne F_0$.

Applying a control chart, we aim at detecting the unknown change point τ as early as possible. The most simple control charts are the so-called *Shewhart charts*, which are based on statistics Z_t being a function only of the most recent observation X_t (or of the most recent sample for a sample-based monitoring). Then Z_t is plotted on a chart against time t with time-invariant lower and upper *control limits* $l < u$. An alarm is triggered at time t for the first time if

$$Z_1, \ldots, Z_{t-1} \in [l;u], \qquad \text{but} \quad Z_t \notin [l;u]. \tag{3.1}$$

An extensive review of Shewhart control charts is given by Montgomery (2009).

Regarding count data monitoring, the so-called *c chart* is particularly relevant where simply $Z_t = X_t$, i.e. the counts are directly plotted on the chart as they arrive in time. More advanced control charts, where $Z_t := f_t(X_1, \ldots, X_t; \delta)$ is an appropriately chosen measurable function of $X_1, \ldots, X_t$ and of a vector δ of design parameters, are considered later in Section 4 in more detail. Applications of the *c* chart to INAR(1) processes (2.1) were considered by Weiß (2007, 2011b), Morais and Pacheco (in press), to NGINAR(1) processes (2.4) by Li, Wang, and Zhu (in press), and to INARCH(1) processes (2.5) by Weiß and Testik (2012).

Remark 3.1 (Change Point Methods) An approach being related to the control chart are tests for a change point within a given time series. For the case of a count data time series stemming from an INARCH(1) model, such change point tests were developed by Franke, Kirch, and Kamgaing (2012), Kang and Lee (2014), Kang and Song (2015), while Torkamani, Niaki, Aminnayeri, and Davoodi (2014), Davoodi, Niaki, and Torkamani (2015) considered an underlying INAR(1) process, also see the references in Hudecová, Hušková, and Meintanis (2015). Note that the main difference between such change point tests and the above control charts is that the first are usually applied in an offline manner, to find the location of the change point withing the available (and static) time series. Online versions of change point tests, where the in-control model is sequentially tested based on the available data at each time, are presented by Hudecová et al. (2015) for the case of the INAR(1) model (2.1), and by Kirch and Kamgaing (2015) for the case of the INARCH(1) model (2.5).

The essential step before starting process monitoring is to find an appropriate chart design, i.e. appropriate values for the control limits $l < u$ in case of the *c* chart. Although sometimes being criticized (Kenett & Pollak, 2012), still, the main approach is to consider appropriately defined mean statistics based on the run length L, i.e. an *average run length (ARL)*, where $L := \min\{t \in \mathbb{N} \mid Z_t \notin [l;u]\}$ is defined as the random number of plotted points until the first alarm is triggered. The most common ARL concepts are as follows (Knoth, 2006): Defining $E_\tau[\cdot]$ as the expectation related to the change point τ,

- the *zero-state ARL* (also *initial-state ARL*) is defined as

$$ARL := E_1[L], \tag{3.2}$$

- the *expected conditional ARL* (also *expected* or *conditional delay*) is defined as

$$ARL^{(\tau)} := E_\tau[L - \tau + 1 \mid L \geq \tau], \tag{3.3}$$

- the *steady-state ARL* is defined as

$$ARL^{(\infty)} := \lim_{\tau \to \infty} ARL^{(\tau)}. \tag{3.4}$$

Obviously, we have $ARL^{(1)} = ARL$. For any of these ARL concepts, we refer to the computed ARL value as the *in-control ARL* (*out-of-control ARL*) if $F_1 = F_0$ ($F_1 \neq F_0$); the in-control ARL is commonly abbreviated by adding the index "0". A popular approach for *chart design* is to choose $l < u$ such that the zero-state ARL_0 reaches a prespecified level (expressing the robustness of the chart against false alarms), and then to evaluate the out-of-control performance based on the steady-state ARL (since the value of the change point is not known but it will satisfy $\tau \gg 1$ in many real applications).

It remains to ask how to compute any of the ARL concepts (3.2)–(3.4) given a certain chart design (this question holds in the same way also for the advanced control charts discussed in Section 4 below). Certainly, in any case where it is possible at all to simulate the considered type of counts data process, ARLs can be approximated based on such simulations with a sufficiently high number of replications (usually at least 10,000). But if $(X_t)_{\mathbb{N}}$ follows a type of discrete Markov model (note that any of the three models (2.1), (2.4), and (2.5) constitutes a discrete Markov chain), then it is often possible to adapt the *Markov chain approach* (MC approach) as first proposed by Brook and Evans (1972). A detailed description for several types of control charts (including the *c* chart for INAR(1) processes), together with corresponding software implementations, is provided by the tutorial by Weiß (2011b).

To conclude this section, let us briefly look at the Phase-I application of control charts, and at the related topic of the effect of estimated parameters on the control charts' performance. To design the control charts for use in Phase II, a model for the in-control behavior of the process is required (which is then used for chart design as outlined before). Since in practice, the true in-control model is hardly known, one has to fit a model to a set of historic data which are believed to stem from the in-control model. There are several issues that have to be considered carefully in this context, see the recent survey by Jones-Farmer, Woodall, Steiner, and Champ (2014). Among others, once a data sample for Phase-I analysis is available, it has to be checked if these data can be assumed at all to stem from a unique model, or if, for instance, the data are contaminated by outliers. In the latter case, such outliers have to be excluded from the data before fitting the in-control model. For this task, control charts are often used (especially Shewhart charts), which is known as the *Phase-I application of control charts*. In Weiß and Testik (2015b), the concrete implementation of the Phase-I analysis for an underlying INAR(1) process is discussed in detail, and the effect of undetected outliers during Phase I on the resulting chart design and performance during Phase II is studied.

Once the available data can be assumed to be "clean", the parameters of the in-control model have to be estimated. The estimated in-control model is then used for chart design for Phase II. Many articles considered the *effect of estimated parameters* on the charts' performance in Phase II (see Jensen, Jones-Farmer, Champ, & Woodall, 2006 for references), where the properties of the used estimators or the sample size play an important role. In the context of autocorrelated count data processes, this topic was considered by Weiß and Testik (2011), Zhang, Nie, He, and Hou (2014), Weiß and Testik (2015b) for the Poisson INAR(1) model and diverse types of control charts.

3.2. Process capability indices

Saying that a process is in control only implies that it is stationary, following a specified model (see above), but it does not imply that the output of the process meets the given quality requirements. Concerning the latter issue, one has to check, for instance, to what extent the given target values and specification limits are met by the process. If the process is not consistent with the given external specifications, adjustments are necessary such that the new in-control model better agrees with the quality requirements. A popular tool for evaluating the actual process capability is *process capability indices*. An introduction to such indices (especially for the variables data case) can be found in the book by Montgomery (2009), the most recent literature survey seems to be the one by Saha and Maiti (2015).

Only few of the works about capability indices refer to attributes data processes. Perakis and Xekalaki (2005) picked up the idea of considering the actual "proportion of conformance": if the *upper specification limit USL* describes, e.g. the maximal acceptable number of non-conformities per produced item, then the probability $P(X > USL)$ is compared to a prespecified acceptable probability level $1 - p_0$. Perakis and Xekalaki (2005) considered an index defined by the quotient

$$C_{PX} := \frac{1 - p_0}{P(X > USL)} \in [1 - p_0;\infty). \tag{3.5}$$

A related approach designed for the specific level $1 - p_0 = 0.0027$ was proposed by Borges and Ho (2001) as

$$C_{BH} := \frac{1}{3} \cdot \Phi^{-1}\left(1 - \frac{1}{2} \cdot P(X > USL)\right) \in [0;\infty), \tag{3.6}$$

where Φ denotes the distribution function of the standard normal distribution $N(0, 1)$.

For practice, a relevant question is how to *estimate* the indices (3.5) and (3.6) from given in-control data (in analogy to the Phase-I analysis discussed before). While Perakis and Xekalaki (2005) considered this task for an underlying i.i.d. process of Poisson counts, Weiß (2012b) extended this work to an underlying Poisson INAR(1) process (2.1), distinguishing between the process capability for the observations or innovations, respectively, from such an INAR(1) process.

4. Advanced control charts

The basic c chart presented in Section 3.1 allows for a continuous monitoring of a serially dependent count data process, but the statistic plotted on the chart at time t, which is simply the count value being observed at time t, does not comprise information about past values of the process (at least not explicitly, beyond the mere effect of autocorrelation). Therefore, the c chart (as any other Shewhart-type chart) is not particularly sensitive to small or moderate changes in the process. For this reason, several types of advanced control charts have been proposed, where the plotted statistic at time t also uses past observations of the process and hence accumulates information about the process for a longer period of time.

4.1. CUSUM charts

The traditional *cumulative sum* (CUSUM) control chart (Page, 1954), being applied directly to the observations X_t of the process, is perhaps the most natural advanced candidate for monitoring autocorrelated processes of counts, because it preserves the discrete nature of the process by only using additions (but no multiplications). Initialized by a starting value $c_0^+ \geq 0$, the *upper-sided CUSUM* is defined by

$$C_0^+ = c_0^+, \qquad C_t^+ = \max(0; X_t - k^+ + C_{t-1}^+) \quad \text{for } t = 1, 2, \ldots \tag{4.1}$$

The starting value is commonly chosen as $c_0^+ = 0$; a value $c_0^+ > 0$ is referred to as a fast initial response (FIR) feature, and it may help to detect an initial out-of-control state more quickly. If k^+ and c_0^+ are taken as integer values, then also $(C_t^+)_{\mathbb{N}_0}$ is integer valued, or, as another example, if $k^+, c_0^+ \in \{0, 1/2, 1, 3/2, \ldots\}$ then so is C_t^+. An alarm is triggered if C_t^+ violates the control limit h^+ (decision interval).

While the upper-sided CUSUM is mainly designed to detect increases in the process mean, the *lower-sided CUSUM*, defined by

$$C_0^- = c_0^-, \qquad C_t^- = \max(0; k^- - X_t + C_{t-1}^-) \quad \text{for } t = 1, 2, \ldots, \tag{4.2}$$

aims at uncovering decreases in the mean. If (C_t^+, C_t^-) are monitored simultaneously, then this chart combination is referred to as a two-sided CUSUM chart. An excellent book with a lot of background information about CUSUM charts is the one by Hawkins and Olwell (1998).

In the context of monitoring autocorrelated counts processes, the upper-sided CUSUM was applied to INAR(1) processes (2.1) by Weiß and Testik (2009),(2011), to NGINAR(1) processes (2.4) by Li et al. (in press), and to INARCH(1) processes (2.5) by Weiß and Testik (2012). The lower-sided and the two-sided version were applied to INAR(1) processes by Yontay, Weiß, Testik, and Bayindir (2013). For performance evaluation, it is important that the CUSUM preserves the discrete range. Therefore, exact run length computations are possible with a type of MC approach (Weiß, 2011b): the one-sided CUSUM requires to consider the bivariate Markov chain $(X_t, C_t^{\pm})$ (Weiß & Testik, 2009), the two-sided CUSUM the trivariate Markov chain (X_t, C_t^+, C_t^-) (Yontay et al., 2013).

Besides the basic CUSUM approach (4.1), also INAR(1) CUSUM charts with additional Winsorization have been considered (Hawkins, 1993; Weiß & Testik, 2011), as well as CUSUM charts for diverse types of residuals from an INAR(1) process (Weiß & Testik, 2015a) and CUSUM charts based on the likelihood ratio of an INARCH(1) process (Weiß & Testik, 2012).

4.2. EWMA charts

Another advanced approach for process monitoring, which is also very popular in applications, is the *exponentially weighted moving average (EWMA)* control chart dating back to Roberts (1959). The standard EWMA recursion defined by

$$Z_t = \lambda \cdot X_t + (1-\lambda) \cdot Z_{t-1} \quad \text{for} \quad t = 1, 2, \ldots, \qquad \text{with} \quad \lambda \in (0;1], \tag{4.3}$$

however, has an important drawback compared to the CUSUM approach of the previous Section 4.1 if applied to count data processes: it does not preserve the discrete range. Quite the contrary, the range of possible values of Z_t changes in time, which rules out, among others, the possibility of an exact ARL computation by the Markov chain approach (remember Section 3.1). Therefore, Gan (1990) suggests to plot rounded values of the statistic (4.3):

$$Q_t = round\big(\lambda \cdot X_t + (1-\lambda) \cdot Q_{t-1}\big) \quad \text{for} \quad t = 1, 2, \ldots, \qquad \text{with} \quad \lambda \in (0;1], \tag{4.4}$$

which is initialized by $Q_0 := q_0 \in \mathbb{N}_0$. q_0 might be chosen as the rounded value of the in-control mean. An alarm is triggered if Q_t violates one of the control limits $0 \leq l \leq u$. Note that the statistics Q_t can take only integer values from $\mathbb{N}_0$.

If the underlying count data process $(X_t)_{\mathbb{N}}$ is a Markov chain, then $(X_t, Q_t)_{\mathbb{N}}$ is a bivariate Markov chain with range $\mathbb{N}_0^2$, so ARLs can be computed again exactly by adapting the MC approach (see Weiß, 2009b for details). In the latter article as well as in Zhang et al. (2014), the particular case of an underlying INAR(1) process (2.1) was considered, while Li et al. (in press) investigated the EWMA approach (4.4) applied to an NGINAR(1) process (2.4).

A possible disadvantage of the rounded EWMA approach (4.4) was presented in Weiß (2011a): especially for small values of λ, which are generally recommended if small mean shifts are to be detected, one may observe some kind of "oversmoothing", i.e. Q_t becomes piecewise constant in time t and rather insensitive to process changes. Therefore, Weiß (2011a) proposed a modification of (4.4), where a refined rounding operation is used: For $s \in \mathbb{N}$, the operation s-round maps x onto the nearest fraction with denominator s. For $s = 1$, we obtain the usual rounding operation, while 2-round rounds onto values in $\{0, 1/2, 1, 3/2, \ldots\}$, for example. The resulting s-EWMA chart follows the recursion

$$Q_t^{(s)} = s - round\big(\lambda \cdot X_t + (1-\lambda) \cdot Q_{t-1}^{(s)}\big) \quad \text{for} \quad t = 1, 2, \ldots, \qquad \text{with} \quad \lambda \in (0;1]. \tag{4.5}$$

If $(X_t)_{\mathbb{N}}$ is a Markov chain (Weiß, 2011a considered the instance of an INAR(1) process (2.1)), then $(X_t, Q_t^{(s)})_{\mathbb{N}}$ again is a discrete Markov chain, now with range $\mathbb{N}_0 \times \mathbb{Q}_{0,s}^+$, where $\mathbb{Q}_{0,s}^+ := \{\frac{r}{s} \mid r \in \mathbb{N}_0\}$ is the set of all non-negative rationals with denominator s. So again, it is possible to adapt the MC approach by Brook and Evans (1972) for an exact ARL computation.

4.3. Jumps chart

The last type of advanced control chart to be presented here is the jumps chart proposed by Weiß (2009c). It considers the "jumps" $J_t := X_t - X_{t-1}$ (Weiß, 2008b), which are particularly sensitive to a reduction of autocorrelation, since this leads to increased jumps. So in view of monitoring changes in the mean and the autocorrelation structure simultaneously, Weiß (2009c) proposed to apply the *combined jumps chart*, where the counts X_t and jumps J_t are plotted simultaneously on a c chart with limits $0 \leq l < u$ and a jumps chart with limits $\mp k$, respectively. If $(X_t)_{\mathbb{N}}$ is a Markov chain (Weiß, 2009c considered the instance of an INAR(1) process (2.1), (Li et al., in press) that of an NGINAR(1) process (2.4)), then $(X_t, J_t)_{\mathbb{N}}$ is a discrete Markov chain with range $\mathbb{N}_0 \times \mathbb{Z}$, so ARLs can be computed exactly by adapting the MC approach.

5. Conclusions

After having been neglected for a long time, there was a rapidly increasing research interest in SPC methods for time-dependent processes of counts during the last few years. The present article provides a comprehensive survey of recent developments in this field in conjunction with a list of relevant references being as complete as possible.

We conclude this article by briefly discussing possible directions for future research in the area of SPC methods for autocorrelated attributes data. Up to now, mainly "well behaved" types of counts data processes have been considered, especially those having a Poisson marginal distribution. But in view of real counts processes as observed, e.g. in epidemiology, future research should also consider phenomena like an excessive number of zeros (zero inflation) or seasonality (the latter leading to a non-stationary but still a "regular" in-control behavior). Also the topic of count data processes having a finite range $\{0, \ldots, n\}$, with a fixed upper limit n reflecting, e.g. the sample size in manufacturing industry or the number of service entities in service industry, would be very relevant in practice, but was considered only casually up to now (Weiß, 2009a; Weiß & Kim, 2013; Weiß & Testik, 2015b). The same applies to multivariate count data processes in Bersimis, Psarakis, and Panaretos (2007, p. 523) . Even more sobering, it seems that the case of serially dependent processes with the *full* set of integers $\mathbb{Z} = \{\ldots, -1, 0, 1, \ldots\}$ as their range (Kim & Park, 2008) has not been discussed so far at all in an SPC context. It is also important to emphasize that INAR(1) processes are related to certain queue length processes (with an infinite number of servers (see Schweer & Wichelhaus, 2015), so control charts for queueing systems as in Chen and Zhou (2015) and the control charts for autocorrelated counts as described in this article might be mutually enriching.

Besides other types of process models, also different approaches for process monitoring and chart design should be considered in future works. These may cover adaptive sampling procedures (e.g. variable sampling intervals) as discussed in Epprecht, Costa, and Mendes (2003), Montgomery (2009), for instance, or the additional use of runs rules as, e.g. in Alwan, Champ, and Maragah (1994), Acosta-Mejia 1999, Koutras, Bersimis, and Maravelakis (2007). Related to the latter approach, the so-called synthetic control charts attracted a lot of research interest during the last years, but recently also drew sound criticism (Knoth, in press). Concerning chart design, it might be interesting to apply economic design principles (Celano, 2011; Montgomery, 2009) in the context of autocorrelated counts, and also the Phase-I analysis for such processes (choice of estimators, effect of parameter estimation, etc.) deserves more attention.

Finally, much more research effort should be put on other types of discrete-valued and serially dependent processes, especially on categorical processes (both ordinal and nominal). There are some works for the special case of serially dependent *binary* attributes, e.g. the Markov Binary CUSUM chart for continuously monitoring a Markov-dependent Bernoulli process as proposed by Mousavi and Reynolds (2009), or the Markov Binomial EWMA chart for monitoring segments taken from such a Markovian Bernoulli process (see Weiß, 2009d). But if product or service quality, for instance, is classified in more than only two categories, then methods for monitoring *non-binary* but serially dependent categorical processes would be required. See Weiß (2012a) and the references therein for a few first approaches in this direction, while a comprehensive treatment of this area is still pending.

Acknowledgements
The author thanks the two referees for carefully reading the manuscript and for their valuable comments, which greatly improved the article.

Funding
The author received no funding for this research.

Author details
Christian H. Weiß[1]
E-mail: weissc@hsu-hh.de
ORCID ID: http://orcid.org/0000-0001-8739-6631
[1] Department of Mathematics and Statistics, Helmut Schmidt University, Postfach 700822, 22008 Hamburg, Germany.

Note
1. While the control limits are chosen according to stochastic properties of the monitored process, see below, the specification limits are determined according to the usability of the produced items: if the considered quality characteristic of a produced item violates the specification limits, it has to be classified as being defective.

References
Acosta-Mejia, C. A. (1999). Improved *p* charts to monitor process quality. *IIE Transactions, 31*, 509–516.
Al-Osh, M. A., & Alzaid, A. A. (1987). First-order integer-valued autoregressive (INAR(1)) process. *Journal of Time Series Analysis, 8*, 261–275.

Alwan, L. C. (1992). Effects of autocorrelation on control chart performance. *Communications in Statistics - Theory and Methods, 21*, 1025–1049.

Alwan, L. C. (1995). The problem of misplaced control limits. *Journal of the Royal Statistical Society C, 44*, 269–278.

Alwan, L. C., Champ, C. W., & Maragah, H. D. (1994). Study of average run lengths for supplementary runs rules in the presence of autocorrelation. *Communications in Statistics - Simulation and Computation, 23*, 373–391.

Alwan, L. C., & Roberts, H. V. (1988). Time series modeling for statistical process control. *Journal of Business & Economic Statistics, 6*, 87–95.

Bersimis, S., Psarakis, S., & Panaretos, J. (2007). Multivariate statistical process control charts: an overview. *Quality and Reliability Engineering International, 23*, 517–543.

Borges, W., & Ho, L. L. (2001). A fraction defective based capability index. *Quality and Reliability Engineering International, 17*, 447–458.

Brook, D., & Evans, D. A. (1972). An approach to the probability distribution of CUSUM run length. *Biometrika, 59*, 539–549.

Celano, G. (2011). On the constrained economic design of control charts: A literature review. *Produção, 21*, 223–234.

Chen, N., & Zhou, S. (2015). CUSUM statistical monitoring of M/M/1 queues and extensions. *Technometrics, 57*, 245–256.

Davoodi, M., Niaki, S. T. A., & Torkamani, E. A. (2015). A maximum likelihood approach to estimate the change point of multistage Poisson count processes. *International Journal of Advanced Manufacturing Technology, 77*, 1443–1464.

Du, J.-G., & Li, Y. (1991). The integer-valued autoregressive (INAR(p)) model. *Journal of Time Series Analysis, 12*, 129–142.

Epprecht, E. K., Costa, A. F. B., & Mendes, F. C. T. (2003). Adaptive control charts for attributes. *IIE Transactions, 35*, 567–582.

Ferland, R., Latour, A., & Oraichi, D. (2006). Integer-valued GARCH processes. *Journal of Time Series Analysis, 27*, 923–942.

Franke, J., Kirch, C., & Kamgaing, J. T. (2012). Changepoints in times series of counts. *Journal of Time Series Analysis, 33*, 757–770.

Gan, F. F. (1990). Monitoring Poisson observations using modified exponentially weighted moving average control charts. *Communications in Statistics - Simulation and Computation, 19*, 103–124.

Hawkins, D. M. (1993). Robustification of cumulative sum charts by Winsorization. *Journal of Quality Technology, 25*, 248–261.

Hawkins, D. M., & Olwell, D. H. (1998). *Cumulative sum charts and charting for quality improvement*. New York, NY: Springer-Verlag.

Heinen, A. (2003). *Modelling time series count data: An autoregressive conditional Poisson model* (CORE Discussion Paper No. 2003-63). Belgium: University of Louvain.

Hudecová, Š, Hušková, M., & Meintanis, S. (2015). Detection of changes in INAR models. In A. Steland, E. Rafajłowicz, & K. Szajowski (Eds.), *Stochastic models, statistics and their applications, Springer proceedings in mathematics & statistics* (Vol. 122, pp. 11–18). Springer.

Jazi, M. A., Jones, G., & Lai, C.-D. (2012). First-order integer valued AR processes with zero inflated Poisson innovations. *Journal of Time Series Analysis, 33*, 954–963.

Jensen, W. A., Jones-Farmer, L. A., Champ, C. W., & Woodall, W. H. (2006). Effects of parameter estimation on control chart properties: a literature review. *Journal of Quality Technology, 32*, 395–409.

Johnson, N. L., Kemp, A. W., & Kotz, S. (2005). *Univariate discrete distributions* (3rd ed.). Hoboken, NJ: Wiley.

Jones-Farmer, L. A., Woodall, W. H., Steiner, S. H., & Champ, C. W. (2014). An overview of phase I analysis for process improvement and monitoring. *Journal of Quality Technology, 46*, 265–280.

Kang, J., & Lee, S. (2014). Parameter change test for Poisson autoregressive models. *Scandinavian Journal of Statistics, 41*, 1136–1152.

Kang, J., & Song, J. (2015). Robust parameter change test for Poisson autoregressive models. *Statistics and Probability Letters, 104*, 14–21.

Kedem, B., & Fokianos, K. (2002). *Regression models for time series analysis*. Hoboken, NJ: Wiley.

Kenett, R. S., & Pollak, M. (2012). On assessing the performance of sequential procedures for detecting a change. *Quality and Reliability Engineering International, 28*, 500–507.

Kim, H.-Y., & Park, Y. (2008). A non-stationary integer-valued autoregressive model. *Statistical Papers, 49*, 485–502.

Kirch, C., & Kamgaing, J. T. (2015). On the use of estimating functions in monitoring time series for change points. *Journal of Statistical Planning and Inference, 161*, 25–49.

Knoth, S. (2006). The art of evaluating monotoring schemes—How to measure the performance of control charts? In H.-J. Lenz & P.-T. Wilrich (Eds.), *Frontiers in statistical quality control 8* (pp. 74–99). Heidelberg: Physica Verlag.

Knoth, S. (in press). The case against the use of synthetic control charts. *Journal of Quality Technology.*

Knoth, S., & Schmid, W. (2004). Control charts for time series: A review. In H. J. Lenz & P. T. Wilrich (Eds.), *Frontiers in statistical quality control 7* (pp. 210–236). Heidelberg: Physica-Verlag.

Koutras, M. V., Bersimis, S., & Maravelakis, P. E. (2007). Statistical process control using Shewhart control charts with supplementary runs rules. *Methodology and Computing in Applied Probability, 9*, 207–224.

Li, C., Wang, D., & Zhu, F. (in press). Effective control charts for monitoring the NGINAR(1) process. *Quality and Reliability Engineering International.*

McKenzie, E. (1985). Some simple models for discrete variate time series. *Water Resources Bulletin, 21*, 645–650.

Montgomery, D. C. (2009). *Introduction to statistical quality control* (6th ed.). New York, NY: Wiley.

Morais, M. C., & Pacheco, A. (in press). On hitting times for Markov time series of counts with applications to quality control. *RevStat.*

Mousavi, S., & Reynolds, Jr., M. R., (2009). A CUSUM chart for monitoring a proportion with autocorrelated binary observations. *Journal of Quality Technology, 41*, 401–414.

Page, E. (1954). Continuous inspection schemes. *Biometrika, 41*, 100–115.

Perakis, M., & Xekalaki, E. (2005). A process capability index for discrete processes. *Journal of Statistical Computation and Simulation, 75*, 175–187.

Psarakis, S., & Papaleonida, G. E. A. (2007). SPC procedures for monitoring autocorrelated processes. *Quality Technology & Quantitative Management, 4*, 501–540.

Ristić, M. M., Bakouch, H. S., & Nastić, A. S. (2009). A new geometric first-order integer-valued autoregressive (NGINAR(1)) process. *Journal of Statistical Planning and Inference, 139*, 2218–2226.

Roberts, S. W. (1959). Control chart tests based on geometric moving averages. *Technometrics, 1*, 239–250.

Saha, M., & Maiti, S. S. (2015). *Trends and practices in process capability studies*. arXiv:1503.06885v1 [stat.AP].

Schweer, S., & Weiß, C. H. (2014). Compound Poisson INAR(1) processes: Stochastic properties and testing for overdispersion. *Computational Statistics & Data Analysis, 77*, 267–284.

Schweer, S., & Wichelhaus, C. (2015). Queueing systems of INAR(1) processes with compound Poisson arrivals. *Stochastic Models, 31*, 618–635.

Steutel, F. W., & van Harn, K. (1979). Discrete analogues of self-decomposability and stability. *Annals of Probability, 7*, 893–899.

Torkamani, E. A., Niaki, S. T. A., Aminnayeri, M., & Davoodi, M. (2014). Estimating the change point of correlated Poisson count processes. *Quality Engineering, 26*, 182–195.

Weiß, C. H. (2007). Controlling correlated processes of Poisson counts. *Quality and Reliability Engineering International, 23*, 741–754.

Weiß, C. H. (2008a). Thinning operations for modelling time series of counts—A survey. *Advances in Statistical Analysis, 92*, 319–341.

Weiß, C. H. (2008b). Serial dependence and regression of Poisson INARMA models. *Journal of Statistical Planning and Inference, 138*, 2975–2990.

Weiß, C. H. (2009a). Monitoring correlated processes with binomial marginals. *Journal of Applied Statistics, 36*, 399–414.

Weiß, C. H. (2009b). EWMA monitoring of correlated processes of Poisson counts. *Quality Technology and Quantitative Management, 6*, 137–153.

Weiß, C. H. (2009c). Controlling jumps in correlated processes of Poisson counts. *Applied Stochastic Models in Business and Industry, 25*, 551–564.

Weiß, C. H. (2009d). Group inspection of dependent binary processes. *Quality Reliability Engineering International, 25*, 151–165.

Weiß, C. H. (2011a). Detecting mean increases in Poisson INAR(1) processes with EWMA control charts. *Journal of Applied Statistics, 38*, 383–398.

Weiß, C. H. (2011b). The Markov chain approach for performance evaluation of control charts—A tutorial. In S. P. Werther (Ed.), *Process control: Problems, techniques and applications* (pp. 205–228). New York, NY: Nova Science.

Weiß, C. H. (2012a). Continuously monitoring categorical processes. *Quality Technology and Quantitative Management, 9*, 171–188.

Weiß, C. H. (2012b). Process capability analysis for serially dependent processes of Poisson counts. *Journal of Statistical Computation and Simulation, 82*, 383–404.

Weiß, C. H., & Kim, H.-Y. (2013). Parameter estimation for binomial AR(1) models with applications in finance and industry. *Statistical Papers, 54*, 563–590.

Weiß, C. H., & Testik, M. C. (2009). CUSUM monitoring of first-order integer-valued autoregressive processes of Poisson counts. *Journal of Quality Technology, 41*, 389–400.

Weiß, C. H., & Testik, M. C. (2011). The Poisson INAR(1) CUSUM chart under overdispersion and estimation error. *IIE Transactions, 43*, 805–818.

Weiß, C. H., & Testik, M. C. (2012). Detection of abrupt changes in count data time series: Cumulative sum derivations for INARCH(1) models. *Journal of Quality Technology, 44*, 249–264.

Weiß, C. H., & Testik, M. C. (2015a). Residuals-based CUSUM charts for Poisson INAR(1) processes. *Journal of Quality Technology, 47*, 30–42.

Weiß, C. H., & Testik, M. C. (2015b). On the phase I analysis for monitoring time-dependent count processes. *IIE Transactions, 47*, 294–306.

Woodall, W. H. (1997). Control charts based on attribute data: Bibliography and review. *Journal of Quality Technology, 29*, 172–183.

Woodall, W. H. (2000). Controversies and contradictions in statistical process control. *Journal of Quality Technology, 32*, 341–350.

Woodall, W. H., & Montgomery, D. C. (2014). Some current directions in the theory and application of statistical process monitoring. *Journal of Quality Technology, 46*, 78–94.

Yontay, P., Weiß, C. H., Testik, M. C., & Bayindir, Z. P. (2013). A two-sided CUSUM chart for first-order integer-valued autoregressive processes of Poisson counts. *Quality and Reliability Engineering International, 29*, 33–42.

Zhang, M., Nie, G., He, Z., & Hou, X. (2014). The Poisson INAR(1) one-sided EWMA chart with estimated parameters. *International Journal of Production Research, 52*, 5415–5431.

Zucchini, W., & MacDonald, I. L. (2009). *Hidden Markov models for time series: An introduction using R*. London: Chapman & Hall/CRC.

11

The properties of the geometric-Poisson exponentially weighted moving control chart with estimated parameters

Aamir Saghir[1,2,*], Zhengyan Lin[2] and Ching-Wen Chen[3]

*Corresponding author: Aamir Saghir, Department of Mathematics, Mirpur University of Science and Technology (MUST) Mirpur, Mirpur, Pakistan

E-mail: aamirstat@yahoo.com; aamir.stat@must.edu.pk

Reviewing editor: Zudi Lu, University of Southampton, UK

Abstract: The geometric-Poisson exponentially weighted moving average (EWMA) chart has been shown to be more effective than the Poisson EWMA chart in monitoring the number of defects in the production processes. In these applications, it is assumed that the process parameters are known or have been accurately estimated. However, in practice, the process parameters are rarely known and must be estimated from reference sample to construct the geometric-Poisson EWMA chart. The performance of the given chart, due to variability in the parameter estimation, might differ from known parameters' case. This article explored the effect of estimated parameters on the conditional and marginal performance of the geometric-Poisson EWMA chart. The run length characteristics are calculated using a Markov chain approach and the effect of estimation on the performance of the given chart is shown to be significant. Recommendations about the proposer choice of sample size, smoothing constant, and dispersion parameter are made. Results of this study highlight the practical implications of estimation error, and to offer advice to practitioners when constructing/analyzing a phase-I sample.

Subjects: Science; Mathematics & Statistics; Statistics & Probability; Statistics; Statistical Computing; Statistics & Computing; Statistical Theory & Methods; Statistics for Business, Finance & Economics

Keywords: the geometric-Poisson chart; parameters estimation; control limits; average run length; marginal performance

ABOUT THE AUTHORS

Aamir Saghir obtained his PhD from Zhejiang University China in 2014. He has been working as a lecturer in Statistics in Mirpur University of Science and Technology (MUST) Mirpur, AJK, Pakistan. His research interests include Statistical Quality Control and application of Statistics.

Zhengyan Lin is currently working as a professor at the Department of Mathematics, Institute of Statistics, Zhejiang University, Hangzhou, China. His research interests include limit theory and industrial applications of statistics.

Ching-Wen Chen is currently working as a professor at the Department of Information Management, National Kaohsiung first university if science and technology, Taiwan. His research interests include industrial statistics and data mining.

PUBLIC INTEREST STATEMENT

The semiconductor industry utilizes expensive production equipment and has rigorous requirements for its production environment, which results in a high production cost. The current paper is an application of statistics in industrial engineering. In these applications, it is assumed that the process parameters are known or have been accurately estimated. However, in practice, the process parameters are rarely known and must be estimated from reference sample to construct the geometric-Poisson EWMA chart. The effect of estimation on the performance of the given chart is shown to be significant. Recommendations regarding sample-size, smoothing constant, and clustering parameter are provided. Finally, a real example is used to highlight the practical implications of estimation error, and to offer advice to practioners when constructing/analyzing a phase-I sample.

1. Introduction

Statistical process control (SPC) is a collection of various tools, which are used to examine a process and improve the quality of its products (see Montgomery, 2009). Among these methods, control charts are known to be more effective tools for process monitoring because they allow practitioners to draw conclusions about the state of the process (in-control or out-of-control). These conclusions about the state of process depend on whether the applied monitoring approach is a phase-I or phase-II method. In phase I, the historical data of the process is used to test the process stability and to estimate the in-control parameters. In phase II, the process is monitored in real time to quickly detect shifts from the baseline established in phase I. Some researchers like Quensberry and Geometric (1995) recommend self-starting charts that bypass phase I and begin monitoring quickly.

Most SPC research has been carried out on developing phase-II control charting methods, where it is assumed that the in-control parameters are known or can be accurately estimated. However, the parameters are rarely known with certainty in practice. Therefore, accurate estimates of the parameters are required to make the statistical performance of the control charts reliable. Also, it is helpful to provide practitioners with phase-I guidelines, such that the effect of estimation error on the performance of control charts in phase II can be better understood (see Woodall & Montgomery, 1999). Along the variable control charts, in the current decade, the performance of estimated control limits of the attribute charts is a burning issue. Jensen, Jones-Farmer, Champ, and Woodall (2006) and Szarka and Woodall (2011) provided a detailed review on the effect of estimation error on control chart performance. Yang, Xie, Kuralmani, and Tusi (2000), Tang and Cheong (2004), Zhang, Peng, Schuh, Megahed, and Woodall (2013) studied the performance of the geometric charts with estimated control limits. They concluded that a large phase-I sample size is required for low in-control proportion of non-conformity. Other related studies on the estimated attributes' control charts performance include Shu, Tsung, and Tsui (2004), Chakraborti and Human (2006), Testik, McCullough, and Borror (2006), Testik (2007), Ozsan, Testik, and Weiβ (2010), Lee, Wang, Xu, Schuh, and Woodall (2013), Chiu and Tsai (2013), Mahmoud and Maravelakis (2013), Saleh, Mahmoud, and Abdel-Salam (2013), Zhang et al. (2013), Saghir and Lin (2013), etc. In fact, the study of the statistical performance of control charts with estimated control limits is a general research issue of importance.

The geometric-Poisson distribution is a natural extension of the Poisson distribution and an adequate model to monitor the number of defects over time in production processes. Chen, Randolph, and Liou (2005) developed CUSUM control charts based on the geometric-Poisson compound distribution. Chen (2012) proposed an exponentially weighted moving average (EWMA) control chart for monitoring the number of defects over time. The geometric-Poisson distribution (compound distribution) was used to develop the proposed chart. The result of the study reveals that the proposed EWMA chart, namely geometric-Poisson EWMA chart, is very effective in monitoring and improving quality in production environment than the usual Poisson EWMA chart. The performance of the geometric-Poisson chart has been investigated by Chen (2012), under the assumption that the parameters of the geometric-Poisson process are known. However, in practice, the parameters are unknown and estimated from the historical data.

The current article investigates the effect of estimation error on the performance of the geometric-Poisson EWMA chart. The run length (RL) properties, such as average run length (ARL), the standard deviation of the run length (SDRL), and percentiles of the RL distribution are analyzed using the Markov chain approach following Saghir and Lin (2013) and Chen (2012). The conditional and marginal performances of the RL metrics of the given chart are evaluated. The conditional analysis allows us to understand the effect of overestimating or underestimating the parameters on the RL performance of the chart. While the marginal performance is useful in providing recommendations regarding minimum sample size, choice of smoothing constant, and dispersion parameter. Because, it considers the distribution of the estimated parameters and thus accounts for the variability introduced through parameters estimations.

The rest of the article is summarized as follows. In Section 2, the geometric-Poisson EWMA control chart with estimated parameters is given. Section 3 describes different performance evaluation measures. Section 4 evaluates the performance of the estimated control limits of EWMA chart under two different conditions. Finally, the conclusion of the study with discussion is made in Section 5.

2. The geometric-Poisson EWMA chart

2.1. The geometric-Poisson EWMA chart with known parameters

Let $Y(t)$ be the random variable of the number of defective items, and $X(t)$ be the random variable of the number of defects that occur up to t, where $t > 0$. According to Chen (2012), the density function of the geometric-Poisson compound distribution with parameters λ (rate) and ρ (dispersion) for any $t > 0$ is

$$P[X(t)=0]=e^{-\lambda t},$$
$$P\left[X(t)=x\right]=\sum_{y=1}^{x}\frac{(\lambda t)^y e^{-\lambda t}}{y!}\cdot\begin{pmatrix} x-1 \\ y-1 \end{pmatrix}\rho^{x-y}(1-\rho)^y,\quad x=1,2,3,\ldots \tag{1}$$

where $\lambda > 0$, $0 < \rho < 1$. The expected value and variance of total defects in a fixed unit $t = 1$, are derived by Chen et al. (2005) and shown as $E[X]=\mu=\frac{\lambda}{1-\rho}$ and $\mathrm{Var}[X]=\sigma^2=\frac{\lambda(1+\rho)}{(1-\rho)^2}$, respectively. Clearly, the variance of the geometric-Poisson distribution is greater than or equal to the mean. If the variance equals the mean, the geometric-Poisson reduces to the Poisson.

Assume that the sequence of $X_1, X_2, \ldots$ forms a repetitive production process of i.i.d. compound geometric random variables with probability mass function defined in Equation 1. To detect the changes from the in-control mean $\mu = \mu_0$ to an out-of-control mean $\mu = \mu_1$, Chen (2012) proposed an EWMA control chart. The EWMA statistic is defined as

$$Z_i=wX_i+(1-w)Z_{i-1},\quad i=1,2,\ldots \tag{2}$$

with $Z_0 = \mu_0$ and $w(0 < w \leq 1)$ being the smoothing constant. Since the EWMA can be viewed as a weighted average of all past and current observations, it is very sensitive to the normality assumption. It is therefore an ideal control chart to monitor individual observations and could effectively detect small and moderate changes in the manufacturing processes. For an in-control process, the mean and variance of the EWMA statistic are

$$E(Z_i)=\mu_0=\frac{\lambda_0}{1-\rho_0}$$
$$\mathrm{Var}(Z_i)=\frac{w}{2-w}\left[1-(1-w)^{2i}\right]\left[\frac{\lambda_0(1+\rho_0)}{(1-\rho_0)^2}\right] \tag{3}$$

The control limits for the EWMA control chart based on the geometric-Poisson compound distribution are defined as:

$$\left.\begin{aligned} LCL &= \frac{\lambda_0}{(1-\rho_0)}-K_l\sqrt{\frac{w}{2-w}\left[1-(1-w)^{2i}\right]\cdot\frac{\lambda_0(1+\rho_0)}{(1-\rho_0)^2}} \\ CL &= \frac{\lambda_0}{1-\rho_0} \\ UCL &= \frac{\lambda_0}{(1-\rho_0)}+K_u\sqrt{\frac{w}{2-w}\left[1-(1-w)^{2i}\right]\cdot\frac{\lambda_0(1+\rho_0)}{(1-\rho_0)^2}} \end{aligned}\right\} \tag{4}$$

For large values of i, the asymptotic limits in Equation 4 reduced to

$$\left.\begin{array}{l} LCL=\frac{\lambda_0}{(1-\rho_0)}-K_l\sqrt{\frac{w}{2-w}\cdot\frac{\lambda_0(1+\rho_0)}{(1-\rho_0)^2}} \\ CL=\frac{\lambda_0}{1-\rho_0} \\ UCL=\frac{\lambda_0}{(1-\rho_0)}+K_u\sqrt{\frac{w}{2-w}\cdot\frac{\lambda_0(1+\rho_0)}{(1-\rho_0)^2}} \end{array}\right\} \quad (5)$$

where K_l and K_u are control chart constants and $\frac{\lambda_0}{(1-\rho_0)}=\mu_0$ is the target mean value of the geometric-Poisson compound process. Chen (2012) provided the values of $K_l = K_u = K$ for various combinations of w, ρ_0, λ_0 and desired in-control ARL using Markov chain approach. The value of the lower control limit LCL should be set to zero, when its computed value is less than zero. This is because the quality characteristic of interest X_i is a compound random variable and therefore the EWMA statistic Z_i in Equation 2 will be non-negative. The choice of the smoothing constant w typically depends on how fast a mean shift of given size should be detected. It is generally accepted that smaller values of w are more effective in rapidly detecting smaller mean shifts and vice versa.

After setting the control limits for the negative binomial EWMA chart, the EWMA statistic given in Equation 2 is plotted against each i. For an in-control process, all of the Z_is should lie inside the control limits, whereas an out-of-control process is signaled by one or more of the Z_is, which exceeds the LCL and UCL. The process will never leave the manufacturing process unless it is modified manually. In other words, the process is at the "absorption" state.

2.2. The geometric-Poisson EWMA chart when parameters are unknown

The control chart constant K for the geometric-Poisson EWMA control statistic can be calculated using the Markov chain approach, if λ_0 and ρ_0 are known. However, when λ_0 and ρ_0 are unknown, their values must be calculated prior to any calculations. The method-of-moments estimators are generally used to estimate λ_0 and ρ_0 from phase-I samples. We have the method-of-moment estimates (see Chen et al., 2005).

$$\hat{\lambda}_0=\frac{2(\bar{X})^2}{S^2+\bar{X}},\ \hat{\rho}_0=\frac{S^2-\bar{X}}{S^2+\bar{X}} \quad (6)$$

Here, $\bar{X}=\sum_{i=}^{m}X_i/m$ and $S^2=\sum_{i=1}^{m}(X_i-\bar{X})^2/m-1$ are the sample mean and variance, respectively, for initial samples of size m. The sample variance S^2 must be greater than sample average $\bar{X}$, so that the value of ρ is positive. Therefore, the average number of non-conformities based on moment estimate is $\hat{\mu}_0=\bar{X}$, and the central limit ensures that its sampling distribution is approximately normal with mean μ_0 and variance μ_0/m. Accordingly, these estimates can be used in any of the RL calculations for the geometric-Poisson EWMA chart.

The estimated control limits for the given EWMA chart based on MM estimates are defined as:

$$\left.\begin{array}{l} \hat{h}_l(m)=LCL=\frac{\hat{\lambda}_0}{(1-\hat{\rho}_0)}-K_l\sqrt{\frac{w}{2-w}\cdot\frac{\hat{\lambda}_0(1+\hat{\rho}_0)}{(1-\hat{\rho}_0)^2}} \\ \hat{h}_c(m)=CL=\frac{\hat{\lambda}_0}{1-\hat{\rho}_0} \\ \hat{h}_u(m)=UCL=\frac{\hat{\lambda}_0}{(1-\hat{\rho}_0)}+K_u\sqrt{\frac{w}{2-w}\cdot\frac{\hat{\lambda}_0(1+\hat{\rho}_0)}{(1-\hat{\rho}_0)^2}} \end{array}\right\} \quad (7)$$

The aim of this article is to determine the effect of the phase-I sample on the geometric-Poisson EWMA chart's performance. Statistical properties of a EWMA control chart are usually evaluated in terms of average run length (ARL), which is the mean of RL distribution. The ARL of a control charting procedure is defined as the expected number of sampling stages until an out-of-control condition is signaled. An effective and efficient control chart can provide optimal ARL. More specifically, the ARL of an optimal control scheme should be large when a process is in control and small when a shift occurs (Chen & Chen, 2007). In following section, we provide some information about our calculations.

3. RL distribution with estimated parameters

Chen (2012) used a Markov chain approach (proposed by Brook & Evans, 1972) to study the RL distribution of a geometric-Poisson EWMA chart. In this section, we will extend the Markov chain approximation for assessing the performance of geometric-Poisson EWMA Chart, when the estimated parameters differ from their actual values. Similar to Zhang et al. (2013), we have considered both conditional and marginal RL properties. We have provided the Markov chain method and the equations used to obtain the RL properties in the coming subsections.

3.1. The Markov chain approach

Suppose D is the number of defects, then the EWMA of D is

$$Z_i = wD_i + (1-w)Z_{i-1}, \quad i = 1, 2, \ldots \tag{8}$$

The corresponding EWMA control scheme would signal, if $Z_i > \hat{h}_u$ or $Z_i < \hat{h}_u$ and a remedy action should be taken. To visualize the transitioning process, the decision interval $[\hat{h}_l, \hat{h}_u]$ is divided into N subintervals as explained in Chen (2012). The transition of Z_i in the interval $[\hat{h}_l, \hat{h}_u]$ is a random walk and the ith subinterval is the ith state, denoted by E_i, and is represented by the mid-point S_i.

If Z_i is within the decision interval, then the process is in-control state and no out-of-control signal would be given. On the other hand, if Z_i moves outside the control limits (above $\hat{h}_u$ or below $\hat{h}_l$, then the process enters the out-of-control status. Thus, the $(N + 1)$st state is absorbing and represents the out-of-control region.

Let P_{ij} denote the probability of transition from state i to state j in one step. Then, the transition probability matrix, $\boldsymbol{P}$, is defined as

$$\boldsymbol{P} = \begin{pmatrix} P_{0,0} & P_{0,1} & \cdots & P_{0,j} & \cdots & P_{0,N+1} \\ P_{1,0} & P_{1,1} & \cdots & P_{1,j} & \cdots & P_{1,N+1} \\ \vdots & \vdots & \vdots & \vdots & \vdots & \vdots \\ P_{i,0} & P_{i,1} & \cdots & P_{i,j} & \cdots & P_{i,N+1} \\ \vdots & \vdots & \vdots & \vdots & \vdots & \vdots \\ 0 & 0 & \cdots & 0 & \cdots & 1 \end{pmatrix} = \begin{pmatrix} \boldsymbol{R} & (\boldsymbol{I}-\boldsymbol{R})\boldsymbol{u} \\ \boldsymbol{0} & 1 \end{pmatrix} \tag{9}$$

Note that all rows sum to unity, and that the last row consists of zeros, except with the last element that is equal to 1 because E_m is an absorbing state. In addition, the matrix $\boldsymbol{R}$ is a $N \times N$ matrix, including the probabilities of moving from one transient state to another, $\boldsymbol{I}$ in an identity matrix of order $N \times N$, $\boldsymbol{u}$ is a $(N \times 1)$ vector of ones, and the $(\boldsymbol{I} - \boldsymbol{R})\boldsymbol{u}$ vector includes the transition probabilities of moving from one transient state to an absorbing state. The transition probabilities for the Markov chain are determined as follows:

$$P_{i,1} = \Pr(E_i \to E_1), P_{i,j} = \Pr(E_i \to E_j), P_{i,N+1} = \Pr(E_i \to E_{N+1}), P_{N+1,i} = 0, P_{N+1,N+1} = 1$$

where

$$P_{ij} = P\left[L_j < Z_t < U_j \mid Z_{t-1} = S_i = \hat{h}_l + \frac{(2i-1)\left(\hat{h}_u - \hat{h}_l\right)}{2N}\right]$$

This transition probability, P_{ij}, can be written as

$$P_{ij} = P\left[\begin{array}{l} \frac{\hat{\lambda}_0}{(1-\hat{p}_0)} - K_l\sqrt{\frac{w}{2-w}\cdot\frac{\hat{\lambda}_0(1+\hat{p}_0)}{(1-\hat{p}_0)^2}} + \frac{(\hat{h}_u-\hat{h}_l)}{2Nw}\{2(j-1)-(1-w)(2i-1)\} \\ < D_i < \frac{\hat{\lambda}_0}{(1-\hat{p}_0)} + K_l\sqrt{\frac{w}{2-w}\cdot\frac{\hat{\lambda}_0(1+\hat{p}_0)}{(1-\hat{p}_0)^2}} + \frac{(\hat{h}_u-\hat{h}_l)}{2Nw}\{2j-(1-w)(2i-1)\} \end{array}\right] \quad i,j=0,1,2,\ldots,N \tag{10}$$

3.2. Conditional performance of the RL distribution

The RL of the geometric-Poisson EWMA chart is the number of steps taken starting from the initial state E_1 to reach the absorbing state E_{N+1}. Using the Markov chain approach, the approximate ARL and SDRL performance measures are computed as follows:

$$\text{ARL} = (\boldsymbol{I} - \boldsymbol{R})^{-1}\boldsymbol{u} \tag{11}$$

$$\text{SDRL} = \left\{\text{ARL} + 2\left[(\boldsymbol{I} - \boldsymbol{R})^{-1} - \boldsymbol{I}\right]\text{ARL} - \text{ARL}^2\right\}^{1/2} \tag{12}$$

where each of ARL and SDRL is a $N \times 1$ vector, including ARLs and SDRLs corresponding to all possible states. Assuming that N is an odd number, then the $((N + 1)/2)$th elements of these vectors correspond to the zero-state ARL and SDRL.

The percentile of the RL distribution is another important performance measure of the control charts. The percentiles of the RL distribution may be determined using the cumulative probability for RL. Let F_r denote the $N \times 1$ cumulative probability vector, where each of the N entries is for one starting value Z_0 and the index $r = 1, 2, \ldots$ represents a value of the RL. Then

$$\boldsymbol{F}_r = \left(\boldsymbol{I} - \boldsymbol{R}^r\right)\boldsymbol{u} \tag{13}$$

The cumulative probability times 100 give the percentile corresponding to the RL value r. For example, the 30th percentile for the case of $Z_0 = \hat{\lambda}_0$ is the smallest value of r for the middle entry of F_r being greater than or equal to 0.3.

3.3. Marginal performance of the RL distribution

The marginal performance can be obtained by integrating the conditional performance measures with respect to the density of parametric space as shown in the following equations:

$$\text{ARL}_{\text{marginal}} = \int_{-\infty}^{\infty} \text{ARL}\, g(\mu)d\mu \tag{14}$$

$$\text{SDRL}_{\text{marginal}} = \int_{-\infty}^{\infty} \text{SDRL}\, g(\mu)d\mu \tag{15}$$

$$\text{Percentile}_{\text{marginal}} = \int_{-\infty}^{\infty} \text{percentile}\, g(\mu)d\mu \tag{16}$$

where the $g(\mu)$ is an approximated normal distribution of random variable μ. The marginal performance measures are weighted averages of the conditional performance over all the values that the estimation may yield for the in-control mean μ_0. These integrals can be solved using a numerical integration procedure. In our calculations, we have followed the approach of Ozsan et al. (2010) and used the Simpson's quadrature method in Matlab.

4. Results and discussion

In this section, we have evaluated the conditional and marginal RL performance of the geometric-Poisson EWMA control chart, when the control limits are estimated. The conditional performance is summarized in Section 4.1 and marginal performance is provided in Section 4.2. All the computations are done in Matlab.

4.1. Conditional performance of the geometric-Poisson EWMA chart

In this section, to investigate the performance of the geometric-Poisson EWMA chart in phase II and observe the impact of parameters estimation, some hypothetical cases of estimated mean are considered. Three different situations are considered in estimating the mean; 25th (underestimation), 50th (nominal), and 75th (overestimation) percentile of the sampling distribution of estimated mean. This is equivalent to the actual in-control process mean being equal to the estimated in-control process mean (nominal), lower (overestimation), or upper (underestimation). Let the true in-control rate parameter be $\lambda_0 = 2$ with *dispersion* rate $\rho = 0.20$. The mean and various percentiles of the sampling distribution of $\hat{\lambda}_0$ are calculated for various samples and provided in Table 1. It is clear that these estimates deviate heavily from the true value $\lambda_0 = 2$ for samples ≤ 500.

The conditional RL performance of the given chart is calculated for various parameters values, smoothing constant, and samples. The estimated conditional RL distribution is provided in Tables 2–4 for some choices. The various amounts of shifts in the average defects in terms of standard deviation of the in-control process are considered i.e. $\mu_a = \mu_0 + \delta\sqrt{\sigma_0}$.

The results of Tables 2–4 indicate that the effect of parameter estimation is significant. The actual in-control ARL_0 is significantly deviated from expected (**500.00**), when the parameters are estimated from *m* initial samples. In the above tables, the nominal case refers to known parameters, so, the performance of the given control chart does not depend on sample size. Overestimation (when parameters assume a value in the 75th percentiles) results in an increase in the number of false alarms and a decrease in the variability of the RL as compared to nominal summaries. On the other hand, underestimation (25th percentiles) results in large EWMA variance than expected one. Therefore, the average run length (ARL_0) is expected to be less than the nominal values and in more frequent false alarms. In case of out-of-control scenario, shift in the average non-conformities due to an assignable cause are detected faster for underestimation case than for the overestimation case, as is well known.

It is apparent that the performance of the geometric-Poisson chart is significantly affected due to estimated parameters, however, the magnitude of the effect decreases as *m* increases. In both over- or underestimation cases, large reference samples, more than 1,000, are required to achieve the desired in-control ARL_0 of 500.00. In most of the quality control applications, 25–30 samples for different control charts is recommended, but this study suggests that more than 1,000 samples are needed for the given chart to monitor quality characteristics. Therefore, the implementation of the geometric-Poisson EWMA chart requires minimum 1,000 reference samples before monitoring.

The choice of smoothing constant or EWMA weight has also significant effect on the performance of chart, as is obvious from Tables 2–4. Considering a sample size constant, it is observed that the actual ARL_0 value is decreased by increase in smoothing constant in case of overestimation, while inverse hold for underestimation. To achieve the desired ARL_0 of 500.00 with a large reference sample size, a large smoothing constant is required. However, when there exists a positive shift, the ARL increases by increase in smoothing constant, as is obvious from Tables 2–4. Therefore, smaller smoothing constant is better than larger one in detecting small shifts of the average non-conformities.

Table 1. The mean and percentiles of the sampling distribution of $\hat{\lambda}_0$ at various phase-I samples, assuming the true values of parameters λ_0 = 2 and ρ = 0.20, respectively

m	The mean and Percentiles of sampling distribution of $\hat{\lambda}_0$			
	Mean	**25th**	**50th**	**75th**
30	1.9710	2.1588	1.9885	1.8825
50	1.9835	2.1150	1.9905	1.9099
100	1.9900	2.0999	1.9920	1.9401
500	1.9970	2.0688	1.9983	1.9657
1000	1.9990	2.0302	1.9995	1.9790
5000	1.9997	2.0255	1.9998	1.9895

Table 2. Conditional RL performance of the geometric-Poisson EWMA chart with $\hat{\lambda}_0$ = 2, $\hat{\rho}$ = 0.20 and $ARL_0 \approx 500$

				25th (overestimation)		50th (nominal)		75th (underestimation)	
M	*w*	*k*	*δ*	**ARL**	**SDRL**	**ARL**	**SDRL**	**ARL**	**SDRL**
30	0.05	2.62	0	**165.10**	**133.40**	**498.78**	**491.88**	**2202.3**	**1991.85**
			0.5	19.30	16.29	26.79	21.05	33.12	27.50
			1.0	8.19	5.83	11.15	9.25	20.70	15.30
			1.5	5.11	3.15	7.41	5.05	10.85	8.11
			2.0	3.25	2.85	5.40	3.98	7.65	5.35
	0.10	2.87	0	**182.56**	**169.88**	**499.60**	**496.77**	**1980.20**	**1784.40**
			0.5	22.10	19.90	27.39	22.74	34.88	28.10
			1.0	8.70	6.50	10.17	8.72	19.20	14.11
			1.5	5.52	3.80	6.48	4.12	10.10	7.50
			2.0	3.65	2.99	4.63	3.50	6.50	4.81
	0.20	3.20	0	**196.05**	**188.56**	**503.69**	**497.23**	**1678.90**	**1480.90**
			0.5	27.56	23.55	33.90	29.76	38.55	33.80
			1.0	8.99	6.85	10.67	8.65	20.10	14.00
			1.5	5.60	3.65	6.26	3.97	9.60	6.95
			2.0	3.80	2.72	4.40	2.98	6.05	4.10
100	0.05	2.62	0	**222.56**	**202.10**	**498.78**	**491.88**	**1705.70**	**1480.19**
			0.5	21.22	17.05	26.79	21.05	31.05	25.85
			1.0	9.05	6.05	11.15	9.25	18.95	13.45
			1.5	5.85	3.80	7.41	5.05	9.10	7.65
			2.0	3.90	2.96	5.40	3.98	6.80	4.85
	0.10	2.87	0	**277.56**	**269.55**	**499.60**	**496.77**	**1115.05**	**985.50**
			0.5	23.11	20.25	27.39	22.74	32.70	26.22
			1.0	9.15	6.99	10.17	8.72	18.15	13.05
			1.5	5.91	3.83	6.48	4.12	9.80	6.10
			2.0	3.98	3.10	4.63	3.50	5.70	4.01
	0.20	3.20	0	**305.50**	**297.55**	**503.69**	**497.23**	**990.50**	**799.00**
			0.5	29.50	24.80	33.90	29.76	35.98	32.50
			1.0	9.65	7.05	10.67	8.65	16.50	12.30
			1.5	6.05	3.79	6.26	3.97	8.12	6.01
			2.0	4.01	2.78	4.40	2.98	5.65	3.70

(Continued)

Table 2. (Continued)

				25th (overestimation)		50th (nominal)		75th (underestimation)	
M	w	*k*	*δ*	ARL	SDRL	ARL	SDRL	ARL	SDRL
1,000	0.05	2.62	0	**295.25**	**288.20**	**498.78**	**491.88**	**650.35**	**595.16**
			0.5	23.15	18.20	26.79	21.05	28.85	23.80
			1.0	10.10	7.55	11.15	9.25	15.88	11.20
			1.5	6.56	4.10	7.41	5.05	8.25	6.15
			2.0	4.40	3.36	5.40	3.98	5.85	4.12
	1.00	2.87	0	**353.15**	**340.15**	**499.60**	**496.77**	**580.10**	**510.05**
			0.5	25.55	21.05	27.39	22.74	30.12	24.12
			1.0	9.85	7.25	10.17	8.72	16.20	11.80
			1.5	6.05	3.99	6.48	4.12	7.80	5.85
			2.0	4.09	3.25	4.63	3.50	5.20	3.25
	2.00	3.20	0	**435.56**	**427.85**	**503.69**	**497.23**	**532.10**	**499.05**
			0.5	31.20	26.05	33.90	29.76	34.25	31.05
			1.0	10.05	7.55	10.67	8.65	13.10	10.45
			1.5	6.12	3.85	6.26	3.97	7.65	5.60
			2.0	4.15	2.86	4.40	2.98	5.10	3.50

Also, the dispersion parameter has significance effect on the performance of the given chart. The larger the *dispersion* parameter, the less effect on the ARL when the *rate* parameter is fixed (see Tables 2 and 4). This is due to the smaller value of *dispersio*n with fixed *rate* parameter that converges to Poisson distribution. However, the more the *rate* parameter with the fixed *dispersion* value, less influenced is the yield on the performance (see Tables 2 and 3). Thus, choice of *dispersion* parameter ρ has also significant effect on the performance of geometric-Poisson EWMA chart.

Table 3. Conditional RL performance of the geometric-Poisson EWMA chart with $\hat{\lambda}_0 = 3$, $\hat{\rho} = 0.20$ and $ARL_0 \approx 500$

				25th (overestimation)		50th (nominal)		75th (underestimation)	
M	w	*A*	*δ*	ARL	SDRL	ARL	SDRL	ARL	SDRL
30	0.05	2.803	0	**265.50**	**220.90**	**503.03**	**491.96**	**1802.3**	**1690.10**
			0.5	20.50	18.29	28.13	22.82	36.52	32.50
			1.0	8.19	7.03	11.56	9.31	25.30	18.30
			1.5	5.11	4.65	7.33	5.85	11.05	9.11
			2.0	3.25	2.90	5.48	3.93	8.55	6.05
	0.10	3.447	0	**310.56**	**285.88**	**500.64**	**494.05**	**1480.20**	**1288.90**
			0.5	21.50	19.90	28.77	21.76	38.40	34.80
			1.0	9.11	7.97	10.48	9.79	26.50	19.80
			1.5	5.85	4.88	6.34	5.95	12.80	10.50
			2.0	3.70	2.99	4.65	3.90	9.01	6.90
	0.20	3.878	0	**378.05**	**318.56**	**500.50**	**497.48**	**1270.50**	**1050.20**
			0.5	27.77	22.50	35.06	30.77	39.55	35.70
			1.0	9.15	8.05	10.87	9.48	27.20	20.88
			1.5	5.80	4.99	6.50	5.38	13.55	11.12
			2.0	3.90	3.01	4.60	4.01	9.50	7.20

(Continued)

Table 3. (Continued)

				25th (overestimation)		50th (nominal)		75th (underestimation)	
M	**w**	**A**	**δ**	**ARL**	**SDRL**	**ARL**	**SDRL**	**ARL**	**SDRL**
100	0.05	2.803	0	**322.20**	**270.66**	**503.03**	**491.96**	**1050.10**	**990.99**
			0.5	23.90	20.15	28.13	22.82	32.30	28.90
			1.0	8.95	7.50	11.56	9.31	20.18	16.10
			1.5	5.88	4.80	7.33	5.85	9.85	7.10
			2.0	4.15	2.70	5.48	3.93	7.05	5.85
	0.10	3.447	0	**390.44**	**318.62**	**500.64**	**494.05**	**995.15**	**785.50**
			0.5	25.30	18.15	28.77	21.76	33.85	29.50
			1.0	9.45	8.50	10.48	9.79	22.30	17.20
			1.5	5.99	4.50	6.34	5.95	10.10	7.70
			2.0	4.23	2.90	4.65	3.90	7.50	6.00
	0.20	3.878	0	**410.80**	**385.72**	**500.50**	**497.48**	**880.10**	**689.33**
			0.5	30.33	25.33	35.06	30.77	37.50	33.28
			1.0	9.76	8.50	10.87	9.48	23.44	17.81
			1.5	5.95	4.92	6.50	5.38	10.35	8.60
			2.0	4.35	3.25	4.60	4.01	7.60	6.19
1,000	0.05	2.803	0	**415.15**	**385.15**	**503.03**	**491.96**	**650.35**	**525.16**
			0.5	25.45	20.90	28.13	22.82	29.85	24.15
			1.0	9.80	8.20	11.56	9.31	15.20	12.10
			1.5	6.15	4.80	7.33	5.85	8.15	6.50
			2.0	4.86	3.10	5.48	3.93	6.10	4.15
	0.10	3.447	0	**425.11**	**392.55**	**500.64**	**494.05**	**560.10**	**495.95**
			0.5	26.30	21.10	28.77	21.76	30.12	25.85
			1.0	9.98	8.45	10.48	9.79	16.60	13.00
			1.5	6.05	4.90	6.34	5.95	8.60	6.80
			2.0	4.25	3.15	4.65	3.90	6.45	4.50
	0.20	3.878	0	**453.85**	**411.50**	**500.50**	**497.48**	**512.10**	**485.85**
			0.5	30.11	27.55	35.06	30.77	36.25	32.10
			1.0	10.05	8.50	10.87	9.48	11.95	9.95
			1.5	6.10	4.99	6.50	5.38	7.05	5.85
			2.0	4.45	3.76	4.60	4.01	4.95	4.20

4.2. Marginal performance of the geometric-Poisson EWMA chart

In practice, it is often not possible to know how the estimated mean compares to the true in-control mean. Therefore, it would also be useful to evaluate the marginal performance of a chart, chart, which considers the distribution of the estimated parameters to take into account the random variability introduced through parameter estimation. The marginal performance for the geometric-Poisson EWMA charts under different parameters values, sample sizes, smoothing constants, and shift magnitudes. The ARL and SDRL for the given chart based on the estimated parameters for in-control $\lambda_0 = 2.0$ and $\rho_0 = 0.20$ and 0.30 values are calculated for in-control and out-of-control situations and given in Table 5. The corresponding ARL for $\lambda_0 = 3$ and $\rho_0 = 0.20$ for different smoothing constant values are given in Figure 1(a–c), which are known as ARL curves. The RL characteristic for any other combination of the parameters could be similarly obtained.

Table 4. Conditional RL performance of the geometric-Poisson EWMA chart with $\hat{\lambda}_0 = 2$, $\hat{\rho} = 0.30$ and $ARL_0 \approx 500$

				25th (overestimation)		50th (nominal)		75th (underestimation)	
m	*w*	*A*	*δ*	ARL	SDRL	ARL	SDRL	ARL	SDRL
30	0.05	2.755	0	**172.45**	**156.80**	**496.79**	**489.79**	**2130.30**	**1851.02**
			0.5	21.50	18.19	27.79	21.95	34.52	28.22
			1.0	9.25	6.53	11.63	8.60	21.10	16.30
			1.5	6.35	4.45	7.34	5.03	11.45	8.65
			2.0	4.45	2.85	5.42	3.04	8.25	5.95
	0.10	3.377	0	**191.76**	**179.08**	**498.23**	**492.63**	**1870.50**	**1624.70**
			0.5	23.50	20.50	28.84	22.12	35.78	29.04
			1.0	9.10	7.30	10.75	6.17	20.50	15.50
			1.5	6.22	4.10	6.47	3.12	10.80	8.10
			2.0	4.15	2.79	1.96	1.18	7.50	5.05
	0.20	3.806	0	**208.14**	**191.50**	**499.56**	**494.15**	**1548.60**	**1380.08**
			0.5	28.16	23.95	36.86	32.75	39.45	34.70
			1.0	9.25	7.35	11.74	8.40	21.40	15.40
			1.5	6.10	3.95	6.34	3.74	10.40	7.50
			2.0	4.20	2.96	3.11	1.76	6.75	4.80
100	0.05	2.755	0	**228.55**	**211.99**	**496.79**	**489.79**	**1635.10**	**1490.99**
			0.5	22.32	17.95	27.79	17.95	32.30	26.90
			1.0	9.05	6.55	11.63	5.60	19.75	14.15
			1.5	6.25	4.05	7.34	3.03	10.50	8.35
			2.0	4.50	3.01	5.42	2.04	7.20	5.05
	0.10	3.377	0	**285.50**	**274.05**	**498.23**	**492.63**	**1205.35**	**1005.80**
			0.5	23.95	20.65	28.84	22.12	33.40	27.32
			1.0	9.15	7.15	10.75	6.17	19.11	13.70
			1.5	6.06	2.55	6.47	3.12	10.35	7.05
			2.0	1.05	0.98	1.96	1.18	6.55	5.35
	0.20	3.806	0	**312.10**	**299.88**	**499.56**	**494.15**	**1010.80**	**859.87**
			0.5	29.95	25.09	36.86	32.75	38.18	34.50
			1.0	10.35	7.45	11.74	8.40	17.80	13.05
			1.5	6.85	3.99	6.34	3.74	9.02	7.41
			2.0	4.55	2.98	3.11	1.76	6.15	4.50
1,000	0.05	2.755	0	**301.25**	**290.60**	**496.79**	**489.79**	**780.45**	**655.87**
			0.5	23.99	18.98	27.79	17.95	30.35	25.40
			1.0	10.76	7.88	11.63	5.60	16.90	10.50
			1.5	7.05	4.74	7.34	3.03	9.95	7.35
			2.0	4.86	3.72	5.42	2.04	7.35	5.12
	0.10	3.377	0	**370.33**	**355.75**	**498.23**	**492.63**	**625.50**	**570.90**
			0.5	26.15	21.56	28.84	22.12	32.50	24.80
			1.0	9.85	6.05	10.75	6.17	16.80	11.80
			1.5	6.05	3.01	6.47	3.12	8.25	5.85
			2.0	1.86	1.06	1.96	1.18	4.20	2.25
	0.20	3.806	0	**445.90**	**437.05**	**499.56**	**494.15**	**560.02**	**509.55**
			0.5	32.20	28.05	36.86	32.75	38.70	33.95
			1.0	10.35	7.85	11.74	8.40	13.75	10.65
			1.5	6.22	3.35	6.34	3.74	8.65	5.60
			2.0	2.96	1.56	3.11	1.76	5.10	2.10

Table 5. Marginal performance summaries of the geometric-Poisson EWMA chart for various sizes, shift magnitude, and parameters

λ_0 = 2.0			Smoothing constant (W)					
			0.05		0.10		0.20	
ρ	m	δ	ARL	SDRL	ARL	SDRL	ARL	SDRL
0.20	30	0	**9985.70**	**9865.30**	**1895.20**	**1650.30**	**565.20**	**532.65**
		0.5	1350.60	1125.30	1012.22	890.25	80.52	62.30
		1.0	25.33	14.26	21.15	12.52	13.50	9.60
		1.5	9.35	6.10	8.50	5.80	7.11	4.52
		2.0	6.10	4.25	5.30	3.90	4.62	3.05
	100	0	**1360.50**	**1280.56**	**1105.20**	**998.56**	**752.20**	**620.56**
		0.5	85.52	71.50	70.15	58.33	45.10	34.56
		1.0	19.45	13.52	17.52	11.23	11.90	8.90
		1.5	8.30	5.85	7.10	4.65	6.40	4.02
		2.0	5.65	4.05	5.25	3.70	4.62	3.25
	1000	0	**630.52**	**590.25**	**562.56**	**542.60**	**520.56**	**505.60**
		0.5	35.50	28.10	30.25	25.36	36.20	31.02
		1.0	16.12	9.40	12.50	8.90	11.20	8.67
		1.5	7.78	5.55	7.05	4.18	6.32	3.97
		2.0	5.05	3.79	4.79	3.58	4.44	3.01
0.30	30	0	**9735.40**	**9525.90**	**1795.70**	**1480.98**	**540.70**	**512.70**
		0.5	1205.45	1075.90	970.70	805.56	82.12	63.50
		1.0	26.80	15.40	22.25	13.12	14.25	9.94
		1.5	9.95	7.15	9.10	6.05	7.90	5.01
		2.0	6.70	4.60	5.70	4.01	4.93	3.70
	100	0	**1228.90**	**1041.55**	**935.70**	**798.06**	**732.50**	**590.56**
		0.5	82.05	67.52	71.75	60.33	44.70	35.76
		1.0	20.75	13.58	18.42	13.23	12.50	9.10
		1.5	9.05	5.95	7.95	4.96	6.92	4.22
		2.0	6.07	4.35	5.95	3.80	4.71	3.42
	1000	0	**615.80**	**565.95**	**543.65**	**542.60**	**516.98**	**503.05**
		0.5	36.10	29.05	31.67	26.22	38.07	32.50
		1.0	16.97	10.25	13.01	9.10	11.82	8.97
		1.5	8.01	5.98	7.80	4.85	6.80	4.01
		2.0	5.85	3.85	4.92	3.70	4.50	3.15

In Table 5 and Figure 1(a–b), the performance metrics are weighted averages over all the values that the estimation may yield for the in-control parameters. To study how large the sample size should be to perform essentially like the known parameter case, the values 30, 100, 1,000, and ∞ for n are evaluated. The infinite sample size ($n = \infty$) corresponds to the known parameters or the nominal case.

Comparing the values in Table 5 and Figure 1(a–b) with their nominal values, it is obvious that estimating control limits can cause both ARL and SDRL to be large or smaller than their desired values, when the process is working in-control. When a shift in the average number of defects occurs, the ARL performance is almost similar for various sample sizes. For a fixed sample size and dispersion parameter, the chart produces a large in-control ARL as the EWMA smoothing constant increases. A smaller sample size increases the in-control ARL. The choice of smoothing constant depends on

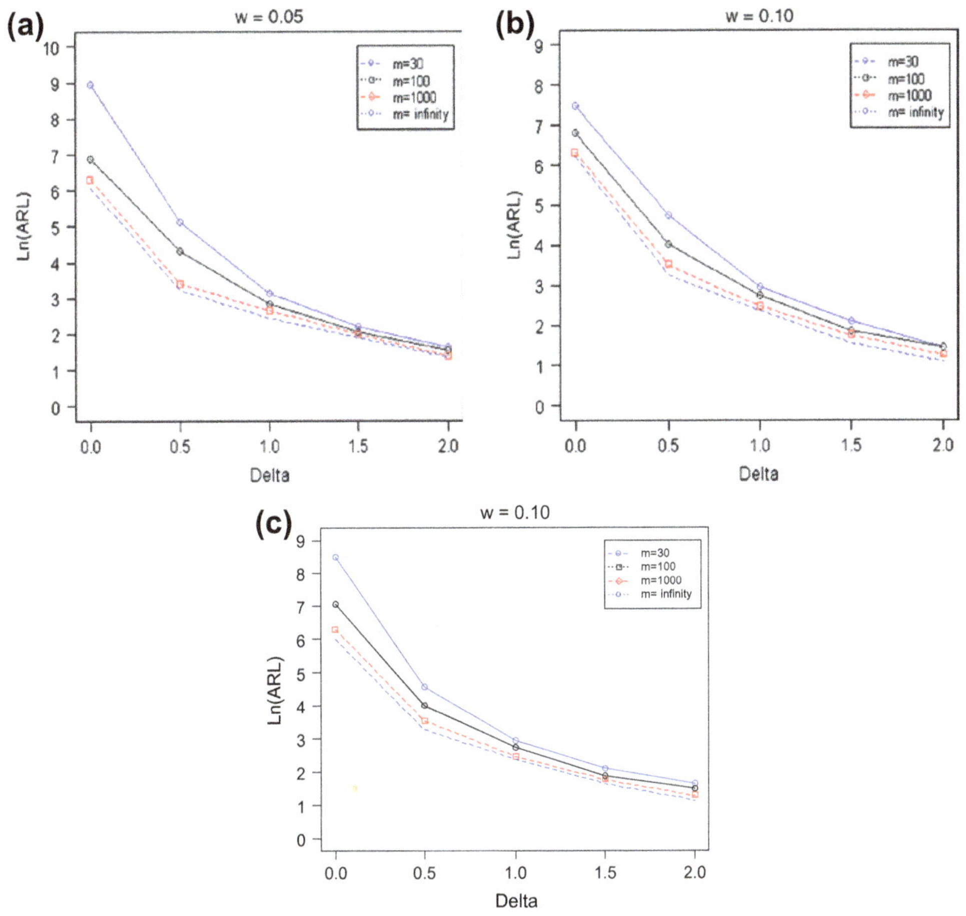

Figure 1. ARL comparison of the estimated control limits for $\lambda_0 = 3, \rho_0 = 0.20$ and (a) $w = 0.05$, (b) $w = 0.10$ and (c) $w = 0.20$ at various samples.

how fast one wants to detect a shift of given size in average non-conformities. Table 4 shows that larger sample size for estimating the parameters of the process would generally require getting fairly close in-control ARL as desired. Also, the larger value of *dispersion* parameter would be required to minimize false alarms. Similar behavior has been observed for other choices of parameters and smoothing constant.

5. Conclusion and discussion

The Poisson distribution is often used to model the count data in all fields. However, the Poisson distribution is not only underlying distribution for counting data. For production processes, the geometric-Poisson EWMA control chart, proposed based on geometric-Poisson compound distribution, is very useful to detect the process variation rapidly to reduce the lost cost. This chart could be used and should be used if small shifts from normal conditions are important to detect quickly.

In real application, actual values of process parameters for designing the given chart are often unknown. In this situation, a typical approach is to conduct a phase-I study, where a reference sample of m observations is obtained and then used for estimating these unknown parameters. However, the performance of the control chart may significantly be different than expected performance if the parameters are not well estimated. This article investigates the performance of the geometric-Poisson EWMA chart, when the process parameters are estimated based on m reference samples. The effect on the RL characteristics, such as ARL and SDRL, has been shown to be significant even with sample sizes as large as 1,000. Furthermore, for smaller EWMA smoothing constants, say 0.05, the chart with estimated parameters produces more false alarm rate, which results into large in-control ARL and SDRL than the chart with known parameters. This study suggest more than 1,000 sample size and smoothing constant greater than 0.05. However, this choice depends on the sensitivity of the chart with respect to detecting changes in wafer quality. The larger value of *dispersion* parameter is better to get the desired in-control ARL and SDRL. The results of the study are very useful for practioners and researchers to design a geometric-Poisson EWMA chart for detecting minor process variations in production processes and improving the process quality in phase-I sample.

Funding
The authors received no direct funding for this research.

Author details
Aamir Saghir[1,2]
E-mail: amirstat@yahoo.com; aamir.stat@must.edu.pk
Zhengyan Lin[2]
E-mail: zlin@zju.edu.cn
Ching-Wen Chen[3]
E-mail: chingwen@nkfust.edu.tw
[1] Department of Mathematics, Mirpur University of Science and Technology (MUST) Mirpur, Mirpur, Pakistan.
[2] Department of Mathematics, Zhejiang University, Hangzhou, P.R. China.
[3] National Kaohsiung First University of Science and Technology, Kaohsiung, Taiwan.

References

Brook, D., & Evans, D. A. (1972). An approach to the probability distribution of CUSUM run length. *Biometrika, 59*, 539–549. http://dx.doi.org/10.1093/biomet/59.3.539

Chakraborti, S., & Human, S. W. (2006). Parameter estimation and performance of the *p*-chart for attributes data. *IEEE Transactions on Reliability, 55*, 559–566. http://dx.doi.org/10.1109/TR.2006.879662

Chen, A., & Chen, Y. K. (2007). Design of EWMA and CUSUM control charts subject to random shift sizes and quality impacts. *IIE Transactions, 39*, 1127–1141. http://dx.doi.org/10.1080/07408170701315321

Chen, C. W. (2012). Using geometric Poisson exponentially weighted moving average control schemes in a compound Poisson production environment. *Computers and Industrial Engineering, 63*, 374–381. http://dx.doi.org/10.1016/j.cie.2012.04.009

Chen, C. W., Randolph, P. H., & Liou, T. S. (2005). Using CUSUM control schemes for monitoring quality levels in compound Poisson production environment: The geometric Poisson process. *Quality Engineering, 17*, 207–217. http://dx.doi.org/10.1081/QEN-200056448

Chiu, J. E., & Tsai, C. H. (2013). Properties and performance of one-sided cumulative count of conforming chart with parameter estimation in high-quality processes. *Journal of Applied Statistics, 40*, 2341–2353. http://dx.doi.org/10.1080/02664763.2013.811479

Jensen, W. A., Jones-Farmer, L. A., Champ, C. W., & Woodall, W. H. (2006). Effects of parameter estimation on control chart properties: A literature review. *Journal of Quality Technology, 38*, 349–364.

Lee, J., Wang, N., Xu, L., Schuh, A., & Woodall, W. H. (2013). The effect of parameter estimation on upper-sided Bernoulli cumulative sum charts. *Quality and Reliability Engineering International, 29*, 639–651. http://dx.doi.org/10.1002/qre.v29.5

Mahmoud, M. A., & Maravelakis, P. E. (2013). The performance of multivariate CUSUM control charts with estimated parameters. *Journal of Statistical Computation and Simulation, 83*, 721–738. http://dx.doi.org/10.1080/00949655.2011.633910

Montgomery, D. C. (2009). *Introduction to statistical quality control* (6th ed.). Hoboken, NJ: Wiley.

Ozsan, G., Testik, M. C., & Weiβ, C. H. (2010). Properties of the exponential EWMA chart with parameter estimation. *Quality and Reliability Engineering International, 26*, 555–569.

Quesenberry, C. P., & Geometric, Q. (1995). Charts for high quality processes. *Journal of Quality Technology, 27*, 304–315.

Saghir, A., & Lin, Z. (2013). The negative Binomial EWMA Chart with estimated control limits. *Quality and Reliability Engineering International*. doi:10.1002/qre.1574

Saleh, N. A., Mahmoud, M. A., & Abdel-Salam, A. S. G. (2013). The performance of the adaptive exponentially weighted moving average control chart with estimated parameters. *Quality and Reliability Engineering International, 29*, 595–606. http://dx.doi.org/10.1002/qre.v29.4

Shu, L., Tsung, F., & Tsui, K. L. (2004). Run-length performance of regression control chart with estimated parameters. *Journal of Quality Technology, 36*, 280–292.

Szarka, J. L., & Woodall, W. H. (2011). A review and perspective on surveillance of Bernoulli processes. *Quality and Reliability Engineering International, 27*, 735–752. http://dx.doi.org/10.1002/qre.v27.6

Tang, L. C., & Cheong, W. T. (2004). Cumulative conformance count chart with sequentially updated parameters. *IIE Transactions, 36*, 841–853. http://dx.doi.org/10.1080/07408170490473024

Testik, M. C. (2007). Conditional and marginal performance of the Poisson CUSUM control chart with parameter estimation. *International Journal of Production Research, 45*, 5621–5638. http://dx.doi.org/10.1080/00207540701325462

Testik, M. C., McCullough, D. B., & Borror, M. C. (2006). The effect of estimated parameters on Poisson EWMA control charts. *Quality Technology and Quantitative Management, 3*, 513–527.

Woodall, W. H., & Montgomery, D. C. (1999). Research issues and ideas in statistical process control. *Journal of Quality Technology, 31*, 376–386.

Yang, Z., Xie, M., Kuralmani, V., & Tusi, K. L. (2000). On the performance of geometric charts with estimated control limits. *Journal of Quality Technology, 34*, 448–458.

Zhang, M., Megahed, F. M., & Woodall, W. H. (2013). Exponential CUSUM charts with estimated control limits. *Quality and Reliability Engineering International, 30*, 275–286.

Zhang, M., Peng, Y., Schuh, A., Megahed, F. M., & Woodall, W. H. (2013). Geometric charts with estimated control limits. *Quality and Reliability Engineering International, 29*, 209–223. http://dx.doi.org/10.1002/qre.1304

12

Response surface designs using the generalized variance inflation factors

Diarmuid O'Driscoll[1] and Donald E. Ramirez[2*]

*Corresponding author: Donald E. Ramirez, Department of Mathematics, University of Virginia, Charlottesville, VA 22904, USA
E-mail: der@virginia.edu

Reviewing editor: Guohua Zou, Chinese Academy of Sciences, China

Abstract: We study response surface designs using the generalized variance inflation factors for subsets as an extension of the variance inflation factors.

Subjects: Mathematics & Statistics; Science; Statistical Computing; Statistical Theory & Methods; Statistics; Statistics & Probability

Keywords: variance inflation factors; generalized variance inflation factors; response surface designs

1. Introduction

We consider a linear regression $\mathbf{Y} = \mathbf{X}\beta + \varepsilon$ with $\mathbf{X}$ a full rank $n \times p$ matrix and $\mathcal{L}(\varepsilon) = N(\mathbf{0}, \sigma^2 \mathbf{I}_n)$. The variance inflation factor *VIF*, Belsley (1986), measures the penalty for adding one non-orthogonal additional explanatory variable to a linear regression model, and they can be computed as a ratio of determinants. The extension of *VIF* to a measure of the penalty for adding a subset of variables to a model is the *generalized variance inflation factor GVIF* of Fox and Monette (1992), which will be used to study response surface designs, in particular, as the penalty for adding the quadratic terms to the model.

2. Variance inflation factors

For our linear model $\mathbf{Y} = \mathbf{X}\beta + \varepsilon$, let $\mathbf{D}_X$ be the diagonal matrix with entries on the diagonal $\mathbf{D}_X[i, i] = (\mathbf{X}'\mathbf{X})_{i,i}^{-1/2}$. When the design has been standardized $\mathbf{X} \to \mathbf{X}D_X$, the *VIF*s are the diagonal entries of the inverse of $\mathbf{S}_X = \mathbf{D}_X(X'\mathbf{X})D_X$. That is, the *VIF*s are the ratios of the actual variances for

ABOUT THE AUTHORS

Diarmuid O'Driscoll is the head of the Mathematics and Computer Studies Department at Mary Immaculate College, Limerick. He was awarded a Travelling Studentship for his MSc at University College Cork in 1977. He has taught at University College Cork, Cork Institute of Technology, University of Virginia, and Frostburg State University. His research interests are in mathematical education, errors in variables regression, and design criteria. In 2014, he was awarded a Teaching Heroes Award by the National Forum for the Enhancement of Teaching and Learning (Ireland).

Donald E. Ramirez is a full professor in the Department of Mathematics at the University of Virginia in Charlottesville, Virginia. He received his PhD in Mathematics from Tulane University in New Orleans, Louisiana. His research is in harmonic analysis and mathematical statistics. His current research interests are in statistical outliers and ridge regression.

PUBLIC INTEREST STATEMENT

Response surface designs are a mainstay in applied statistics. The variance inflation factors *VIF* are a measure of collinearity for a single variable in a linear regression model. The generalization to subsets of variables is the generalized variance inflation factor *GVIF*. This research introduces *GVIF* as a penalty measure for extending a linear response model to a response surface with the included quadratic terms. The methodology is demonstrated with case studies, and, in particular, it is shown that using *GVIF*, the *H*310 design can be improved for the standard global optimality criteria of *A*, *D*, and *E*.

the explanatory variables to the "ideal" variances had the columns of $\mathbf{X}$ been orthogonal. Note that we follow Stewart (1987) and do not necessarily center the explanatory variables.

For our linear model $\mathbf{Y} = \mathbf{X}\beta + \varepsilon$, view $\mathbf{X} = [\mathbf{X}_{[p]}, \mathbf{x}_p]$ with $\mathbf{x}_p$ the pth column of $\mathbf{X}$ and $\mathbf{X}_{[p]}$ the matrix formed by the remaining columns. The variance inflation factor VIF_p measures the effect of adding column $\mathbf{x}_p$ to $\mathbf{X}_{[p]}$. For notational convenience, we demonstrate VIF_p with the last column p. An ideal column would be orthogonal to the previous columns with the entries in the off-diagonal elements of the pth row and p^{th} column of $\mathbf{X}'\mathbf{X}$ all zeros. Denote by $\mathbf{M}_p$ the idealized moment matrix

$$\mathbf{M}_p = \begin{bmatrix} \mathbf{X}'_{[p]}\mathbf{X}_{[p]} & \mathbf{0}_{p-1} \\ \mathbf{0}'_{p-1} & \mathbf{x}'_p\mathbf{x}_p \end{bmatrix}.$$

The VIFs are the diagonal entries of $\mathbf{S}_X^{-1} = \mathbf{D}_X^{-1}(\mathbf{X}'\mathbf{X})^{-1}\mathbf{D}_X^{-1}$. It remains to note that the inverse, $\mathbf{S}_X^{-1}$, can be computed using cofactors $\mathbf{C}_{i,j}$. In particular,

$$\begin{aligned} VIF_p &= [\mathbf{S}_X^{-1}]_{p,p} = [\mathbf{D}_X^{-1}(\mathbf{X}'\mathbf{X})^{-1}\mathbf{D}_X^{-1}]_{p,p} \\ &= (\mathbf{x}'_p\mathbf{x}_p)^{1/2}\frac{\det(\mathbf{C}_{p,p})}{\det(\mathbf{X}'\mathbf{X})}(\mathbf{x}'_p\mathbf{x}_p)^{1/2} = \frac{\det(\mathbf{M}_p)}{\det(\mathbf{X}'\mathbf{X})} \end{aligned} \quad (1)$$

the ratio of the determinant of the idealized moment matrix $\mathbf{M}_p$ to the determinant of the moment matrix $\mathbf{X}'\mathbf{X}$. This definition extends naturally to subsets and is discussed in the next section.

For an alternate view of the how collinearities in the explanatory variables inflate the model variances of the regression coefficients when compared to a fictitious orthogonal reference design, consider the formula for the model variance

$$Var_M(\hat{\beta}_j) = \frac{\sigma^2}{\sum_{i=1}^{n}(x_{ij} - \bar{x}_j)^2}\frac{1}{1 - R_j^2}$$

where R_j^2 is the square of the multiple correlation from the regression of the jth column of $\mathbf{X} = [x_{ij}]$ on the remaining columns as in Liao and Valliant (2012). The first term $\sigma^2 / \sum(x_{ij} - \bar{x}_j)^2$ is the model variance for $\hat{\beta}_j$ had the jth explanatory variable been orthogonal to the remaining variables. The second term $1/(1 - R_j^2)$ is a standard definition of the jth VIF as in Thiel (1971).

3. Generalized variance inflation factors

In this section, we introduce the $GVIF$s as an extension of the classical variance inflation factors VIF from Equation 1. For the linear model $\mathbf{Y} = \mathbf{X}\beta + \varepsilon$, view $\mathbf{X} = [\mathbf{X}_1, \mathbf{X}_2]$ partitioned with $\mathbf{X}_1$ of dimension $n \times r$ usually consisting of the lower order terms and $\mathbf{X}_2$ of dimension $n \times s$ usually consisting of the higher order terms. The idealized moment matrix for the (r, s) partitioning of $\mathbf{X}$ is

$$\mathbf{M}_{(r,s)} = \begin{bmatrix} \mathbf{X}'_1\mathbf{X}_1 & \mathbf{0}_{r\times s} \\ \mathbf{0}_{s\times r} & \mathbf{X}'_2\mathbf{X}_2 \end{bmatrix}.$$

Following Equation 1, to measure the effect of adding $\mathbf{X}_2$ to the design $\mathbf{X}_1$, that is for $\mathbf{X}_2|\mathbf{X}_1$, we define the *generalized variance inflation factor* as

$$GVIF(\mathbf{X}_2|\mathbf{X}_1) = \frac{\det(\mathbf{M}_{(r,s)})}{\det(\mathbf{X}'\mathbf{X})} = \frac{\det(\mathbf{X}'_1\mathbf{X}_1)\det(\mathbf{X}'_2\mathbf{X}_2)}{\det(\mathbf{X}'\mathbf{X})} \quad (2)$$

as in Equation 10 of Fox and Monette (1992), who compared the sizes of the joint confidence regions for β for partitioned designs and noted when $\mathbf{X} = [\mathbf{X}_{[p]}, \mathbf{x}_p]$ that $GVIF[\mathbf{x}_p|\mathbf{X}_{[p]}] = VIF_p$. Equation 2 is in the spirit of the efficiency comparisons in linear inferences introduced in Theorems 4 and 5 of Jensen and Ramirez (1993). A similar measure of collinearity is mentioned in Note 2 in Wichers (1975),

Theorem 1 of Berk (1977), and Garcia, Garcia, and Soto (2011). For the simple linear regression model with $p = 2$, Equation 2 gives $VIF = \frac{1}{1-\rho^2}$ with ρ the correlation coefficient as required. Fox and Monette (1992) suggested that $\mathbf{X}_1$ contains the variables which are of "simultaneous interest," while $\mathbf{X}_2$ contains additional variables selected by the investigator. We will set $\mathbf{X}_1$ for the constant and main effects and set $\mathbf{X}_2$ the (optional) quadratic terms with values from $\mathbf{X}_1$.

Willan and Watts (1978) measured the effect of collinearity using the ratio of the volume of the actual joint confidence region for $\widehat{\boldsymbol{\beta}}$ to the volume of the joint confidence region in the fictitious orthogonal reference design. Their ratio is in the spirit of $GVIF$ as $\det(\mathbf{X}'\mathbf{X})$ is inversely proportional to the square of the volume of the joint confidence region for $\widehat{\boldsymbol{\beta}}$. They also introduced a measure of relative predictability and they note: "The existence of near linear relations in the independent variables of the actual data reduces the overall predictive efficiency by this factor." For a simple case study, consider the simple linear regression model with $n = 4$, $\mathbf{x}_1 = [-2, -1, 1, 2]'$, and $\mathbf{y} = [4, 1, 1, 4]'$. The 95% prediction interval for $x_1 = 0$ is 2.5 ± 10.20. If the model also includes $\mathbf{x}_2 = [-2.001, -1.001, 1.001, 2.001]'$, then the 95% prediction interval for $(x_1, x_2) = (0, 0)$ is 2.5 ± 46.02 demonstrating the loss of predictive efficiency due to the collinearity introduced by $\mathbf{x}_2$.

For the (r, s) partition of $\mathbf{X} = [\mathbf{X}_1, \mathbf{X}_2]$ with $\mathbf{X}_1$ of dimension $n \times r$ and $\mathbf{X}_2$ of dimension $n \times s$, set

$$\mathbf{D}_{(r,s)} = \begin{bmatrix} (\mathbf{X}_1'\mathbf{X}_1)^{-1/2} & \mathbf{0} \\ \mathbf{0} & (\mathbf{X}_2'\mathbf{X}_2)^{-1/2} \end{bmatrix},$$

and denote the *canonical moment matrix* as

$$\begin{aligned} \mathbf{R} &= \mathbf{D}_{(r,s)}(\mathbf{X}'\mathbf{X})D_{(r,s)} \\ &= \begin{bmatrix} \mathbf{I}_{r\times r} & (\mathbf{X}_1'\mathbf{X}_1)^{-1/2}(\mathbf{X}_1'\mathbf{X}_2)(\mathbf{X}_2'\mathbf{X}_2)^{-1/2} \\ (\mathbf{X}_2'\mathbf{X}_2)^{-1/2}(\mathbf{X}_2'\mathbf{X}_1)(\mathbf{X}_1'\mathbf{X}_1)^{-1/2} & \mathbf{I}_{s\times s} \end{bmatrix}; \end{aligned} \quad (3)$$

with determinant

$$\det(\mathbf{R}) = \frac{\det(\mathbf{X}'\mathbf{X})}{\det(\mathbf{X}_1'\mathbf{X}_1)\det(\mathbf{X}_2'\mathbf{X}_2)} = \frac{1}{GVIF(\mathbf{X}_2|\mathbf{X}_1)};$$

equivalently,

$$\det(\mathbf{R}) = \det(\mathbf{I}_{r\times r} - \mathbf{B}_{r\times s}\mathbf{B}_{s\times r}') = \det(\mathbf{I}_{s\times s} - \mathbf{B}_{s\times r}'\mathbf{B}_{r\times s})$$

where $\mathbf{B}_{r\times s} = (\mathbf{X}_1'\mathbf{X}_1)^{-1/2}(\mathbf{X}_1'\mathbf{X}_2)(\mathbf{X}_2'\mathbf{X}_2)^{-1/2}$.

In the case $\{r = p - 1, s = 1\}$, $\mathbf{X}_2 = \mathbf{x}_p$ is a $n \times 1$ vector and the partitioned design $\mathbf{X} = [\mathbf{X}_1, \mathbf{x}_p]$ has $\det(\mathbf{R}) = 1 - [\mathbf{x}_p'\mathbf{X}_1(\mathbf{X}_1'\mathbf{X}_1)^{-1}\mathbf{X}_1'\mathbf{x}_p]/\left(\mathbf{x}_p'\mathbf{x}_p\right)$. From standard facts for the inverse of a partitioned matrix, for example, Myers (1990, p. 459), $VIF_p = [\mathbf{R}^{-1}]_{p,p} = [\mathbf{D}_{(p-1,1)}^{-1}(\mathbf{X}'\mathbf{X})^{-1}\mathbf{D}_{(p-1,1)}^{-1}]_{p,p}$ can be computed directly as

$$\begin{aligned} \left(\mathbf{x}_p'\mathbf{x}_p\right)^{1/2}(\mathbf{X}'\mathbf{X})_{p,p}^{-1}\left(\mathbf{x}_p'\mathbf{x}_p\right)^{1/2} &= \frac{\mathbf{x}_p'\mathbf{x}_p}{\mathbf{x}_p'\mathbf{x}_p - \mathbf{x}_p'\mathbf{X}_1(\mathbf{X}_1'\mathbf{X}_1)^{-1}\mathbf{X}_1'\mathbf{x}_p} \\ &= \frac{1}{1 - [\mathbf{x}_p'\mathbf{X}_1(\mathbf{X}_1'\mathbf{X}_1)^{-1}\mathbf{X}_1'\mathbf{x}_p]/\left(\mathbf{x}_p'\mathbf{x}_p\right)} \\ &= \frac{1}{\det(\mathbf{R})} = GVIF(\mathbf{X}_2|\mathbf{X}_1). \end{aligned}$$

We study the eigenvalue structure of $\mathbf{M}_{(r,s)}$ in Appendix 1. Let $\{\lambda_1 \geq \lambda_2 \geq \ldots \geq \lambda_{\min(r,s)} \geq 0\}$ be the non-negative singular values of $(\mathbf{X}_1'\mathbf{X}_1)^{-1/2}(\mathbf{X}_1'\mathbf{X}_2)(\mathbf{X}_2'\mathbf{X}_2)^{-1/2}$. It is shown in Appendix 1 that an alternative formulation for *GVIF* is

$$GVIF(\mathbf{X}_2|\mathbf{X}_1) = \prod_{i=1}^{\min(r,s)} (1 - \lambda_i^2)^{-1}. \tag{4}$$

4. Quadratic model with $p = 3$

For the partitioning $\mathbf{X} = [\mathbf{X}_r|\mathbf{X}_s]$, the canonical moment matrix, Equation 3, has the identity matrices $\mathbf{I}_r, \mathbf{I}_s$ down the diagonal and off-diagonal array $(\mathbf{X}_1'\mathbf{X}_1)^{-1/2}\mathbf{X}_1'\mathbf{X}_2(\mathbf{X}_2'\mathbf{X}_2)^{-1/2}$. For the quadratic model $y = \beta_0 + \beta_1 x + \beta_2 x^2$ and partitioning $\mathbf{X} = [\mathbf{1}, \mathbf{x}|\mathbf{x}^2]$, we have

$$\mathbf{R} = \begin{bmatrix} 1 & 0 & \rho_1 \\ 0 & 1 & \rho_2 \\ \rho_1 & \rho_2 & 1 \end{bmatrix}.$$

From Equation 4, $GVIF(\mathbf{x}^2|\mathbf{1}, x) = (1 - \lambda^2)^{-1}$ where $\lambda = \sqrt{\rho_1^2 + \rho_2^2}$ is the unique positive singular value of $[\rho_1, \rho_2]'$. Denote

$$\gamma_X^2 = \rho_1^2 + \rho_2^2$$

as the *canonical index* with $GVIF(\mathbf{x}^2|\mathbf{1}, \mathbf{x}) = \frac{1}{1-\gamma_X^2} = \frac{1}{\det(\mathbf{R})}$. Surprisingly, many higher order designs also have the off-diagonal entry of the canonical moment matrix with a unique positive singular value with $GVIF(\mathbf{X}_2|\mathbf{X}_1) = \frac{1}{1-\gamma_X^2}$ with the collinearity between the lower order terms and the upper order terms as a function of the canonical index γ_X^2.

5. Central composite and factorial designs for quadratic models ($p = 6$)

In this section, we compare the central composite design (CCD) $\mathbf{X}$ of Box and Wilson (1951) and the factorial design $\mathbf{Z}$. The design points are shown in Table B1 of Appendix 2. Both designs are 9×6 and use the quadratic response model

$$y = \beta_0 + \beta_1 x_1 + \beta_2 x_2 + \beta_{11} x_1^2 + \beta_{22} x_2^2 + \beta_{12} x_1 x_2 + \varepsilon.$$

The CCD traditionally uses the value $a = \sqrt{2}$ in four entries, while the factorial design uses the value $a = 1$. To study the difference in the designs with these different values, we computed the *GVIF* to compare the "orthogonality" between the lower order terms $\mathbf{X}_1$ of dimension 9×3 and the higher order quadratic terms $\mathbf{X}_2$ of dimension 9×3. The off-diagonal $\mathbf{B}_{3\times3}$ entry of $\mathbf{R}$ from Equation 3 in Section 3 has the form

$$\mathbf{B}_{3\times3} = \begin{bmatrix} \rho_1 & \rho_2 & 0 \\ 0 & 0 & 0 \\ 0 & 0 & 0 \end{bmatrix}$$

Table 1. CCD with parameter a, canonical index γ_X^2, and *GVIF*

a	γ_X^2	$GVIF(\mathbf{X}_2|\mathbf{X}_1)$
0.000	0.4444	1.800
1.000	0.8000	5.000
1.500	0.8858	8.758
$\sqrt{2}$	0.8889	9.000
1.750	0.8514	6.729
∞	0.4444	1.800

with $\rho_1 = \rho_2 = \frac{2}{3}\frac{2+a^2}{\sqrt{8+2a^4}}$, canonical index $\gamma_X^2 = \rho_1^2 + \rho_2^2$ and $GVIF(\mathbf{X}_2|\mathbf{X}_1) = \frac{1}{1-\gamma_X^2}$ as in the quadratic model case with $p = 3$ shown in the Section 4. For Table 1, if $a = 1$, then $\rho_1 = \rho_2 = 2/\sqrt{10}$, $\gamma_X^2 = 8/10$, and $GVIF(\mathbf{X}_2|\mathbf{X}_1) = 5$. Surprisingly, the classical choice of $a = \sqrt{2}$ gives the largest value for $GVIF(\mathbf{X}_2|\mathbf{X}_1)$, that is the worst value, indicating the greatest collinearity between the lower and higher order terms, as noted in O'Driscoll and Ramirez (in press).

6. Larger designs ($p = 10$)

We consider the quadratic response surface designs for

$$y = \beta_0 + \beta_1 x_1 + \beta_2 x_2 + \beta_3 x_3 + \\ \beta_{11}x_1^2 + \beta_{22}x_2^2 + \beta_{33}x_3^2 + \beta_{12}x_1x_2 + \beta_{13}x_1x_3 + \beta_{23}x_2x_3 + \varepsilon \quad (5)$$

with n responses and with $\mathbf{X}$ partitioned into $[\mathbf{X}_1|X_2]$ with $\mathbf{X}_1$ the four lower order terms ($r = 4$) and $\mathbf{X}_2$ the six quadratic terms ($s = 6$). Four popular designs are given in Appendix 2. They are the hybrid designs (*H*310 and *H*311*B*) of Roquemore (1976) Tables B2 and B3, the Box and Behnken (1960) (*BBD*) design Table B4, and the CCD of Box and Wilson (1951) Table B5.

For each design, we compute the 10×10 canonical moment matrix. It is striking that, for all these designs, the off-diagonal 4×6 array in **R** has only one non-zero singular value with its square the canonical index γ_X^2. It follows that $GVIF(\mathbf{X}_2|\mathbf{X}_1) = \frac{1}{1-\gamma_X^2}$.

Table 2 reports that the design *H*310 is the most conditioned with respect to the *GVIF* with the least amount of collinearity between the lower and higher order terms.

7. More complicated designs with ordered singular values

Let $\mathbf{X}$ be the minimal design of Box and Draper (1974) BDD with $n = 11$ from Table B6, and let $\mathbf{Z}$ be the small composite design of Hartley (1959) *SCD* with $n = 11$ from Table B7 for the quadratic response surface model ($r = 4$ and $s = 6$) as in Equation (5). Let $\boldsymbol{\alpha} = \{\alpha_1 \geq \ldots \geq \alpha_r \geq 0\}$ and $\beta = \{\beta_1 \geq \ldots \geq \beta_r \geq 0\}$ be the non-negative singular values of the off-diagonal array for $\mathbf{R}_X$ and $\mathbf{R}_Z$, respectively. As $\alpha_i \leq \beta_i (1 \leq i \leq r)$ (Table 3), it follows that $GVIF(\mathbf{X}_2|\mathbf{X}_1) \leq GVIF(\mathbf{Z}_2|\mathbf{Z}_1)$ showing less collinearity between the lower and higher order terms for the BDD design.

Table 2. Hybrid designs *H*310, *H*311*B*, Box and Behnken *BBD*, *and* CCD

Design	*n*	*r*	γ_X^2	$GVIF(X_2|X_1)$
*H*310	11	4	0.8199	5.553
*H*311*B*	11	4	0.9091	11.00
BBD	13	4	0.9231	13.00
CCD	15	4	0.9333	15.00

Table 3. Singular values for off-diagonal array of R for BDD and SCD with *GVIF*

	BDD		SCD
	0.9017		0.9535
$\alpha =$	0.3424	$\beta =$	0.6325
	0.3424		0.6325
	0.1877		0.6325
GVIF =	7.114		50.93

8. An improved H310 design

When the diagonal matrix $\Lambda_{r \times s}$ in Equation 6 in Appendix 1 has only one non-zero entry, we have denoted the square of this value the *canonical index*. We extend this definition to the case when $(\mathbf{X}_1'\mathbf{X}_1)^{-1/2}(\mathbf{X}_1'\mathbf{X}_2)(\mathbf{X}_2'\mathbf{X}_2)^{-1/2}$ has multiple positive singular values. The Frobenious norm for a rectangular matrix $\mathbf{A}_{r \times s}$ is defined by $||\mathbf{A}||_F^2 = \sum_{i=1}^{r}\sum_{j=1}^{s} a_{ij}^2 = trace(\mathbf{A}'\mathbf{A})$. For a design matrix $\mathbf{X}$, we extend the definition of the *canonical index* with $\gamma_{\mathbf{X}}^2 = ||\Lambda_{r \times s}||_F^2$. Alternatively, $\gamma_{\mathbf{X}}^2 = trace((\mathbf{X}_2'\mathbf{X}_2)^{-1}(\mathbf{X}_2'\mathbf{X}_1)(\mathbf{X}_1'\mathbf{X}_1)^{-1}(\mathbf{X}_1'\mathbf{X}_2))$ as in Equation 7.

We examine, in detail, the *H*310 design matrix $\mathbf{X}_{11 \times 10}$, Table B2 in Appendix 2, with our attention to the value of -0.1360 in row 2 for $\mathbf{x}_3$. In succession, we will replace the values $\{1.1736, 0.6386, -0.9273, 1.0000, 1.2906, -0.1360\}$ by a free parameter and use $\gamma_{\mathbf{X}}^2$ to determine an optimal value. For example, replacing the four entries which are 1.1736 with c_1, we calculate the minimum value for $\gamma_{\mathbf{X}}^2 = 0.8199$ with $c_1 = 1.1768$ denoted c_{min} in Table 4. These values are within the four digit accuracy of the data. We performed a similar calculation with c_2 using the four entries which are 0.6386; with c_3 with the four entries which are -0.9273; with c_4 with the eight entries which are 1; and with c_5 with the single entry 1.2906. The original design has $\gamma_{\mathbf{X}}^2 = 0.8199$. The entries in the *H*310 design are given to four significant digits. With this precision, the original design is nearly optimal with respect to the canonical index $\gamma_{\mathbf{X}}^2$ for the first five entries in Table 4. The sixth entry of $c_6 = -0.1360$ was not optimal with $\gamma_{\mathbf{X}}^2 = 0.8181$ with $c_{min} = -0.01264$, a magnitude value smaller.

Denote the "improved" *H*310 design as the *H*310 design with the value of $c_6 = -0.01264$. The "improved" *H*310 also has a unique positive singular value for the off-diagonal of $\mathbf{R}$ with its square the canonical index $\gamma_{\mathbf{X}}^2$. All of the standard design criteria favor the "improved" *H*310 design over the *H*310 design, which was originally constructed based on the rotatability criterion to maintain equal variances for predicted responses for points that have the same distance from the design center. As usual $A(\mathbf{X}) = tr((\mathbf{X}'\mathbf{X})^{-1})$, $D(\mathbf{X}) = \det((\mathbf{X}'\mathbf{X})^{-1})$, and $E(\mathbf{X}) = \max\{\text{eigenvalues of } (\mathbf{X}'\mathbf{X})^{-1}\}$. The small relative changes Δ in the design criteria are shown in Table 5 in Column 4.

The abnormality of the second row in *H*310 has been noted in Jensen (1998) who showed that the design is least sensitive to the second row of $\mathbf{X}$, the row containing the value $c_6 = -0.1360$.

Table 4. Optimal values c_{min} for $\gamma_{\mathbf{X}}^2$

Design entries						c_{min}	$\gamma_{\mathbf{X}}^2$
c_1	0.6386	-0.9273	1	1.2906	−0.1360	1.1768	0.8199
1.1736	c_2	-0.9273	1	1.2906	−0.1360	0.6356	0.8199
1.1736	0.6386	c_3	1	1.2906	−0.1360	−0.9303	0.8199
1.1736	0.6386	−0.9273	c_4	1.2906	−0.1360	0.9975	0.8199
1.1736	0.6386	−0.9273	1	c_5	−0.1360	1.2880	0.8199
1.1736	0.6386	−0.9273	1	1.2906	c_6	−0.0126	0.8181

Table 5. Design criteria for the "improved" *H*310 with Δ the relative change

	$c_6 = -0.1360$	$c_6 = -0.01264$	Δ
A	2.697	2.688	0.33%
D	0.1552×10^{-6}	0.1542×10^{-6}	0.64%
E	0.8611	0.8512	1.15%
$\gamma_{\mathbf{X}}^2$	0.8199	0.8181	0.22%

9. Conclusions

The *VIF* measure the penalty for adding a non-orthogonal variable to a linear regression. The *VIF* can be computed as a ratio of determinant as in Equation 1. A similar ratio criterion was studied by Fox and Monette (1992) to measure the effect of adding a subset of new variables to a design and they dubbed it the generalized variance inflation factor *GVIF*, Equation 2. We have noted the relationship between *GVIF* and the singular values of the off-diagonal array in the canonical moment matrix and have used *GFIV* to study standard quadratic response designs. The *H*310 design of Roquemorer (1976) was shown not to be optimal with respect to *GFIV* and an "improved" *H*310 design was introduced which was favored over *H*310 using the standard design criteria *A*, *D*, and *E*.

Funding
The authors received no direct funding for this research.

Author details
Diarmuid O'Driscoll[1]
E-mail: diarmuid.odriscoll@mic.ul.ie
Donald E. Ramirez[2]
E-mail: der@virginia.edu
[1] Department of Mathematics and Computer Studies, Mary Immaculate College, Limerick, Ireland.
[2] Department of Mathematics, University of Virginia, Charlottesville, VA, 22904, USA.

References

Belsley, D. A. (1986). Centering, the constant, first-differencing, and assessing conditioning. In E. Kuh & D. A. Belsley (Eds.), *Model reliability* (pp. 117–153). Cambridge: MIT Press.

Berk, K. (1977). Tolerance and condition in regression computations. *Journal of the American Statistical Association, 72*, 863–866.

Box, G. E. P., & Behnken, D. W. (1960). Some new three-level designs for the study of quantitative variables. *Technometrics, 2*, 455–475.

Box, M. J., & Draper, N. R. (1974). On minimum-point second order design. *Technometrics, 16*, 613–616.

Box, G. E. P., & Wilson, K. B. (1951). On the experimental attainment of optimum conditions. *Journal of the Royal Statistical Society, Series B, 13*, 1–45.

Eaton, M. L. (1983). *Multivariate statistics*. New York, NY: Wiley.

Fox, J., & Monette, G. (1992). Generalized collinearity diagnostics. *Journal of the American Statistical Association, 87*, 178–183.

Garcia, C. B., Garcia, J., & Soto, J. (2011). The raise method: An alternative procedure to estimate the parameters in presence of collinearity. *Quality and Quantity, 45*, 403–423.

Hartley, H. O. (1959). Smallest composite design for quadratic response surfaces. *Biometrics, 15*, 611–624.

Jensen, D. R. (1998). Principal predictors and efficiency in small second-order designs. *Biometrical Journal, 40*, 183–203.

Jensen, D. R., & Ramirez, D. E. (1993). Efficiency comparisons in linear inference. *Journal of Statistical Planning and Inference, 37*, 51–68.

Liao, D., & Valliant, R. (2012). Variance inflation in the analysis of complex survey data. *Survey Methodology, 38*, 53–62.

Myers, R. (1990). *Classical and modern regression with applications* (2nd ed.). Boston, MA: PWS-Kent.

O'Driscoll, D., & Ramirez, D. E. (in press). *Revisiting some design criteria* (under review).

Roquemorer, K. G. (1976). Hybrid designs for quadratic response surfaces. *Technometrics, 18*, 419–423.

Stewart, G. W. (1987). Collinearity and least squares regression. *Statistical Science, 2*, 68–84.

Thiel, H. (1971). *Principles of econometrics*. New York, NY: Wiley.

Wichers, R. (1975). The detection of multicollinearity: A comment. *Review of Economics and Statistics, 57*, 366–368.

Willan, A. R., & Watts, D. G. (1978). Meaningful multicollinearity measures. *Technometrics, 20*, 407–412.

Appendix 1

We study the eigenvalue structure of $\mathbf{M}_{(r,s)}$. Let $\{\lambda_1 \geq \lambda_2 \geq \ldots \geq \lambda_{\min(r,s)} \geq 0\}$ be the non-negative singular values of $(\mathbf{X}_1'\mathbf{X}_1)^{-1/2}(\mathbf{X}_1'\mathbf{X}_2)(\mathbf{X}_2'\mathbf{X}_2)^{-1/2}$.

As with the canonical correlation coefficients Eaton (1983), write the off-diagonal rectangular array $\mathbf{B}_{r\times s}$ of $\mathbf{R}$ as $\mathbf{P}\Lambda Q'$with $\mathbf{P}$ and $\mathbf{Q}$ orthogonal matrices and $\Lambda_{r\times s}$ the rectangular diagonal matrix with the non-negative singular values down the diagonal. Set

$$\mathbf{L} = \begin{bmatrix} \mathbf{P}_{r\times r} & \mathbf{0}_{r\times s} \\ \mathbf{0}_{s\times r} & \mathbf{Q}_{s\times s} \end{bmatrix}.$$

For notational convenience, we assume $r \leq s$. The matrix $\mathbf{L}$ is orthogonal and transforms $\mathbf{R} \to \mathbf{L}'\mathbf{R}\mathbf{L}$ into diagonal matrices:

$$\begin{bmatrix} \mathbf{I}_r & \Lambda_{r\times s} \\ \Lambda'_{s\times r} & \mathbf{I}_s \end{bmatrix} = \begin{bmatrix} \mathbf{I}_r & [\mathbf{SV}_{r\times r}|0_{r\times(s-r)}] \\ [\mathbf{SV}_{r\times r}|0_{r\times(s-r)}]' & \mathbf{I}_s \end{bmatrix} \tag{A1}$$

with $\Lambda_{rxs} = [SV_{rxr}|0_{rx(s-r)}]$ where SV_{rxr} is the diagonal matrix of the non-negative singular values. Since $\mathbf{L}$ is orthogonal, this transformation has not changed the eigenvalues. To compute the determinant of $\mathbf{R}$, convert the matrix in Equation 6 into an upper diagonal matrix by Gauss Elimination on Λ'_{sxr}. This changes r of the 1's on the diagonal in rows $r+1$ to $r+r$ into $1-\lambda_i^2$, and thus $\det(\mathbf{R}) = \prod_{i=1}^{\min(r,s)}(1-\lambda_i^2)$ with

$$GVIF(\mathbf{X}_2|\mathbf{X}_1) = \prod_{i=1}^{\min(r,s)} \frac{1}{1-\lambda_i^2}.$$

The singular values of $\mathbf{R}_{12} = (\mathbf{X}_1'\mathbf{X}_1)^{-1/2}(\mathbf{X}_1'\mathbf{X}_2)(\mathbf{X}_2'\mathbf{X}_2)^{-1/2}$ are the non-negative square roots of the eigenvalues of $\Lambda'\Lambda$ denoted by

$$\begin{aligned} eigvals(\Lambda'\Lambda) &= eigvals((\mathbf{Q}'\mathbf{R}_{12}'\mathbf{P})(P'\mathbf{R}_{12}\mathbf{Q}))) \\ &= eigvals((\mathbf{X}_1'\mathbf{X}_1)^{-1})(\mathbf{X}_1'\mathbf{X}_2)(\mathbf{X}_2'\mathbf{X}_2)^{-1}(\mathbf{X}_2'\mathbf{X}_1)). \end{aligned} \tag{A2}$$

If the trace of the inverse of the matrix in Equation 6 is required, then we note that

$$\begin{bmatrix} \mathbf{I}_r & \Lambda_{rxs} \\ \Lambda'_{sxr} & \mathbf{I}_s \end{bmatrix}^{-1} = \begin{bmatrix} (\mathbf{I}_r - \Lambda_{rxs}\Lambda'_{sxr})^{-1} & -\Lambda_{rxs}(I_s - \Lambda'_{sxr}\Lambda_{rxs})^{-1} \\ -\Lambda'_{sxr}(I_r - \Lambda_{rxs}\Lambda'_{sxr})^{-1} & (\mathbf{I}_s - \Lambda'_{sxr}\Lambda_{rxs})^{-1} \end{bmatrix}$$

with trace given by $tr((\mathbf{L}'\mathbf{R}L)^{-1}) = |r-s| + 2\sum_{i=1}^{\min(r,s)} \frac{1}{1-\lambda_i^2}$.

Appendix 2

Table B1. The lower order matrix for the CCD with center run with $a = \sqrt{2}$, $n = 9$ and the lower order matrix for the factorial design with center run $n = 9$

Central composite design			Factorial design		
1	**x_1**	**x_2**	**1**	**x_1**	**x_2**
1	1	1	1	1	1
1	1	-1	1	1	-1
1	-1	1	1	-1	1
1	-1	-1	1	-1	-1
1	a	0	1	1	0
1	$-a$	0	1	-1	0
1	0	a	1	0	1
1	0	$-a$	1	0	-1
1	0	0	1	0	0

Table B2. The lower order matrix for the hybrid (*H*310) design of Roquemore (1976) with center run, *n* = 11

1	x_1	x_2	x_3
1	0	0	1.2906
1	0	0	-0.1360
1	1	1	0.6386
1	1	-1	0.6386
1	-1	1	0.6386
1	-1	-1	0.6386
1	1.1736	0	-0.9273
1	-1.1736	0	-0.9273
1	0	1.1736	-0.9273
1	0	-1.1736	-0.9273
1	0	0	0

Table B3. The lower order matrix for the hybrid (*H*311*B*) design of Roquemore (1976) with center run, *n* = 11

1	x_1	x_2	x_3
1	0	0	2.4495
1	0	0	-2.4495
1	2.1063	0.7507	1
1	2.1063	-0.7507	-1
1	-2.1063	-0.7507	1
1	-2.1063	0.7507	-1
1	0.7507	2.1063	-1
1	0.7507	-2.1063	1
1	-0.7507	-2.1063	-1
1	-0.7507	2.1063	1
1	0	0	0

Table B4. The lower order matrix for the Box and Behnken (1960) design (*BBD*) with center run, *n* = 13

1	x_1	x_2	x_3
1	1	1	0
1	1	-1	0
1	1	0	1
1	1	0	-1
1	-1	1	0
1	-1	-1	0
1	-1	0	1
1	-1	0	-1
1	0	1	1
1	0	1	-1
1	0	-1	1
1	0	-1	-1
1	0	0	0

Table B5. The lower order matrix for the Box and Wilson (1951) CCD for $\alpha = 1.732$ with center run, $n = 15$

1	x_1	x_2	x_3
1	1	1	1
1	1	-1	-1
1	-1	1	-1
1	-1	-1	1
1	1.732	0	0
1	-1.732	0	0
1	0	1.732	0
1	0	-1.732	0
1	0	0	1.732
1	0	0	-1.732
1	0	0	0

Table B6. The lower order matrix for the Box and Draper (1974) minimal design (BDD) with center run, $n = 11$

1	x_1	x_2	x_3
1	1	-1	-1
1	-1	1	-1
1	-1	-1	1
1	-1	-1	-1
1	0.1925	0.1925	-1
1	0.1925	-1	0.1925
1	-1	0.1925	0.1925
1	-0.2912	1	1
1	1	-0.2912	1
1	1	1	-0.2912
1	0	0	0

Table B7. The Lower order matrix for the small composite design of Hartley (1959) (*SCD*) for $\alpha = 1.732$ with center run, $n = 11$

1	x_1	x_2	x_3
1	1	1	1
1	1	1	-1
1	1	-1	1
1	1	-1	-1
1	-1	1	1
1	-1	1	-1
1	-1	-1	1
1	-1	-1	-1
1	1.732	0	0
1	-1.732	0	0
1	0	1.732	0
1	0	-1.732	0
1	0	0	1.732
1	0	0	-1.732
1	0	0	0

13

Sequential Monte Carlo methods for filtering of unobservable components of multidimensional diffusion Markov processes

Ellida M. Khazen[1*]

*Corresponding author: Ellida M. Khazen, 13395 Coppermine Rd. Apartment 410, Herndon, VA 20171, USA
E-mail: ellida_khazen@yahoo.com
Reviewing editor: Zudi Lu, University of Southampton, USA

Abstract: The problem of filtering of unobservable components $x(t)$ of a multidimensional continuous diffusion Markov process $z(t) = (x(t), y(t))$, given the observations of the (multidimensional) process $y(t)$ taken at discrete consecutive times with small time steps, is analytically investigated. On the base of that investigation the new algorithms for simulation of unobservable components, $x(t)$, and the new algorithms of nonlinear filtering with the use of sequential Monte Carlo methods, or particle filters, are developed and suggested. The analytical investigation of observed quadratic variations is also developed. The new closed-form analytical formulae are obtained, which characterize dispersions of deviations of the observed quadratic variations and the accuracy of some estimates for $x(t)$. As an illustrative example, estimation of volatility (for the problems of financial mathematics) is considered. The obtained new algorithms extend the range of applications of sequential Monte Carlo methods, or particle filters, beyond the hidden Markov models and improve their performance.

Subjects: Mathematical Statistics; Mathematics & Statistics; Probability; Probability Theory & Applications; Science; Statistics; Statistics & Probability; Stochastic Models & Processes

ABOUT THE AUTHOR

Ellida M. Khazen graduated from Department of Mathematics of Lomonosov Moscow State University (Moscow, Russia, USSR) with Honor Diploma in 1959 (Master Degree). She received her PhD Degree in Mathematics from Department of Mathematics of Lomonosov Moscow State University in 1962. She received her Doctor Degree in Applied Mathematics in 1994. She was working as Senior Scientist Researcher Mathematician at Moscow Scientific Research Institute of Device Automation in 1962–1996, and also as a Visiting Lecturer at Department of Mathematics of Lomonosov Moscow State University and as a Visiting Lecturer at Moscow Institute of Radio Engineering, Electronics and Automation (MIREA). She has published two scientific monographs (in Russian and in English), chapters in books (in Russian), and more than 50 scientific research papers in the areas of the theory of turbulence onset, the theory of random processes, filtering and signal detection, optimal statistical decisions and statistical sequential analysis, informational estimation of risks, and the theory of optimal control.

PUBLIC INTEREST STATEMENT

The problems of filtering of an unobservable process of interest, $x(t)$, or estimation of the value of some function $f(x(t))$ (or functional) based on observations of the other random process $y(t)$ related to $x(t)$ arise in many areas: signal detection and filtering in radio physical devices; design of control and intelligent systems; financial mathematics; physical chemistry. In the last two decades, a great deal of works has been devoted to development and investigation of Monte Carlo methods for filtering, or particle filtering, for hidden Markov models where $x(t)$ is a Markov process in itself. In the present paper, the new algorithms for the solution of nonlinear filtering problems (with the use of Monte Carlo calculations) are derived and obtained in explicit and closed analytical form *for the general case* where $(x(t), y(t))$ represents a multidimensional continuous diffusion Markov process. The new, effective methods for solution of nonlinear filtering difficult problems are obtained.

Keywords: nonlinear filtering; multidimensional diffusion Markov process; particle filters; sequential Monte Carlo methods; simulation; quadratic variation; volatility

MSC codes: 60G35; 60J60; 62M05; 93E11

1. Introduction

In the last two decades, a great deal of works has been devoted to the development and the investigation of particle filters, or sequential Monte Carlo algorithms, for filtering of an unobservable process $x(t) \stackrel{def}{=} (x_1(t), \dots, x_m(t))$ given observations of the other process $y(t) \stackrel{def}{=} (y_{m+1}(t), \dots, y_p(t))$, taken at discrete times $t_k, t_0 < t_1 < t_2 < \dots < t_k < \dots$, with small time steps $t_k - t_{k-1} \stackrel{def}{=} \Delta t_k$ (see, e.g. survey (Crisan & Doucet, 2002), collection (Doucet, Freitas, & Gordon, 2001), works (Carvalho, Del Moral, Monin, & Salut, 1997; Del Moral, 1998; Doucet, Godsill, & Andrieu, 2000), recent complete survey (Doucet & Johansen, 2011), survey (Del Moral & Doucet, 2014), and references wherein). The observations $y(t_k)$ are being obtained consecutively, and an estimate for $x(t_k)$ should be updated at each time t_k. It is assumed that the whole process $z(t) \stackrel{def}{=} (x(t), y(t)) \stackrel{def}{=} (z_1(t), \dots, z_p(t))$ is a Markov process.

In order to solve the problem of filtering of $x(t_n)$, given the sequence of observations$y(t)|_0^n \stackrel{def}{=} (y(t_0), y(t_1), \dots, y(t_n))$, with the use of Monte Carlo methods, the samples of the random sequences $x^i(t)|_0^n \stackrel{def}{=} (x^i(t_0), x^i(t_1), \dots, x^i(t_n))$ (from the distributions of $x(t_k)$ when $y(t)|_0^k$ and $x(t)|_0^{k-1}$ are given) should have been simulated numerically for $i = 1, \dots, N$.

In general case, first of all, the problem arises: how to obtain an explicit and compact analytical expression (exactly or approximately) for the conditional probability density $P\left(x(t_k)|x(t)|_0^{k-1}, y(t)|_0^k\right)$ in order to simulate samples of $x(t_k)$ when $y(t)|_0^k$ and $x(t)|_0^{k-1}$ are given.

Suppose that such sample sequences $x^i(t)|_0^k$ are being simulated, for $i = 1, \dots, N$. The joint probability density for the given observations $y(t)|_0^n$ and the sequence $x^i(t)|_0^n$ could be computed:

$$P_i(n) \stackrel{def}{=} P\left(x^i(t)|_0^n, y(t)|_0^n\right) = P_0(x^i(t_0), y(t_0)) \prod_{k=1}^{n} P(x^i(t_k), y(t_k)|x^i(t_{k-1}), y(t_{k-1})), \tag{1}$$

where $P(x(t_k), y(t_k)|x(t_{k-1}), y(t_{k-1}))$ represents the transition probability density of the Markov process $(x(t), y(t))$, and $P_0(x_0, y_0)$ is the probability density for the initial joint distribution of the initial value $x(t_0)$ and the initial observation $y(t_0)$.

In a large number of works devoted to particle filters the various algorithms of resampling were introduced in order to obtain the most a posteriori probable samples. For the problem of searching for the maximum a posteriori probable sample, the following algorithm of resampling can be suggested: introduce the values $W_i(n) \stackrel{def}{=} P_i(n) / \sum_{j=1}^{N} P_j(n)$, for $i = 1, \dots, N$. Then $W_i(n)$ can be considered as a weight of the sample sequence $x^i(t)|_0^n$ (as well as of the sample point $x^i(t_n)$), which characterizes the a posteriori probability (or the importance) of the sample sequence $x^i(t)|_0^n$ in comparison with the other samples $x^j(t)|_0^n$ given $y(t)|_0^n$. Due to the Markov property of the process$z(t)$, the recursive formulae for $P_i(n)$ and $W_i(n)$ can be written in general form; we have

$$P_i(n+1) = P_i(n) P\left(x^i(t_{n+1}), y(t_{n+1})|x^i(t_n), y(t_n)\right). \tag{2}$$

Then, it is possible to calculate recursively the new values $P_i(n+1)$ and $W_i(n+1)$, when the new measurement $y(t_{n+1})$ is obtained and the new sample point $x^i(t_{n+1})$ is augmented to the sequence $x^i(t)|_0^n$, so that $x^i(t)|_0^{n+1} = (x^i(t_0), \dots, x^i(t_n), x^i(t_{n+1}))$. It is easy to see, that it is not necessary to keep in memory all the sequence $x^i(t)|_0^n$, but only the last point $x^i(t_n)$. Then the samples $x^i(t)|_0^n$ with small weights $W_i(n)$ could be deleted, but the "more important" samples $x^j(t)|_0^n$ should be continued with a few "offsprings", so that the number of all sample sequences under consideration is equal to N (exactly or approximately). The point $x^{j^*}(t_n)$ and the sequence $x^{j^*}(t)|_0^n$ that have the maximum weight $W_{j^*}(n)$ correspond to the point $x^j(t_n)$ and to the sample sequence that is maximally a posteriori probable given the observations $y(t)|_0^n$, among all the considered sample points $x^j(t_n)$ and sample sequences $x^j(t)|_0^n$. Then the value of $x^{j^*}(t_n)$ can be taken as the sought estimate of $x(t_n)$. On the base of the Laws of Large Numbers, it could be proved that this Monte Carlo estimate converges to the maximum a posteriori probability estimate for $x(t_n)$ if N increases (see also Miguez, Crisan, & Djuric, 2013).

In most of works the additional assumption is accepted that the process $x(t)$ is a Markov process in itself, and that the conditional probability density $P(y(t_k)|y(t_{k-1}),x(t_k),x(t_{k-1}))$ (for the observable process $y(t)$) can be presented in explicit and simple analytical form. Such cases are often referred to as hidden Markov models. Then, in many works, the sample sequences $x^j(t)|_0^n$ are being simulated as trajectories of the Markov process $x(t)$ in itself, since such a simulation can be easily done. The joint probability density, corresponding to the constructed sample $x^j(t)|_0^n$ and to the observations $y(t)|_0^n$, is equal to

$$P(x^j(t)|_0^n, y(t)|_0^n) = P_0\left(x^j(t_0), y(t_0)\right) \prod_{k=1}^{n} P(x^j(t_k)|x^j(t_{k-1}))P\left(y(t_k)|y(t_{k-1}), x^j(t_k), x^j(t_{k-1})\right) = P_j(n). \tag{3}$$

Then the weights $W_j(n)$ (or the oth.er similar weights) could be easily calculated, and the above procedure of sampling and resampling could be realized, in order to obtain the estimate for $x(t_n)$. In some cases it would be better to simulate samples $x^j(t_k)$ with the use of the conditional probability density $P(x(t_k)|x(t_{k-1}),y(t_k),y(t_{k-1}))$, which includes the observed value $y(t_k)$. But this could have required a large amount of computations if we do not have an explicit and compact analytical expression for that conditional probability density. Meantime, if the process $x(t)$ is being simulated just as a Markov process in itself, the sample sequences $x^j(t)|_0^n$ represent samples from the a priori distribution for $x(t_k)$ that can be far apart from the a posteriori distribution for $x(t_k)$ given $y(t)|_0^k$. Then resampling with the use of weights and significant increase in the number N are needed in order to find the most a posteriori probable values of $x(t_n)$ using the algorithm described above. Nevertheless, for hidden Markov models the particle filters with such a simple simulation of the sample sequences $x^j(t)|_0^n$, with resampling based on consideration of weights, but with large N, have been implemented and proved to be useful in some applications (Carvalho et al., 1997; Thrun, Fox, Burgard, & Dellaert, 2001).

In general case when $x(t)$ represents only some components of a multidimensional Markov process $(x(t), y(t))$ (and $x(t)$ is not a Markov process in itself), it was not shown in the literature how to simulate sample sequences $x^i(t)|_0^n$ when the values of the other components, $y(t)|_0^n$, are given (i.e. how to simulate sample sequences $x^i(t)|_0^n$ without a formidable amount of computations at each time t_k).

Note that the difficulties in obtaining the samples $x^i(t)|_0^n$ that correspond to the distributions $P(x(t_k)|x(t)|_0^{k-1}, y(t)|_0^k)$ have led to the introduction and the use of some "proposal sampling distributions" (which can be simulated easier) and "auxiliary particle filters" in many works (see, e.g. collection (Doucet, De Freitas, & Gordon, 2001) and survey (Doucet & Johansen, 2011)). But the "optimal proposal sampling distribution" would be equal to the true distribution $P(x(t_k)|x(t)|_0^{k-1}, y(t)|_0^k)$ if the latter could have been found in closed form. In the present work the precise, explicit and compact analytical formulae and algorithms for simulation of the sample sequences $x^i(t)|_0^n$, when $y(t)|_0^n$ is given, and for recursive calculations of $P_i(n)$, weights, and estimates are obtained for the general case of a multidimensional continuous diffusion Markov process $(x(t), y(t))$.

For the problem of estimation of a function $f(x(t_n))$ or $\varphi(x(t)|_0^n)$, the new estimates in explicit and closed analytical form are obtained in the following Section 2.2.

Moreover, in the quotients $P_i(n)/\sum_{j=1}^{N} P_j(n)$ the "scale" of all the $P_j(n)$ is cancelled, and the information that could characterize the smallness of the a posteriori probability of all the generated samples $x^j(t)|_0^n$ is lost. In the present work, the new sequential Monte Carlo algorithms are derived that include some tests (developed in Section 2.4) in order to discard the samples of low a posteriori probability before the calculation of all the weights is done. The implementation of the suggested tests guarantees that the samples that remain under consideration belong to the domain where the a posteriori probability density is localized.

In some important cases of hidden Markov models, in the particle filters that were suggested and studied in the literature, where the samples $x^i(t)|_0^n$ (with $i = 1, ..., N$) are being simulated as the trajectories of the Markov process $x(t)$ in itself, it is possible that the large part of those N generated samples $x^i(t)|_0^n$ would not be localized in the "vicinity" of the "true" realization of the trajectory, $x^{tr}(t)|_0^n$, because they are simulated with the use of the a priori transition probability density for $x(t)$, and they could be of low a posteriori probability given $y(t)|_0^n$. For example, if the "level of intensity of noise" in observations $y(t)$ is small, the chance to generate randomly the sample $x(t)|_0^n$ that belongs to the domain where the a posteriori probability density (given $y(t)|_0^n$) is concentrated might be as small as the chance to catch randomly a needle in a haystack. Thus, although theoretically in those cases the Monte Carlo estimates converge as $N \to \infty$, the required number N increases if the "level of intensity of noise" decreases. Meanwhile, a large amount of calculations would be wasted for processing of those samples $x^i(t)|_0^n$ of low a posteriori probability. In the present paper, in the following Section 2.6, we shall show (for the problem of nonlinear filtering of a signal) that with the use of the new algorithm (17) of simulation $x^i(t)|_0^n$ given $y(t)|_0^n$ and with some appropriate change of the mathematical model of description of the process of observations (which could be justified in applications), the new algorithm (17) will generate at once the samples $x^i(t)|_0^n$ that are localized in the domain that is a posteriori probable given $y(t)|_0^n$, for all $i = 1, ..., N$.

The special case when the diffusion coefficients of the observable process $y(t)$ depend on unobservable components $x(t)$ is considered. In that case "observed quadratic variations" of the process $y(t)$ can be introduced (when observations $y(t_k)$ are taken in discrete time), and they contain a lot of information about $x(t)$. In the present paper, the analytical formulae that characterize the observed quadratic variations are obtained in explicit and closed form. On the base of these results, the algorithms of filtering and estimation with the use of the observed quadratic variations are developed.

Theoretically, the derivative of the process of quadratic variation could be incorporated into the set of observed, known processes. If that derivative process were known, then with the use of the new algorithm of simulation (17), new estimates (26) and algorithms developed in the present paper, the filtering problem would be effectively solved, as shown in Section 2.5. But in practice we cannot assume that this derivative process is straightforwardly observed. However, if additional observations of $y(t)$ are available, namely, $y(t_0 + s\delta t)$, with $\delta t \ll \Delta t$, $s = 1, 2, \dots$, then the estimates $\hat{u}(t_k)$ for the derivative $u(t_k)$ can be obtained and used for the filtering, as shown in Section 2.5.

The similar systems arise if some additional precisely known observations are available, for example, measurements of some function $H(x(t), t)$. In that case the solution of the filtering problem, described in Section 2.5, could be implemented.

The obtained new algorithms improve performance of particle filters, or sequential Monte Carlo methods, and extend the range of their applications beyond the hidden Markov models.

The implementation of particle filters, as it is usual for Monte Carlo methods, requires a lot of repeating computations, but it became accessible and useful in some applications, since the speed of computations and the capacity of computers have increased dramatically in the last two decades. Note that the speed can be increased further with the use of parallel computing. The computational cost of implementation of a general scheme of particle filtering on some existing processing units that allow parallel programming was considered e.g. in the paper by Hendeby, Karlsson, and Gustafsson (2010).

2. Derivation of new recursive algorithms for Monte Carlo simulation and filtering of unobservable components of multidimensional diffusion Markov processes

2.1. Simulation of trajectories of unobservable components. Analytical investigation of a multidimensional diffusion Markov process observed at discrete times

Consider the multidimensional diffusion Markov process $z(t) \overset{def}{=} (x(t), y(t)) \overset{def}{=} (z_1(t), \dots, z_p(t))$. The components $(z_1(t), \dots, z_m(t)) \overset{def}{=} x(t)$ are unobservable, but the other components $(z_{m+1}(t), \dots, z_p(t)) \overset{def}{=} y(t)$ are available for observations, which are being taken at discrete times t_k.

A diffusion Markov process with continuous trajectories $z(t)$ can be characterized by its drift and diffusion coefficients:

$$E\left\{ \left(z_j(t+\Delta t) - z_j(t)\right) | z(t) = z \right\} = A_j(z,t)\Delta t + o(\Delta t), i,j = 1, \dots, p, \tag{4}$$

$$E\left\{ ((z_i(t+\Delta t) - z_i(t))\left(z_j(t+\Delta t) - z_j(t)\right) | z(t) = z \right\} = b_{ij}(z,t)\Delta t + o(\Delta t), \tag{5}$$

where Δt is a small time step. Denote $\Delta z(t) \overset{def}{=} z(t+\Delta t) - z(t) \overset{def}{=} \left(\Delta z_1(t), \dots, \Delta z_p(t)\right)$.

From the assumption that the trajectories of the process $z(t)$ are continuous functions of t it follows (Kolmogorov, 1931/1986, 1933/1986) that

$$\lim_{\Delta t \to 0} \frac{1}{\Delta t} E\left\{ \Delta z_{i_1}(t) \dots \Delta z_{i_r}(t) | z(t) = z \right\} = 0, with\ r > 2. \tag{6}$$

The matrix $\| b_{ij} \|$ is symmetric and nonnegative definite. Therefore, in general case, that matrix can be represented with the use of its "square root" in the form $\| b_{ij} \| = \| a_{ij} \| \| a_{ij} \|^T$, where $\| a_{ij} \|^T$ stands for transpose matrix $\| a_{ij} \|$.

Then the process $z(t)$ could be constructed as a solution of the system of stochastic differential equations:

$$dz_i(t) = A_i(z,t)dt + \sum_{j=1}^{p} a_{ij}(z,t)dw_j(t), i = 1, \dots, p, t > t_0, \tag{7}$$

where $w_j(t)$ are independent Wiener processes. The initial condition at $t = t_0$ is given as $z(t_0) = z_0$, where z_0 is a random variable independent of all $w_j(t)$, with given probability density $P_0(z_0)$. The system (7) should be interpreted as the system of stochastic integral equations:

$$z_i(t) = \int_{t_0}^{t} A_i(z(s),s)ds + \sum_{j=1}^{p} \int_{t_0}^{t} a_{ij}(z(s),s)dw_{j(}s), i = 1, \dots, p, \tag{8}$$

with stochastic integrals in the sense of Ito (Doob, 1953). It is assumed that the drift coefficients $A_i(z, t)$ satisfy the Lipschitz condition

$$\| A_i(z^{(1)},t) - A_i(z^{(2)},t) \| \leq K \| z^{(1)} - z^{(2)} \|, with\ K = Const. \tag{9}$$

It is also assumed that the diffusion coefficients $b_{ij}(z, t)$ are continuous and differentiable functions of z and t. The Lipschitz conditions (9) guarantee that trajectories $z(t)$ do not have finite escapes, i.e. $|z(t)|$ does not tend to infinity when t tends to some finite moments of time.

The system of integral equation (8) with any given continuous trajectory $w(t) \overset{def}{=} (w_1(t), \dots, w_p(t))$ can be solved with the use of successive approximations, which converge and define the continuous trajectory $z(t)$. Thus, the trajectories of the process $z(t)$ are continuous with the probability 1.

The diffusion Markov process $z(t)$ can be also constructed as the limit solution of the system of finite difference equations:

$$z_i(t_k) - z_i(t_{k-1}) = A_i(z(t_{k-1}), t_{k-1})\Delta t_k + \sum_{j=1}^{p} a_{ij}(z(t_{k-1}), t_{k-1})\eta_j(t_k)\sqrt{\Delta t_k}, \tag{10}$$

where the random impacts $\eta_j(t_k)$ are independent random variables (for all $j = 1, \dots, p, k = 1, 2, \dots$), with $E\eta_j(t_k) = 0$ and $E(\eta_j(t_k))^2 = 1$; $\Delta t_k \overset{def}{=} t_k - t_{k-1}$; $i = 1, \dots, p$. In particular, the increments of Wiener processes, $\Delta w_j(t_k) \overset{def}{=} w_j(t_k) - w_j(t_{k-1})$, can be used in the finite difference equations (10) instead of $\eta_j(t_k)\sqrt{\Delta t_k}$:

$$z_i(t_k) - z_i(t_{k-1}) = A_i(z(t_{k-1}), t_{k-1})\Delta t_k + \sum_{j=1}^{p} a_{ij}(z(t_{k-1}), t_{k-1})\Delta w_j(t_k), \tag{11}$$

Here again it is assumed that the Lipschitz conditions (9) are satisfied, and that the functions $A_i(z, t)$ and $b_{ij}(z, t)$ are differentiable with respect to z. The construction of diffusion Markov processes by the passage to the limit in the scheme of finite difference equations when time steps Δt_k tend to zero was first introduced and investigated by Academician S.N. Bernstein in 1934, 1938 years. The works (Bernstein, 1934/1964, 1938/1964) were published at first in French. They are republished in *Collected works of S.N. Bernstein, vol.4* (in Russian), Publishing House Nauka (Academy of Science of USSR, Moscow, 1964). The results, obtained in these works (Bernstein, 1934/1964, 1938/1964), define analytically all the joint probability distributions for the limit process $z(t)$, for all the random variables $z(t_{i_1}), z(t_{i_2}), \dots, z(t_{i_r})$, for any r and t_{i_r}. By the Theorem of A.N. Kolmogorov on extension (or continuation) of measures in function space (Kolmogorov, 1933/1950), those multidimensional distributions can be continued to the measure on the σ-algebra that corresponds to the process $z(t)$ with continuous time t. The convergence of the linear broken lines, which represent interpolation of the solutions of finite difference stochastic equations (10) or (11), to the continuous trajectories of diffusion Markov processes was also proved in (Kushner, 1971). Thus, the diffusion Markov process $z(t)$ is well defined by the system of finite difference equations (which could be more general, but similar to (10)) and by the passage to the limit from those Markov processes with discrete time (Bernstein, 1934/1964, 1938/1964). Analytical methods and solutions for some problems of filtering and estimation, based on recursive finite difference stochastic equations, with the passage to the limit (as the time steps tend to zero), similarly to the scheme of S.N. Bernstein, were developed and investigated in (Khazen, 1968, 1971, 1977; Chapter 3; Khazen, 2009)

In the following Section 2.2, the goal is to obtain the estimates of $x(t_n)$ or $f(x(t_n))$ with the use of Monte Carlo calculation of some integrals that can be interpreted as mathematical expectations. We should notice the following properties that are important in order to achieve that goal. Denote $P_\Delta(z(t_{i_1}), \dots, z(t_{i_r}))$ the joint probability density for the random process $z_\Delta(t)$ that represents the solution of the finite difference equations (10) or (11) with small time steps $\Delta t_k = \Delta$ (and with piecewise constant or piecewise linear interpolation on the small time intervals $(t_0 + k\Delta, t_0 + (k+1)\Delta)$, $k = 0, 1, 2, \dots$). Denote $P(z(t_{i_1}), \dots, z(t_{i_r}))$ the joint probability density for the limit diffusion Markov continuous process $z(t)$ obtained as $\Delta \to 0$. It was established in the works by Bernstein (1934/1964, 1938/1964) that $P_\Delta(z(t_{i_1}), \dots, z(t_{i_r})) \to P(z(t_{i_1}), \dots, z(t_{i_r}))$ as $\Delta \to 0$. Hence, the contribution to the error of the Monte Carlo estimates of the integrals $\int f(x(t_n))P(x(t)|_0^n, y(t)|_0^n)dx(t)|_0^n$ and $\int P(x(t)|_0^n, y(t)|_0^n dx(t)|_0^n$ (which are considered below in (23), (24)), caused by approximation of the limit diffusion process $z(t)$ by its pre-limit finite difference model (11), tends to zero as $\Delta \to 0$. For the scheme (11), it was also proved in (Kushner, 1971, Chapter 10) that trajectories $z_\Delta(t)$ converge in mean square to the trajectories of the limit diffusion process $z(t)$ as $\Delta \to 0$, so that $E\left\{ \left[z(t) - z_\Delta(t)\right]^2 \right\} \to 0$. Consequently, $\left|E\{f(z(t))\} - E\{f(z_\Delta(t))\}\right| \to 0$, if the function $f(z)$ satisfies Lipschitz condition. Meanwhile, in the problems of filtering the best estimate of $f(x(t_n))$ (by the criterion of minimum mean square error) is $\widehat{f(x(t_n)} = E\{f(x(t_n))|y(t)|_0^n\}$, and, in general case, the value of the mean square error of filtering, $E\left\{ \left[f(x(t_n)) - \widehat{f(x(t_n))}\right]^2 \right\}$, remains finite, positive, limited from below, even if the error of calculation of that conditional expectation tends to zero. Thus, the accuracy of the estimate of $x(t_n)$ or $f(x(t_n))$ could not be noticeably improved even if the accuracy of approximation of the limit diffusion process $z(t)$ by its finite difference model (11) increased as Δ decreased. Therefore, it is justified to use the finite difference model (11) for description and simulation of the considered process $z(t)$, in order to obtain the estimates of $x(t_n)$ or $f(x(t_n))$ that provide solutions to the considered filtering problems (with feasible precision) in case if the observations $y(t_k)$ are being taken in discrete consecutive times t_k with small time steps.

In the present work, we are interested to describe analytically the conditional probability density $P(x(t_k)|x(t)|_0^{k-1}, y(t)|_0^k)$. Consider the local increment Δz of the Markov process $z(t)$ when the value $z(t) = z$ is given, with small time step Δt; denote$\Delta z \overset{def}{=} z(t+\Delta t) - z(t) \overset{def}{=} (\Delta x, \Delta y)$. We have the relations (4)–(6) for the moments of Δz given $z(t) = z$. Then the characteristic function for the random value Δz can be presented in the form:

$$F(u|z,t) \overset{def}{=} E\left\{ exp\left(i\sum_{k=1}^{p} u_k \Delta z_k \right)|z(t) = z \right\} \overset{def}{=} \int exp\left(i\sum_{k=1}^{p} u_k \Delta z_k \right) P(\Delta z|z,t) d\Delta z$$
$$= exp\left\{ \Delta t \sum_{k=1}^{p} A_k(z,t)(iu_k) + \frac{1}{2}\Delta t \sum_{k,l=1}^{p} b_{kl}(z,t)(iu_k)(iu_l) + o(\Delta t) \right\}, \tag{12}$$

where $u = \left(u_1, \ldots, u_p\right)$ is a real vector. The last equality follows from the known expression for the characteristic function of a random variable with the use of its moments (see, for example, course of the Theory of Probability (Gnedenko, 1976). The inverse transformation provides the representation

$$P(\Delta z|z,t) \overset{def}{=} (2\pi)^{-p} \int exp\left(-i\sum_{k=1}^{p} u_k \Delta z_k \right) F(u|z,t) du = (2\pi\Delta t)^{-\frac{p}{2}} (Det \parallel b_{ij} \parallel)^{-1/2}$$
$$exp\left\{ -\frac{1}{2\Delta t} \sum_{i,j=1}^{p} r_{ij}(z,t)(\Delta z_i - A_i(z,t)\Delta t)(\Delta z_j - A_j(z,t)\Delta t) \right\}, \tag{13}$$

where $\parallel r_{ij} \parallel \overset{def}{=} \parallel b_{ij} \parallel^{-1}$ is the inverse matrix of $\parallel b_{ij} \parallel$ (with $1 \le i,j \le p$). The above expressions (12), (13) show that the local increment Δz (in the small, but finite time interval Δt) of the multidimensional diffusion Markov process $z(t)$ with continuous trajectories, when $z(t) = z$ is given, can be considered as a multidimensional Gaussian random variable.

Using the Theorem on Normal Correlation, we find the following expressions for the first and second moments of the conditional probability distribution of the increments Δx_α (with ($\alpha = 1, \ldots, m$) provided that the increments Δy_ρ (with $\rho = (m+1), \ldots, p$) and the value $z(t) = z$ are given (Khazen, 2009, Chapter 3, Sections 3.1.2 and 3.3.1, pages 79–81, 101–106):

$$E\{\Delta x_\alpha|\Delta y, z\} = A_\alpha(z,t)\Delta t + b_{\alpha\sigma}(z,t)c_{\sigma\rho}(z,t)\left(\Delta y_\rho - A_\rho(z,t)\Delta t\right) + o(\Delta t), \tag{14}$$

$$E\left\{ \Delta x_\alpha \Delta x_\beta|\Delta y, z\right\} = \left(b_{\alpha\beta}(z,t) - b_{\alpha\sigma}(z,t)c_{\sigma\rho}(z,t)b_{\rho\beta}(z,t)\right)\Delta t + o(\Delta t). \tag{15}$$

Hereinafter the summation is being made over repeated indices, with $1 \le \alpha, \beta \le m, (m+1) \le \sigma, \rho \le p$; the matrix $\| c_{\sigma\rho} \|$ represents the inverse or pseudo-inverse (Moore–Penrose) matrix of the matrix of the diffusion coefficients of the observable components $y_\rho(t)$, so that $\| c_{\sigma\rho} \| \overset{def}{=} \| b_{\sigma\rho} \|^+$.

For the probability density of the increments Δy, we obtain the following expression:

$$P(\Delta y | z(t) = z) = C(z,t) exp\left\{ -\frac{1}{2\Delta t} c_{\sigma\rho}(z,t)\left(\Delta y_\rho - A_\rho(z,t)\Delta t\right)(\Delta y_\sigma - A_\sigma(z,t)\Delta t)\right\}, \quad (16)$$

where $C(z, t)$ is the normalization factor.

Note, that in the theory of filtering of diffusion Markov processes with observations made in continuous time t, i.e. if $y(s)$ is supposed to be known exactly on the time interval $t_0 \le s \le t$, in such case it is assumed that the diffusion coefficients $b_{\sigma\rho}$, with $(m+1) \le \sigma, \rho \le p$, do not depend on unobservable components $x(t)$. In the contrary case (as it was pointed out by this author (Khazen, 1977; Chapter 3; Khazen, 2009, Chapter 3), since the diffusion coefficients $b_{\sigma\rho}(x(t), y(t), t)$ could be (at least, theoretically) precisely restored on the base of a single realization of the observed trajectory $y(s)$ on the small time interval $t - \delta \le s \le t + \delta$ (no matter how small the taken value of δ is), the functions $b_{\sigma\rho}(x(t), y(t), t)$ could have been incorporated into the set of observable functions. In the problem at hand, when filtering of the process $x(t)$ should be obtained on the base of observations $y(t_k)$, taken at discrete times t_k with small, but finite time steps Δt_k, that restriction may be cancelled. We shall consider further (in Sections 2.4, 2.5) how to incorporate some estimates of the values of $b_{\sigma\rho}(x(t_k), y(t_k), t_k)$ given $y(t)|_0^k$ in order to improve filtering of $x(t_n)$.

The matrix $\| b_{\alpha\beta} - b_{\alpha\sigma} c_{\sigma\rho} b_{\rho\beta} \|$ is symmetric and nonnegative definite, and it can be presented in the following form:

$$\| b_{\alpha\beta} - b_{\alpha\sigma} c_{\sigma\rho} b_{\rho\beta} \| =^{def} \| g_{\alpha\beta} \| \, \| g_{\gamma\delta} \|^T, \text{ with } 1 \le \alpha, \beta, \gamma, \delta \le m,$$

Then the samples $x^j(t_k)$, when $x^j(t)|_0^{k-1}$ and $y(t)|_0^k$ are given, can be simulated as

$$\begin{aligned} x_\alpha^j(t_k) &= x_\alpha^j(t_{k-1}) + A_\alpha\left(x^j(t_{k-1}), y(t_{k-1})\right)\Delta t_k + g_{\alpha\beta}(x^j(t_{k-1}), y(t_{k-1}))\xi_\beta^j(k)\sqrt{\Delta t_k} \\ &+ b_{\alpha\sigma}\left(x^j(t_{k-1}), y(t_{k-1})\right) c_{\sigma\rho}(x^j(t_{k-1}), y(t_{k-1}))\left(y_\rho(t_k) - y_\rho(t_{k-1}) - A_\rho\left(x^j(t_{k-1}), y(t_{k-1})\right)\Delta t_k\right) \end{aligned} \quad (17)$$

where $\xi_\beta^j(k)$ are independent samples from Gaussian standard distribution, with $E\xi_\beta^j(k) = 0$, $E(\xi_\beta^j(k))^2 = 1$ (for all $\beta = 1, \ldots, m$, $j = 1, \ldots, N$, $k = 1, 2, \ldots$); $\Delta t_k = t_k - t_{k-1}$.

We shall denote shortly $b_{\alpha\sigma}\left(x^j(t_{k-1}), y(t_{k-1})\right) \overset{def}{=} b_{\alpha\sigma}^j(t_{k-1})$, $c_{\sigma\rho}\left(x^j(t_{k-1}), y(t_{k-1})\right) \overset{def}{=} c_{\sigma\rho}^j(t_{k-1})$, $A_\sigma\left(x^j(t_{k-1}), y(t_{k-1})\right) \overset{def}{=} A_\sigma^j(t_{k-1})$, $c_{\sigma\rho}(x(t_{k-1}), y(t_{k-1})) \overset{def}{=} c_{\sigma\rho}(t_{k-1})$.

Note, that the above expressions (14)–(17) show that if all the diffusion coefficients $b_{\alpha\rho}$ vanish (with $1 \le \alpha \le m$, $(m+1) \le \rho \le p$) and if $A_\alpha(z, t)$, $b_{\alpha\beta}(z, t)$ do not depend on $y(t)$, then $x(t)|_1^n$ will be simulated without the use of $y(t)|_1^n$. Only the initial measurement $y(t_0)$ will be taken into account: the initial sample point $x^j(t_0)$ should be simulated as a random variable with the probability density

$$P_0(x(t_0) | y(t_0) = y_0) = P_0(x(t_0), y_0) / \int P_0(x(t_0), y_0) dx(t_0). \quad (18)$$

This particular case do not correspond to all the hidden Markov models: The process $x(t)$ could be a Markov process in itself, but it is still possible, that some coefficients $b_{\alpha\rho}$ are not equal to zero, for example, in case, if there are some "white noises", which enter simultaneously into equations for $x(t)$ and into "random disturbances or random errors" of the observable process $y(t)$.

In general case of a multidimensional continuous Markov diffusion process $(x(t), y(t))$, the influence of the observed data $y(t)|_1^n$ is manifested in the expressions (17), which describe simulation of the random samples $x^j(t_k)$.

The values $P_i(n)$ and $W_i(n)$ can be easily calculated:

$$P_i(n) = P_0\left(x^i(t_0), y(t_0)\right)\prod_{k=1}^{n} P(\Delta x^i(t_k)|\Delta y(t_k), x^i(t_{k-1}), y(t_{k-1}))P\left(\Delta y(t_k)|x^i(t_{k-1}), y(t_{k-1})\right), \tag{19}$$

where the conditional probability densities for the increments, $\Delta x^i(t_k) \overset{def}{=} x^i(t_k) - x^i(t_{k-1})$, $\Delta y(t_k) \overset{def}{=} y(t_k) - y(t_{k-1})$, are Gaussian densities with moments determined by (14), (15), and (16). And the above procedure of sampling and resampling, which corresponds to a search for the point $x(t_n)$ that provides maximum to the a posteriori probability density, could be implemented. In the next Sections 2.2–2.4, the results of the further analytical investigation and development of the algorithms are presented.

2.2. Estimating of a function $f(x(t_n))$

Consider the problem of estimation of some functions $f(x(t_n))$ or $\varphi(x((t)|_0^n))$ given $y(t)|_0^n$. The general expression for the estimate can be written in the following form:

$$\widehat{f(x(t_n))} = E\{f(x(t_n))|y(t)|_0^n\} = \frac{\int f(x(t_n))P(x(t)|_0^n, y(t)|_0^n)dx(t)|_0^n}{\int P(x(t)|_0^n, y(t)|_0^n)dx(t)|_0^n}. \tag{20}$$

It is assumed that for the considered functions $f(x)$ or $\varphi(x(t)|_0^n)$ this conditional expectation exists. In case of a Markov process $(x(t_n), y(t_n))$, we can write:

$$\begin{aligned} P(x(t)|_0^n, y(t)|_0^n) &= P_0(x(t_0)|y(t_0))P_0(y(t_0))\prod_{k=1}^{n} P(x(t_k)|y(t_k), y(t_{k-1}), x(t_{k-1})) \\ &\times \prod_{s=1}^{n} P(y(t_s)|y(t_{s-1}), x(t_{s-1})). \end{aligned} \tag{21}$$

In our case of a multidimensional continuous diffusion Markov process, we have derived the explicit analytical expressions (19), (16). Hence, in this case, we obtain

$$\begin{aligned} &\prod_{s=1}^{n} P(y(t_s)|y(t_{s-1}), x(t_{s-1})) = C(x(t)|_0^{n-1}, y(t)|_0^{n-1}) \\ &exp\left\{ -\frac{1}{2}\sum_{s=1}^{n}\frac{1}{\Delta t_s}c_{\sigma\rho}(t_{s-1})[\Delta y_\sigma(t_s)\Delta y_\rho(t_s) - 2\Delta y_\sigma(t_s)A_\rho(t_{s-1})\Delta t_s + A_\sigma(t_{s-1})A_\rho(t_{s-1})(\Delta t_s)^2]\right\}, \end{aligned} \tag{22}$$

where $C(x(t)|_0^{n-1}, y(t)|_0^{n-1})$ represents the known normalization factor. Denote shortly $C_i(n-1) \overset{def}{=} C(x^i(t)|_0^{n-1}, y(t)|_0^{n-1})$.

If N samples $x^i(t)|_0^n$, with $i = 1, \ldots, N$ and $N \gg 1$, have been independently taken from the distribution

$$P_0(x(t_0)|y(t_0))\prod_{k=1}^{n} P(x(t_k)|y(t_k), y(t_{k-1}), x(t_{k-1})),$$

then, in accordance with basics of Monte Carlo methods, the following integrals can be interpreted as mathematical expectations. Then we obtain

$$\int f(x(t_n))P(x(t)|_0^n, y(t)|_0^n)dx(t)|_0^n \approx P_0(y(t_0)\frac{1}{N}\sum_{i=1}^{N} f(x^i(t_n))C_i(n-1)$$
$$\times \exp\left\{-\frac{1}{2}\sum_{s=1}^{n}\frac{1}{\Delta t_s}c^i_{\sigma\rho}(t_{s-1})[\Delta y_\sigma(t_s)\Delta y_\rho(t_s) - 2\Delta y_\sigma(t_s)A^i_\rho(t_{s-1})\Delta t_s + A^i_\sigma(t_{s-1})A^i_\rho(t_{s-1})(\Delta t_s)^2]\right\}, \quad (23)$$

and similarly

$$\int P(x(t)|_0^n, y(t)|_0^n)dx(t)|_0^n \approx P_0(y(t_0)\frac{1}{N}\sum_{i=1}^{N} C_i(n-1)$$
$$\times \exp\left\{-\frac{1}{2}\sum_{s=1}^{n}\frac{1}{\Delta t_s}c^i_{\sigma\rho}(t_{s-1})[\Delta y_\sigma(t_s)\Delta y_\rho(t_s) - 2\Delta y_\sigma(t_s)A^i_\rho(t_{s-1})\Delta t_s + A^i_\sigma(t_{s-1})A^i_\rho(t_{s-1})(\Delta t_s)^2]\right\}. \quad (24)$$

Due to the Laws of Large Numbers, the accuracy of those approximate estimates increases as N increases.

Denote shortly

$$\Upsilon_i(n) = C_i(n-1) \quad (25)$$
$$\exp\left\{-\frac{1}{2}\sum_{s=1}^{n}\frac{1}{\Delta t_s}c^i_{\sigma\rho}(t_{s-1})[\Delta y_\sigma(t_s)\Delta y_\rho(t_s) - 2\Delta y_\sigma(t_s)A^i_\rho(t_{s-1})\Delta t_s + A^i_\sigma(t_{s-1})A^i_\rho(t_{s-1})(\Delta t_s)^2]\right\}.$$

Then the sought estimate $\widehat{f(x(t_n))}$ can be written in the following form:

$$\widehat{f(x(t_n))} = \sum_{i=1}^{N} f(x^i(t_n))\,\tilde{W}_i(n), \text{where } \tilde{W}_i(n) = \frac{\Upsilon_i(n)}{\sum_{j=1}^{N}\Upsilon_j(n)}. \quad (26)$$

In most of applications, we can assume that there exists the mathematical expectation of the random value $\left|f(x^i(t_n))\right|\Upsilon_i(n)$, where $y(t)|_0^n$ is fixed and $x^i(t)|_0^n$ represent the results of independent random simulations with the use of (17); i.e. $E\{\left|f\left(x^i(t_n)\right)\right|\Upsilon_i(n)\} < \infty$. This assumption is equivalent to the following: $E\left\{\left|f(x(t_n))\right| |y(t)|_0^n\right\} < \infty$. Denote $\zeta_i = f(x^i(t_n))\Upsilon_i(n)$. The random values ζ_i are independent, and they have one and the same probability distribution, and $E\{|\zeta_i|\} < \infty$. Then the Strong Law of Large Numbers (the Theorem of Kolmogorov) guarantees that the sums (i.e. the arithmetic means $\frac{1}{N}\sum_{i=1}^{N}\zeta_i$) in right-hand side of the formulae (23), (24) converge with probability one (i.e. almost surely) to the integrals in left-hand side if $N \to \infty$.

Note that in case if the diffusion coefficients $b_{\sigma\rho}$ (for $m < \sigma, \rho \le p$) of the observable process $y(t)$ do not depend on unobservable process $x(t)$, then $C_i(n-1)$ and $c^i_{\sigma\rho}(t_{s-1})$ do not depend on i, and cofactors $C_i(n-1)\exp\left\{-\frac{1}{2}\sum_{s=1}^{n}\frac{1}{\Delta t_s}c^i_{\sigma\rho}(t_{s-1})\Delta y_\sigma(t_s)\Delta y_\rho(t_s)\right\}$ should be cancelled in the numerator and denominator of the formula (26). In that case we can put:

$$\Upsilon_i(n) = \exp\left\{-\frac{1}{2}\sum_{s=1}^{n} c_{\sigma\rho}(t_{s-1})[-2A^i_\rho(t_{s-1})\Delta y_\sigma(t_s) + A^i_\sigma(t_{s-1})A^i_\rho(t_{s-1})\Delta t_s]\right\}. \quad (27)$$

Note that the calculations of the values $Y_i(n)$ (25) or (27) can be implemented recurrently since they are determined by the values of the recursively accumulated sums.

In case of hidden Markov models where the unobservable process $x(t)$ is a Markov process in itself, and the observable process $y(t)$ is described by a stochastic differential equation (SDE):

$$dy(t) = h(x(t), t)dt + \sigma dw(t), \tag{28}$$

where $h(x, t)$ is a given nonlinear function with respect to $x, \sigma = Const$, and $w(t)$ is a standard Wiener process, the above estimate $\widehat{f(x(t_n))}$ takes the form, which is similar to the estimates for $f(x(t_n))$ that were earlier constructed and studied for this particular case in the literature. The novelty of the estimates (23)–(27) and the simulation (17), obtained in the present paper, is that the simulation of unobservable components and estimates $\widehat{f(x(t_n))}$ are obtained in explicit and closed analytical form for the general case of partially observable multidimensional continuous diffusion Markov processes, and the obtained Monte Carlo estimate (26) converges with probability one to the sought posterior expectation of $f(x(t_n))$ given $y(t)|_0^n$, if $N \to \infty$.

Now we shall demonstrate that the Monte Carlo estimates (23)–(26), (27) for $f(x(t_n))$ (obtained in the present paper) hold true also in the case if the sample sequences $x^i(t)|_0^n$ are being generated with the use of some branching sampling procedures.

Consider the following branching resampling. As it was already pointed out in the Introduction, it is purposeful to discard the samples $x^i(t)|_0^n$ that have negligibly small weights $W_i(n)$ or $\tilde{W}_i(n)$, in order to decrease the amount of all the computations.

Suppose that in the end of each time interval $(T_k, T_{k+1}]$ (with $k = 0, 1, 2, \ldots; T_0 = t_0; T_{k+1} = T_k + m_k\Delta t$, and m_k and N_k are given numbers) each sample sequence $x^i(t)|_0^n$ that still existed at the time T_{k+1} is being continued with N_{k+1} "offsprings". At the initial moment of time, T_0, there are N_0 independently taken initial sample points $x^i(t_0)$, $i = 1, \ldots, N_0$. When $t_n = T_{k+1}$ and $t_{n+1} = T_{k+1} + \Delta t$, sample points $x^{(i,j)}(t_{n+1})$ are being augmented to the sequence $x^i(t)|_0^n$, with $j = 1, \ldots, N_{k+1}$, in order to construct the "offsprings". The sample points $x^{(i,j)}(t_{n+1})$ are being independently taken from the distribution $P(x(t_{n+1})|x^i(t)|_0^n, y(t)|_0^{n+1})$. Then those sample sequences (with "offsprings") will be continued until the next time of branching, T_{k+2}, except for some sample sequences that are discarded before T_{k+2} since their weights become "too small" (for example, smaller than some chosen threshold W_{cr}); and so on. For the simplicity of discussion, we can renumber again all the current sample sequences (that still exist at the time t_n) as $x^i(t)|_0^n$. Consider at first the case when all the sample sequences are being continued without discarding. Then their number is growing, and at the time T_{k+1} it is equal to $N_0 \times N_1 \times \ldots \times N_k$ (with $k > 0$).

Consider the estimate

$$\overline{f(x(t_n))} = E\{f(x(t_n))|y(t)|_0^n\},$$

under condition that the above branching sampling procedure is being implemented. Denote $(t_{r(n)}, t_{(r+1)(n)}]$ the time interval (between the moments of branching) that contains t_n, so that $t_{r(n)} = T_{k(n)} < t_n \leq T_{k(n)+1} = t_{(r+1)(n)}$.

From the Markov property of the random process $(x(t), y(t))$ and the factorization (21), (22), and from the "tower property" of conditional expectations:

$$E\{f(x(t_n))|y(t)|_0^n\} = E\{E\{f(x(t_n)|y(t)|_0^n, x(t)|_0^{r(n)}\}\} =$$
$$E\left\{E\left\{E\{f(x(t_n)|y(t)|_0^n, x(t)|_0^{r(n)}\}|x(t)|_0^{(r-1)(n)}\right\}\right\} = \ldots,$$

it follows that the formulae (23)–(26), (27) hold true for the Monte Carlo estimate of integral $\overline{f(x(t_n))}$ in the case if the sample sequences $x^i(t)|_0^n$ are generated with the use of the above branching procedure.

We can consider the Monte Carlo estimates for the integrals (23), (24) for each i-th "tree" of the branching sample sequences, which begins at the point $x^i(t_0)$, with $i = 1, \ldots, N_0$. Denote these random values as $\xi_i(n)$, with $i = 1, \dots, N_0$. Then the random values $\xi_i(n)$ are independent of each other, and they have one and the same probability distribution, and $E\{|\xi_i|\} < \infty$. Then they obey the Strong Law of Large Numbers, so that $\frac{1}{N_0}\sum_{i=1}^{N_0} \xi_i$ converges with probability one to the sought integral as $N_0 \to \infty$.

But many of the exponential weights $\tilde{W}_i(n)$ are rapidly decreasing with increase in the number n of time steps. Therefore, it is possible to discard some highly a posteriori improbable sample sequences $x^i(t)|_0^n$, which do not provide noticeable contribution into the estimates $\widehat{f(x(t_n))}$. The numbers m_k, N_k, and the threshold W_{cr} can be adjusted (in practical implementation for a particular class of applications) in order to decrease the amount of all the calculations and make them feasible, and at the same time to keep the current sample points $x^i(t_n)$ in the domain where the a posteriori probability density is localized. In practical implementation, it can be also purposeful to begin branching not at the fixed moment of time T_k but at some current moment of time t_n when the number of all existed sample sequences (or sample points) becomes less than some given fraction of N.

It is worth noting that in the following Section 2.4 the tests (47), (48) are obtained that allow detection and rejection of samples $x^i(t_n)$ of low a posteriori probability to be done without the use of the weights $W_i(n)$ or $\tilde{W}_i(n)$ (for $i = 1, \dots, N$), i.e. independently for each sample sequence $x^i(t)|_0^n$. Thus, in the above algorithm we can use such tests instead of the comparison of the weight $\tilde{W}_i(n)$ with the threshold W_{cr}. Then the algorithm described above can be effectively implemented with the use of parallel computing, since the simulation and continuation (the branching resampling) of each sample sequence $x^i(t)|_0^m$ can be done independently of the other samples $x^k(t)|_0^m$ with $k \neq i$. The values $Y_i(n)$ determined by (25) or (27) will be also calculated recurrently and independently for each sample sequence $x^i(t)|_0^n$. Thus, that large amount of calculations could be effectively performed with the use of parallel computing. And only on the other stage all the obtained accumulated values $Y_i(n)$ (with $i = 1, \dots, N$) will be used for the calculation of the weights $\tilde{W}_i(n)$, which are needed for the calculation of the sought estimate $\widehat{f(x(t_n))}$.

The new Monte Carlo algorithm of estimation of $f(x(t_n))$ or $\varphi(x(t)|_0^n)$ (presented above) is derived straightforwardly from Bayes formula (20). The estimates are constructed with the use of the samples $x^i(t_n)$ or $x^i(t)|_0^n$ that are simulated by (17), and their weights $\widehat{f(x(t_n))}$ are defined by (23)–(26), (27).

For the particular case of hidden Markov models, another algorithm with random branching resampling for "particle filtering" of unobservable Markov "signal" $x(t_n)$ was developed, that is the Sequential Importance Resampling (SIR) algorithm; it is studied in the works (Crisan & Doucet, 2002; Del Moral, 1998), for the processes with discrete time. In that SIR algorithm with random branching resampling, the a posteriori probability distribution for $x(t_n)$ is approximated by some "cloud of random particles" $x^i(t_n)$ with weights $w_i(n) = 1/N, i = 1, \dots, N$. With the use of our new algorithm of simulation of unobservable components $x(t_n)$ given $y(t)|_0^n$ (see Section 2.1, (17)) and the new closed form analytical expressions (14)–(17), (23)–(27), that SIR algorithm with random branching can be generalized for the general case of filtering of unobservable components $x(t)$ of a multidimensional diffusion Markov process $(x(t), y(t))$, in the following way.

In the generalized algorithm that we are proposing the sample sequence $x^i(t)|_0^m$ should be chosen for continuation at the time of branching $T_m = t_{r(m)}$ (and kept up to the next time of branching, T_{m+1}) with probability $p_i(m)$ (which will be determined below) in each of N independent attempts to continue their ensemble, so that the expectation of the random number $K_i(m)$ of its "offsprings" is equal to $E\{K_i(m)|x^1(t_{r(m)}), \dots, x^{K(m)}(t_{r(m)}), y(t)|_0^{r(m)}\} = p_i(m)N$, with $i = 1, \dots, K(m)$, where $K(m)$ denotes the number of all the sample points $x^i(t_{r(m)})$ that still existed at the time $t_{r(m)}$. We can suggest that the times of branching $T_{m+1} = T_m + M\Delta t$, with $M \geq 1, m = 0, 1, 2, \dots$. (Note, that $M = 1$ in the standard SIR algorithms). The probability $p_i(m)$ depends on all the sample points $x^1(t_m), \dots, x^{K(m)}(t_m)$ or sample sequences $x^i(t)|_0^m$. Such a procedure of random resampling is similar to the standard SIR algorithm, but we have to determine the probability $p_i(m)$ for the general case of multidimensional diffusion Markov processes $(x(t), y(t))$.

For the general case, with the use of analytical expressions (16)–(27), we derive the following expression:

$$p_i(m) = \frac{Q_i(m) exp\{-R_i(m)\}}{\sum_{j=1}^{K(m)} Q_j(m) exp\left\{-R_j(m)\right\}},$$

where

$$R_i(m) = \frac{1}{2}\sum_{s=0}^{M-1} \frac{1}{\Delta t} c^i_{\sigma\rho}(t_{r(m-1)+s})[\Delta y_\sigma(t_{r(m-1)+s+1})\Delta y_\rho(t_{r(m-1)+s+1})$$
$$- 2\Delta y_\sigma(t_{r(m-1)+s+1})A^i_\rho(t_{r(m-1)+s})\Delta t + A^i_\sigma(t_{r(m-1)+s})A^i_\rho(t_{r(m-1)+s})(\Delta t)^2],$$

and $Q_i(m)$ denotes the normalization factor of the probability density $Q_i(m) exp\{-R_i(m)\}$. Here the notations are the same as earlier in Sections 2.1, 2.2.

In the case if the diffusion coefficients $b_{\sigma\rho}$ of the observed process $y(t)$ do not depend on the unobservable process $x(t)$, the above expression can be written in the following more concise form:

$$p_i(m) = \frac{exp\{-V_i(m)\}}{\sum_{j=1}^{K(m)} \exp\left\{-V_j(m)\right\}},$$

where

$$V_i(m) = \frac{1}{2}\sum_{s=0}^{M-1} c_{\sigma\rho}(t_{r(m-1)+s})[-2\Delta y_\sigma(t_{r(m-1)+s+1})A^i_\rho(t_{r(m-1)+s}) + A^i_\sigma(t_{r(m-1)+s})A^i_\rho(t_{r(m-1)+s})\Delta t].$$

In case of hidden Markov models (28), which can be considered as a particular case of a multidimensional diffusion Markov process $(x(t), y(t))$, from the above expression follows:

$$p_i(m) = \frac{exp\{-\Gamma_i(m)\}}{\sum_{j=1}^{K(m)} \exp\left\{-\Gamma_j(m)\right\}},$$

where

$$\Gamma_i(m) = \frac{1}{2\sigma^2}\sum_{s=0}^{M-1}[-2h(x^i(t_{r(m-1)+s}))\Delta y(t_{r(m-1)+s+1}) + h^2(x^i(t_{r(m-1)+s}))\Delta t]$$

For the simplicity of notation, the above formula is written for the case of one-dimensional processes $x(t)$ and $y(t)$, which satisfy (28). The last expression for $p_i(m)$ (for the case $M = 1$) is in agreement with the determination of the probabilities of continuation of the samples $x^i(t_n)$ presented in (Crisan & Doucet, 2002; Del Moral, 1998) for the case of hidden Markov models.

With the use of the above procedure of random branching and with our algorithm (17) of simulation of $x(t_n)$ given $y(t)|_0^n$, we can generalize the SIR algorithm of particle filtering with random branching, which was developed for hidden Markov models, to the general case of multidimensional diffusion Markov processes $(x(t), y(t))$.

Thus, in the present paper a few different possible versions of algorithms of particle filters are provided for filtering of unobservable components $x(t_n)$ given $y(t)|_0^n$, which are justified theoretically. The new algorithm developed above, with branching sampling and recurrent calculation of the values (25) or (27), appears to be preferable for implementation with the use of parallel computing. The further practical implementation and comparison for various applications could be achieved in the future works.

2.3. Analytical investigation of observed quadratic variations

Consider the case when some diffusion coefficients $b_{\sigma\rho}$ of the observable process $y(t)$ depend on the unobservable process $x(t)$. We assume throughout this paper that the diffusion coefficients $b_{ij}(z,t)$ are continuous and differentiable functions with respect to z. For simplicity of notation, consider at first the case when $y(t)$ is one-dimensional. For example, consider the following model that plays an important role in financial mathematics (Hull, 2000):

$$dx = A(x,t)dt + \sigma_1(x,t)dw_1(t), dy = \left(\mu - \frac{x^2}{2}\right)dt + \sigma_2 x dw_2(t). \tag{29}$$

Here $w_1(t)$, $w_2(t)$ are independent Wiener processes, μ and σ_2 are known constants, and the functions $A(x, t)$ and $\sigma_1(x, t)$ are continuous and differentiable with respect to x. In the famous Black–Scholes models (Hull, 2000), the observable process $y(t)$ in (29) represents the natural logarithm of a stock price, $S(t)$, so that $y(t) \stackrel{def}{=} \log S(t)$, while the process $|x(t)|$ corresponds to volatility, which is to be estimated given the path $y(s)$, $t_0 \le s \le t$. If the process $y(s)$ were available for observations continuously over the time interval $t_0 \le s \le t$, the value $|x(t)|$ could have been restored precisely, at least theoretically.

It is well known that a diffusion Markov process $z(t)$ is not differentiable at any time t. It can be proved that its quadratic variations

$$\sum_{k=1}^{M}[z_i(t_k) - z_i(t_{k-1})]\left[z_j(t_k) - z_j(t_{k-1})\right] \underset{max\{\Delta t_k\}\to 0}{\to} \int_{t-\delta}^{t+\delta} b_{ij}(z(s),s)ds, \tag{30}$$

if time steps Δt_k tend to zero, while $M \to \infty$; here $(t-\delta) = t_0 < t_1 < \ldots < t_k < \ldots < t_M = (t+\delta)$, with given $\delta > 0$, and the value of δ can be arbitrarily small. Note that the relation (30) will be also obtained as a consequence of the new analytical estimates that are developed below.

For the system (29), $b_{22}(x(t),y(t),t) = (\sigma_2)^2 x^2(t)$, and from (30) it follows that the value of $x^2(t)$ could be restored with any desirable accuracy. Then $|x(t)| = +\sqrt{x^2(t)}$, and the filtering problem would be solved exactly.

But the observations $y(t_k)$ are being taken at discrete times t_k with small, but finite time steps Δt_k, so that it is impossible to observe precisely the values of $b_{\sigma\rho}(x(t_k), y(t_k), t_k)$. In general case, we can calculate the "observed quadratic variations",

$$V_{\sigma\rho}^{(obs)}(t_k) \stackrel{def}{=} \sum_{s=0}^{M}[y_\sigma(t_{k-s}) - y_\sigma(t_{k-s-1})]\left[y_\rho(t_{k-s}) - y_\rho(t_{k-s-1})\right], \tag{31}$$

and consider it as an estimate for $b_{\sigma\rho}(x(t_k), y(t_k), t_k)$. The similar estimates for volatilities were introduced and considered in (Hull, 2000, chapter 15). It is more convenient to use the recursive averaging instead of the moving averaging (31). Then

$$V_{\sigma\rho}^{(obs)}(t_k) \stackrel{def}{=} e^{-\lambda\Delta t_k} V_{\sigma\rho}^{(obs)}(t_{k-1}) + [y_\sigma(t_k) - y_\sigma(t_{k-1})]\left[y_\rho(t_k) - y_\rho(t_{k-1})\right], \tag{32}$$

with $\lambda > 0$. In the limit, if all Δt_k decreased and tended to zero, we would have obtained

$$V_{\sigma\rho}^{(obs)}(t) \underset{max\{\Delta t_k\}\to 0}{\to} \int_{t_0}^{t} e^{-\lambda(t-s)} b_{\sigma\rho}(x(s), y(s), s)ds. \tag{33}$$

We can incorporate the observed quadratic variations $V_{\sigma\rho}^{(obs)}(t_k)$ into the set of observed data. We shall show further, in Section 2.4, how to use that "observed quadratic variations" in order to reject at once the samples $x^i(t)|_0^k$ that are highly improbable given $y(t)|_0^k$.

The recursive formula (32) implies that

$$V_{\sigma\rho}^{(obs)}(t_k) = \sum_{s=1}^{k} e^{-\lambda(t_k-t_s)} [y_\sigma(t_s) - y_\sigma(t_{s-1})][y_\rho(t_s) - y_\rho(t_{s-1})]. \tag{34}$$

We may assume that the considered realization of the process $z(t)$, $(x(t_0), y(t_0)), (x(t_1), y(t_1)), \ldots, (x(t_s), y(t_s)), \ldots$, is being obtained consecutively in accordance with finite difference stochastic equations (11). The value of observed quadratic variation $V_{\sigma\rho}^{(obs)}(t_k)$ is being accumulated in parallel, along with that realization of $z(t)$, so that the random increment $\Delta y_\rho(t_s) \overset{def}{=} y_\rho(t_s) - y_\rho(t_{s-1})$ will be produced after $(x(t_{s-1}), y(t_{s-1}))$ is realized. We are interested to describe the properties of the observed quadratic variations $V_{\sigma\rho}^{(obs)}(t_k)$ under condition that the realization of unobservable components, $x(t)|_0^k$, is fixed, although it is unknown, and that the next measurement $y(t_s)$ will arrive after the previous measurement $y(t_{s-1})$ is already given.

The conditional expectation

$$E\left\{\Delta y_\sigma(t_s)\Delta y_\rho(t_s) | x(t_{s-1}), y(t_{s-1})\right\} = b_{\sigma\rho}(x(t_{s-1}), y(t_{s-1}), t_{s-1})\Delta t_s.$$

In general case, as it was demonstrated above, the increment $\Delta y = (\Delta y_{m+1}, \ldots, \Delta y_p)$ in small time step Δt, given $z(t) = z$, can be described as the Gaussian multidimensional random variable with the probability density (16). The increments $\Delta y(t_k)$ are independent from the past history before t_{k-1} if $z(t_{k-1})$ is given. The following properties of Gaussian distributions will be useful in the sequel: If ξ is a Gaussian random variable with $E\xi = 0$, then $E\xi^4 = 3(E\xi^2)^2$. If $(\xi_1, \xi_2, \xi_3, \xi_4)$ is a four-dimensional Gaussian random variable with $E\{\xi_i\} = 0$ (for $i = 1, \ldots, 4$), then

$$E\{\xi_1\xi_2\xi_3\xi_4\} = E\{\xi_1\xi_2\}E\{\xi_3\xi_4\} + E\{\xi_1\xi_3\}E\{\xi_2\xi_4\} + E\{\xi_1\xi_4\}E\{\xi_2\xi_3\}.$$

In general case of multidimensional process $y(t)$, we obtain

$$E\left\{(\Delta y_\sigma(t_k)\Delta y_\rho(t_k) - b_{\sigma\rho}(t_k)\Delta t_k)^2 | x(t_{k-1}), y(t_{k-1})\right\}$$
$$= [b_{\sigma\sigma}(x(t_{k-1}), y(t_{k-1}), t_{k-1})\Delta t_k][b_{\rho\rho}(x(t_{k-1}), y(t_{k-1}), t_{k-1})\Delta t_k] + [b_{\sigma\rho}(x(t_{k-1}), y(t_{k-1}), t_{k-1})\Delta t_k]^2.$$

Denote shortly $b_{\sigma\rho}(x(t_k), y(t_k), t_k) \overset{def}{=} b_{\sigma\rho}(t_k)$. Introduce the following value:

$$\sum_{s=1}^{k} e^{-\lambda(t_k-t_s)} b_{\sigma\rho}(t_{s-1})(t_s - t_{s-1}) \overset{def}{=} \tilde{b}_{\sigma\rho}(t_k). \tag{35}$$

Formula (35) can be written in recursive form, similar to (32):

$$\tilde{b}_{\sigma\rho}(t_k) = e^{-\lambda(t_k-t_{k-1})} \tilde{b}_{\sigma\rho}(t_{k-1}) + b_{\sigma\rho}(t_{k-1})(t_k - t_{k-1}). \tag{36}$$

Consider deviation

$$\psi_{\sigma\rho}(t_k) \overset{def}{=} V_{\sigma\rho}^{(obs)}(t_k) - \tilde{b}_{\sigma\rho}(t_k). \tag{37}$$

For simplicity of discussion suppose that all the time steps $\Delta t_s = \Delta t$., we obtain: $E\psi_{\sigma\rho}(t_k) = o(\Delta t)$. That notation means that this small value is proportional to a value which decreases faster than Δt if Δt decreases.

The increments $\Delta y_\rho(t_s)$ and $\Delta y_\rho(t_q)$ with $t_q < t_s$ are independent when $(x(t_{s-1}), y(t_{s-1}))$ is given. We obtain that the dispersion of the deviation $\psi_{\sigma\rho}(t_k)$ (37) (under condition that $x(t)|_0^k$ is fixed) is equal to

$$Var\left\{\psi_{\sigma\rho}(t_k)\right\} \overset{\text{def}}{=} E\left\{\left(\psi_{\sigma\rho}(t_k)\right)^2 |x(t)|_0^k\right\}$$
$$= \sum_{s=1}^{k} e^{-2\lambda(t_k-t_s)} E\left\{\left(b_{\sigma\sigma}(t_{s-i}) b_{\rho\rho}(t_{s-1}) + \left[b_{\sigma\rho}(t_{s-1})\right]^2\right)|x(t)|_0^{s-1}\right\}(t_s - t_{s-1})^2. \tag{38}$$

In case if the diffusion coefficients of the observable process, $b_{\sigma\rho}(z,t)$,are the functions of x and t only, so that $b_{\sigma\rho}(z(t),t) = b_{\sigma\rho}(x(t),t)$ (with $(m+1) \le \sigma, \rho \le p$)we obtain

$$Var\left\{\psi_{\sigma\rho}(t_k)\right\} = \sum_{s=1}^{k} e^{-2\lambda(t_k-t_s)}\left(b_{\sigma\sigma}(t_{s-1}) b_{\rho\rho}(t_{s-1}) + \left[b_{\sigma\rho}(t_{s-1})\right]^2\right)(t_s - t_{s-1})^2$$
$$\overset{\text{def}}{=} \tilde{d}_{\sigma\rho}(x(t)|_0^k, t_k) \overset{\text{def}}{=} \tilde{d}_{\sigma\rho}(t_k). \tag{39}$$

The value (39) also can be calculated recursively, similarly to (36). *Note, that the value $\tilde{d}_{\sigma\rho}(t_k)$ is a value proportional to Δt.* If $\sigma = \rho$, the above expression takes more concise form:

$$\tilde{d}_{\sigma\rho}(t_k) = 2\sum_{s=1}^{k} e^{-2\lambda(t_k-t_s)}[b_{\rho\rho}(t_{s-1})]^2 (t_s - t_{s-1})^2. \tag{40}$$

In general case, if $b_{\sigma\rho}(z(t),t)$ depend on $y(t)$, we can find the estimates $E\left\{b_{\sigma\rho}(t_s)|x(t)|_0^{s-1}\right\} < K_{\sigma\rho} < \infty$, where $K_{\sigma\rho}$ may be determined either as some functions of $x(t)|_0^{s-1}$ or as some constants $K_{\sigma\rho} \le K_1$. Then for the variance (38), we obtain the following estimates:

$$Var\left\{\psi_{\sigma\rho}(t_k)\right\} < 2\sum_{s=1}^{k} e^{-2\lambda(t_k-t_s)}(K_{\sigma\rho}(x(t)|_0^{s-1}))^2(\Delta t_s)^2,$$

$$Var\left\{\psi_{\sigma\rho}(t_k)\right\} < (K_1)^2\left(\frac{1}{\lambda}\right)\Delta t. \tag{41}$$

The process $z(t)$ is considered on the finite time interval $t_0 \le s \le T$, so that the values of $K_{\sigma\rho}(x(t)|_0^{s-1})$ are limited. Thus, the variance of the deviation (37) is a value proportional to Δt, and if $\Delta t \to 0$ the observed quadratic variation tends to the limit (33). We shall demonstrate below, that the probability distribution for the deviation $\psi_{\rho\rho}$ or $\psi_{\sigma\rho}$ tends to the Gaussian one as $\Delta t \to 0$. It follows that the deviations converge to zero if $\Delta t \to 0$.

Besides, we obtained new analytical formulae (38)-(41), which describe the dispersion of the deviation (37) of the observed quadratic variation when observations are taken in discrete times, with small time steps.

For the system (28), the general analytical formulae (38)-(40), obtained in the present paper, provide characterization of the accuracy of the estimate of volatility $|x(t_k)|$, which is constructed with the use of the observed quadratic variation. If this accuracy satisfies requirements, the problem of estimation is solved. In that case the value of the given time step Δt can be considered as “small enough”.

In general case, if the value of $x(t_k)$ cannot be uniquely recovered only on the base of observed quadratic variations, the further filtering may be needed.

Finally, we are going to demonstrate that the probability density of the random value $\psi_{\rho\rho}(t_k)$ can be approximately described as a Gaussian one. The deviation $\psi_{\rho\rho}(t_k)$ contains the sum of squares of the increments $\Delta y_\rho(t_s)$, which are independent of the past history before t_{s-1} given $y(t_{s-1})$, $x(t_{s-1})$. The increments $\Delta y_\rho(t_s)$ (given $y(t_{s-1})$, $x(t_{s-1})$) can be described as the Gaussian random variables with probability density (16). The sum of squares of Gaussian random variables is not a Gaussian random variable. However, the probability density for the random deviation $\psi_{\rho\rho}(t_k)$ can be approximately described as the Gaussian probability density. Consider the case when $b_{\sigma\rho}(z(t),t)$ (with $(m+1) \le \rho, \sigma \le p$)are the functions of $x(t)$ and t only. Then the variance of deviation $\psi_{\rho\rho}(t_k)$ is equal to $\tilde{d}_{\rho\rho}(t_k)$ (40), which represents a value proportional to Δt (here the time step Δt can be small, but Δt is finite).

Note, that for a Gaussian random variable ξ with $E\xi = 0$ the following relations hold true: $E\left\{\xi^{2r}\right\} = Q(r)\left(E\left\{\xi^2\right\}\right)^r$, where $Q(r)$ is a constant, and for the odd moments $E\left\{\xi^{2r+1}\right\} = 0$, with $r = 1, 2, 3, \ldots$. Those relations can be useful in estimating of the moments of $\Delta y_\rho(t_s)$ when $(y(t_{s-1}), x(t_{s-1}))$ is given. Then the consideration of the higher moments of the deviation proves that $E\left(\psi_{\rho\rho}(t_k)\right)^q = o(\Delta t)$, if $q > 2$. Then the characteristic function of the random deviation $\psi_{\rho\rho}(t_k)$ takes the form similar to (12), which implies the Gaussian expression for its inverse transformation, similar to (13). Hence, the probability density of that deviation can be approximately described as a Gaussian probability density. The smaller Δt the higher the precision of that approximation, although the deviation itself tends to zero if $\Delta t \to 0$. But in practice the time step should not be chosen too small since in our mathematical model $y(t)$ is considered as a diffusion random process, which is not differentiable at any time t. That is not the case in practice where we have a smooth trajectory $\tilde{y}(t)$, and the quadratic variation of $\tilde{y}(t)$ is equal to zero. Diffusion approximation can be accepted when the time steps $\Delta t_s \ge \tau_{cor}$, where τ_{cor} represents the characteristic time span of decay of correlation, so that correlation between the random values $\Delta\tilde{y}(t_s)$ and $\Delta\tilde{y}(t_{s-1})$ (given $x(t_{s-1}), \tilde{y}(t_{s-1})$) would be negligibly small (Khazen, 1968, 1971, 1977; Chapters 2, 3; Khazen, 2009).

2.4. *Detection and rejection of highly a posteriori improbable samples*

In the process of resampling, the a posteriori improbable samples $x(t)_0^n$ are being rejected when their weights become small. Besides, it is possible to discard some highly a posteriori improbable samples before all the weights $W_i(n)$ or $\tilde{W}_i(n)$ are calculated. It can be done with the use of the following tests that are based on analysis of the obtained analytical expressions (19), (21), (22).

The measurements $y(t)|_0^k$ are being obtained under condition that there is an unobservable realization of the process $x(t)$, $x^{tr}(t)|_0^n$, which is unknown. (Here the superscript "tr" stands for "true"). The first and the second moments of the random variables $\Delta y_\rho(t_s)$ given $x^{tr}(t)|_0^k$ are equal to

$$E\left\{\Delta y_\rho(t_s)|x^{tr}(t_{s-1}), y(t_{s-1})\right\} = A_\rho(x^{tr}(t_{s-1}), y(t_{s-1}), t_{s-1})\Delta t_s \overset{\text{def}}{=} A_\rho^{tr}(t_{s-1})\Delta t_s, \tag{42}$$

$$E\left\{\Delta y_\sigma(t_s)\Delta y_\rho(t_s)|x^{tr}(t_{s-1}), y(t_{s-1})\right\} = b_{\sigma\rho}(x^{tr}(t_{s-1}), y(t_{s-1}), t_{s-1})\Delta t_s$$
$$+ o(\Delta t_s) \overset{\text{def}}{=} b_{\sigma\rho}^{tr}(t_{s-1})\Delta t_s + o(\Delta t_s)$$

For simplicity of discussion we can assume that all the time steps are equal to Δt.

Consider at first the case when the diffusion coefficients $b_{\sigma\rho}$ of the observable process $y(t)$ do not depend on $x(t)$; then $b_{\sigma\rho}^i = b_{\sigma\rho}^{tr} = b_{\sigma\rho}$.

The value $P_i(k)$ (19) contains the following cofactor:

$$Cexp\left\{-\frac{1}{2}\sum_{s=1}^{k}\frac{1}{\Delta t_s}c_{\sigma\rho}(t_{s-1})[\Delta y_\sigma(t_s)\Delta y_\rho(t_s)-2\Delta y_\sigma(t_s)A^i_\rho(t_{s-1})\Delta t_s+A^i_\sigma(t_{s-1})A^i_\rho(t_{s-1})(\Delta t_s)^2]\right\}, \tag{43}$$

where the normalization factor C does not depend on $x^i(t)|_0^k$ in that case.

Consider the hypothesis H_i that the sample $x^i(t)|_0^k$ is situated in the vicinity of the true realization $x^{tr}(t)|_0^k$, and the "closeness" means that $|A^i_\rho(t_{s-1})-A^{tr}_\rho(t_{s-1})|\le D_\rho(x^i(t_{s-1}),y(t_{s-1}),t_{s-1})\stackrel{\text{def}}{=}D^i_\rho(t_{s-1})$. Here the function $D_\rho(x,y,t)$ can be determined on the basis of some preliminary analysis of the considered dynamical system, for example, we can put $D_\rho(x,y,t)=\left|\frac{\partial A_\rho(x,y,t)}{\partial x_\alpha}\right|R_\alpha$, where R_α are constants, $1\le\alpha\le m$. In order to distinguish between two hypotheses, H_i and its negation $\bar{H}_i$, consider the following value:

$$Q^i(t_k)\stackrel{\text{def}}{=}\sum_{s=1}^{k}c_{\sigma\rho}(t_{s-1})A^i_\rho(t_{s-1})\Delta y_\sigma(t_s). \tag{44}$$

Note, that $Q^i(t_k)$ can be recursively calculated. Denote

$$G^i(t_k)\stackrel{\text{def}}{=}\sum_{s=1}^{k}c_{\sigma\rho}(t_{s-1})A^i_\sigma(t_{s-1})A^i_\rho(t_{s-1})\Delta t_s. \tag{45}$$

The larger the difference $\left|Q^i(t_k)-G^i(t_k)\right|$ the farther the sample $x^i(t)|_0^k$ can be from $x^{tr}(t)|_0^k$ Denote $\Delta\tilde{y}_\sigma(t_s)=\Delta y_\sigma(t_s)-A^{tr}_\sigma(t_{s-1})\Delta t_s$, $U^i(t_k)\stackrel{\text{def}}{=}\sum_{s=1}^{k}c_{\sigma\rho}(t_{s-1})A^i_\rho(t_{s-1})\,\Delta\tilde{y}_\sigma(t_s)$.

Consider the variance of the random value $U^i(t_k)$ when the realization $x^{tr}(t)|_0^k$ is given. We obtain

$$Var\{U^i(t_k)\}=\sum_{s=1}^{k}E\{c_{\sigma\rho}(t_{s-1})b_{\sigma\eta}(t_{s-1})c_{\eta\tau}(t_{s-1})A^i_\rho(t_{s-1})A^i_\tau(t_{s-1})|x^{tr}(t)|_0^s\}\,\Delta t_s.$$

If the matrix $\|b_{\sigma\rho}\|$ is not degenerated, then $b_{\sigma\eta}c_{\eta\tau}=\delta_{\sigma\tau}$, where $\delta_{\rho\tau}$ is Kronecker symbol, $\delta_{\sigma\tau}=1$ if $\sigma=\tau$ and $\delta_{\sigma\tau}=0$ if $\sigma\ne\tau$. In that case, the above expression for $Var\{U^i(t_k)\}$can be written in more concise form:

$$Var\{U^i(t_k)\}=\sum_{s=1}^{k}E\{c_{\sigma\rho}(t_{s-1})A^i_\rho(t_{s-1})A^i_\sigma(t_{s-1})|x^{tr}(t)|_0^s\}\,\Delta t_s \tag{46}$$

In order to reject the samples $x(t)|_0^k$, which are highly improbable given $y(t)|_0^k$, we can use approximately the following test. If the considered dynamic system is stable, the following value can be used as an approximate estimate for the above expression (46), if the hypothesis H_i is true and the measurements $y(t)|_0^k$ are obtained:

$$\tilde{V}ar\{U^i(t_k)\}=\sum_{s=1}^{k}c_{\sigma\rho}(t_{s-1})A^i_\rho(t_{s-1})A^i_\sigma(t_{s-1})\Delta t_s.$$

The probability of the following event:

$$\left|Q^i(t_k)-G^i(t_k)\right|>[\sum_{s=1}^{k}c_{\sigma\rho}(t_{s-1})A^i_\rho(t_{s-1})D^i_\sigma(t_{s-1})\Delta t_s+3\sqrt{\tilde{V}ar\left\{U^i(t_k)\right\}}] \tag{47}$$

is small under condition that H_i is true. Hence, if the inequality (47) is satisfied for the considered sample $x^i(t)|_0^k$, the hypothesis $H_i$should be rejected, and that sample $x^i(t)|_0^k$ should be discarded. If all the considered sample sequences are discarded, the new sample points $x(t_k)$ should be generated as initial points, i.e. as samples from the initial distribution $P_0(x(t_k)|y(t_k))$ (18). For the samples $x^i(t)|_0^k$ that remain under consideration the values $P_i(k)$ and $W_i(k)$ or $\tilde{W}_i(k)$ will be recursively computed. The samples that were not discarded can be continued with a few "offsprings". Due to the test (47) the samples $x^i(t)|_0^k$ that are highly a posteriori improbable (when the sequence of observations $y(t)|_0^k$ is given) will be rejected at once, before the weights $W_j(k)$ or $\tilde{W}_j(k)$ are computed.

Consider the case when some diffusion coefficients depend on $x(t)$. In some cases, as it was shown above, the unobservable value of $x(t)$ can be restored with the use of the observed quadratic variations. In other cases, if $x(t)$ cannot be uniquely recovered, it is purposeful to use observed quadratic variations $V_{\rho\rho}^{(obs)}(t_k)$ in order to reject the samples $x^i(t)|_0^k$ that are highly a posteriori improbable. That test is similar to the above test (47). Assume that if hypothesis H_i is true, the sample $x^i(t)|_0^k$ is "close" to the true realization $x^{tr}(t)|_0^k$, and the "closeness" means that

$$\left|b_{\rho\rho}\left(x^i(t_s),y(t_s),t_s\right) - b_{\rho\rho}(x^{tr}(t_s),y(t_s),t_s)\right| \le B_\rho(x^i(t_s),y(t_s),t_s) \stackrel{\text{def}}{=} B_\rho^i(t_s).$$

Here the function $B_\rho(x, y, t)$ can be determined as $B_\rho(x,y,t) = \left|\frac{\partial b_{\rho\rho}(x,y,t)}{\partial x_\alpha}\right| R_\alpha$, where R_α are constants. Denote shortly (similarly to (35), (36), (38),(39)):

$$\tilde{b}_{\sigma\rho}^i(t_k) \stackrel{\text{def}}{=} \tilde{b}_{\sigma\rho}(x^i(t)|_0^k, y(t)|_0^k, t_k) \quad \tilde{d}_{\sigma\rho}^i(t_k) \stackrel{\text{def}}{=} \tilde{d}_{\sigma\rho}(x^i(t)|_0^k, y(t)|_0^k, t_k),$$

$$\tilde{B}_\rho^i(t_k) \stackrel{\text{def}}{=} \tilde{B}_\rho\left(x^i(t)|_0^k, y(t)|_0^k, t_k\right) \stackrel{\text{def}}{=} \sum_{s=1}^{k} e^{-\lambda(t_k-t_s)} B_\rho\left(x^i(t_s), y(t_s), t_s\right)(t_s - t_{s-1})$$

$$\tilde{B}_\rho^i(t_k) = e^{-\lambda(t_k-t_{k-1})}\,\tilde{B}_\rho^i(t_{k-1}) + B_\rho^i(t_k)(t_k - t_{k-1})$$

Consider the case when $b_{\sigma\rho}(z(t),t) = b_{\sigma\rho}(x(t),t)$ (with $(m+1) \le \sigma, \rho \le p$). Then the probability of the following event

$$\left|V_{\rho\rho}^{(obs)}(t_k) - \tilde{b}_{\rho\rho}^i(t_k)\right| > \left[\tilde{B}_\rho^i(t_k) + 3\sqrt{b_{\rho\rho}^i(t_k)}\right] \tag{48}$$

is small if the hypothesis H_i is true. Hence, if the inequality (48) is satisfied at least for one number ρ, with $(m+1) \le \rho \le p$, then the hypothesis H_i should be rejected, and the sample $x^i(t)|_0^k$ should be discarded at once, before the calculation of all the weights $W_j(k)$ or $\tilde{W}_j(k)$ is done.

In case if $b_{\sigma\rho}(z(t),t) = b_{\sigma\rho}(x(t),y(t),t)$, with $E\{\left(b_{\sigma\rho}\right)^2 | x^{tr}(t)|_0^k\} \le K_1^2 = Const$, the following test can be used instead of (48):

$$\left|V_{\rho\rho}^{(obs)}(t_k) - \tilde{b}_{\rho\rho}^I(t_k)\right| > \left[\tilde{B}_\rho^i(t_k) + 3K_1\sqrt{\frac{1}{\lambda}\Delta t}\right] \tag{49}$$

2.5. Incorporation of the derivative of the process of quadratic variation into the set of observed data. Systems with some additional precise observations

In case of continuous time of observation, the process of quadratic variation can be included into the set of observed processes, as it was demonstrated in Section 2.3.

For simplicity of notation, consider the system of one-dimensional processes $x(t)$, $y(t)$:

$$dx = m(x,t)dt + \sigma_1(x,t)dw_1(t), \tag{50}$$

$$dy = h(x,t)dt + \sigma_2(x,t)dw_2(t), \tag{51}$$

where $w_1(t)$, $w_2(t)$ are independent Wiener processes, the functions $m(x, t)$, $h(x, t)$ satisfy Lipschitz condition, and the functions $h(x,t), \sigma_1(x,t)$, $\sigma_2(x, t)$ are twice differentiable with respect to x. The process

$$V(t) = \int_{t_0}^{t} e^{-\alpha(t-s)} \sigma_2^2(x(s), s) ds \tag{52}$$

can be considered theoretically as an observed process, given the trajectory $y(s)$for $t_0 \leq s \leq t$. Here $\alpha = Const \geq 0$.The process $V(t)$ obeys the ordinary differential equation

$$\frac{dV}{dt} = -\alpha V + \sigma_2^2(x(t), t).$$

For continuous time of observation and multidimensional diffusion processes $(x(t), y(t))$, in some cases the possible values of $x(t)$ are precisely determined by the equalities similar to (52) (and, consequently, $x(t)$ can be restored precisely, at least theoretically), and in other cases the trajectory $x(s)$ must be localized in some "layer", where those equalities are satisfied, given the observed path of the process of quadratic variation, $V(s)$, for $t_0 \leq s \leq t$.

Theoretically, it is possible to consider the derivative $\frac{dV(t)}{dt} \stackrel{def}{=} u(t)$ as an observed process, and incorporate it into the set of observed components: $(y(t), u(t))$. For the system (50), (51), the process $u(t)$ satisfies the following stochastic differential equation (in the sense of Ito):

$$du = -\alpha u dt + 2\sigma_2(x,t)\frac{\partial \sigma_2}{\partial x}\left[m(x,t)dt + \sigma_1(x,t)dw_1(t)\right] + \left[\sigma_2 \frac{\partial^2 \sigma_2}{\partial x^2} + \left(\frac{\partial \sigma_2}{\partial x}\right)^2\right]\sigma_1^2(x,t)dt. \tag{53}$$

The above SDE (53) could be augmented to the system of SDE (50)–(51) for $x(t)$, $y(t)$.

But in practice we cannot assume that the process $u(t)$ (or, equivalently, the process $\sigma_2^2(x(t), t)$) is straightforwardly observed and known precisely. Practically, the process $V(t_k)$ is not precisely given. Even in case if the deviations of "observed quadratic variation" $V^{(obs)}(t_k)$ from $V(t_k)$ are small, in obtaining the estimates for derivative, $\frac{dV(t)}{dt}|_{t=t_k} \stackrel{def}{=} u(t_k)$, the sought values could be inevitably corrupted by errors.

However, if additional observations of $y(t)$ are available, namely, $y(t_0 + s\delta t)$, with $\delta t \ll \Delta t$, $s = 1, 2, \ldots$, then the estimates $\hat{u}(t_k)$ for the values of $u(t_k)$ can be improved. Suppose that there is a special digital device that takes measurements of $y(t_s')$ at times $t_s' = t_0 + s\delta t$, with $t_{k-1} < t_s' \leq t_k$, and use them in order to calculate the estimate $\hat{u}\left(t_k\right)$. For the moment, consider the case if we had $\hat{u}\left(t_k\right) = u(t_k)$. Then the problem of filtering (based on observations $y(t_k)$, $u(t_k)$ taken at discrete times t_k, with small time steps Δt_k) could be solved effectively with the use of the new algorithm of simulation (17) and the estimates (26) and algorithms suggested in the present paper. For the system (50), (51), and (53), we obtain in (17): $b_{11} - b_{1\sigma} c_{\sigma\rho} b_{\rho 1} \equiv 0$. That means that the sample paths $x^i(t)|_1^n$ would be deterministically continued given the observations $u(t)|_1^n$ and the initial values $x^i(t_1)$, with $i = 1, \ldots, N$. It is purposeful to choose the probability distribution of the initial random samples $x^i(t_1)$ in such way that $E\left\{\sigma_2^2(x^i(t_1), t_1)\right\} = u(t_1)$, with some random scattering. Meanwhile, the weights $\tilde{W}_i(n)$ are being defined by (25), (26) with the use of observations $y(t)|_1^n$. Thus, the a posteriori distribution is being determined not only by $u(t)|_1^n$ but also by $y(t)|_1^n$. Then, as it was shown in Section 2.2, the estimate (26) converges with probability one (due to the Strong Law of Large Number, as $N \to \infty$) to the sought a posteriori expectation that provides solution to the filtering problem. That algorithm could be implemented with the use of parallel computing.

The random errors of the obtained estimates $\hat{u}(t_k)$ can be taken into account in the following way. The random deviations $\hat{u}(t_k) - u(t_k) \stackrel{def}{=} \varepsilon(t_k)$ can be approximately described as independent of each other (for different k) Gaussian random values. Then $\hat{u}(t_k) = u(t_k) + \varepsilon(t_k)$ can be considered as measurements (taken at discrete times t_k) of the process $\hat{u}(t_k)$ that satisfies the following SDE:

$$d\hat{u} = -\alpha\,\hat{u}\,dt + 2\sigma_2(x,t)\frac{\partial \sigma_2}{\partial x}\left[m(x,t)dt + \sigma_1(x,t)dw_1(t)\right] + \left[\sigma_2\frac{\partial^2 \sigma_2}{\partial x^2} + \left(\frac{\partial \sigma_2}{\partial x}\right)^2\right]\sigma_1^2(x,t)dt + \sigma_3 dw_3(t), \quad (54)$$

where $w_3(t)$ is an independent Wiener process, and $\sigma_3^2(t_k)$ is proportional to the variance of $\varepsilon(t_k)$, i.e. $E(\varepsilon^2(t_k))$. Thus, σ_3 can be determined as some function of x, t. Then the solution of the filtering problem for the system (50), (51), (54) can be obtained in the similar way as above.

Note that the systems that are similar to (50), (51), (53) arise if the measurements of some given function $H(x(t), t)$ are known without errors. Here the function $H(x, t)$ is supposed to be twice differentiable with respect to x. Then the solution of the filtering problem (given observations of $y(t)$ and $H(x(t), t)$ at discrete times $t_k = t_0 + k\Delta t$, with small time step Δt) can be obtained with the use of the algorithm of simulation (17) and the estimates (26), similarly to the solution described above for the system (50), (51), (53).

2.6. The problem of nonlinear filtering of a signal. Change of the mathematical model that describes the process of observations

For simplicity of notations, consider one-dimensional processes $x(t)$, $y(t)$. In most of works devoted to nonlinear filtering problems the following hidden Markov model is accepted for description of a signal process $x(t)$ and an observed process $y(t)$:

$$dx = m(x,t)dt + \sigma_1(x,t)dw_1(t), \quad (55)$$

$$dy = h(x(t),t)dt + \sigma_2 dw_2(t), \quad (56)$$

where $w_1(t)$, $w_2(t)$ are independent Wiener processes, $\sigma_2 \equiv Const$, the functions $m(x, t)$, $h(x,t)$, $\sigma_1(x, t)$ satisfy Lipschitz condition, and the functions $h(x, t)$, $\sigma_1(x, t)$ are twice differentiable with respect to x.

In many applications, it would be possible to describe an observed process $Y(t)$ as follows:

$$Y(t) = h(x(t),t) + u(t), \quad (57)$$

where $u(t)$ represents the errors of measurements. The "noise process" $u(t)$ can be considered as a Gaussian Markov process, which can be described by the following stochastic differential equation:

$$du = -\beta u dt + \sigma_3 dw_3(t), \quad (58)$$

where β and σ_3 are constant, and $w_3(t)$ is an independent Wiener process. Then $E\{u\} = 0$, the variance $E\left\{u^2\right\} = \sigma_3^2/2\beta$, and the correlation function $E\{u(t)u(t+\tau)\} = \left(\sigma_3^2/2\beta\right)e^{-\beta\tau}$.

Denote $z_1(t) \overset{def}{=} x(t)$, $z_2(t) \overset{def}{=} u(t)$, $z_3(t) \overset{def}{=} Y(t)$.

Then the model (55), (57), (58) of the signal process $x(t)$ and the observed process $Y(t)$ can be described as follows:

$$dz_1 = m(z_1,t)dt + \sigma_1(z_1,t)dw_1(t),\ dz_2 = -\beta z_2 dt + \sigma_3 dw_3(t),$$
$$dz_3 = \frac{\partial h(z_1,t)}{\partial t}dt + \frac{\partial h}{\partial z_1}\left[m(z_1,t)dt + \sigma_1(z_1,t)dw_1\right] + \frac{1}{2}\frac{\partial^2 h}{\partial z_1^2}\sigma_1^2(z_1,t)dt - \beta z_2 dt + \sigma_3 dw_3. \quad (59)$$

Here $z_3(t)$ represents the observed process, and $z_1(t)$, $z_2(t)$ are unobservable components of the process $z(t) =^{def} (z_1(t), z_2(t), z_3(t))$. It is easy to see that the diffusion coefficients $b_{11} = \sigma_1^2(z_1, t)$, $b_{13} = \frac{\partial h}{\partial z_1}\sigma_1^2(z_1, t)$, $b_{12} = 0$, $b_{22} = \sigma_3^2$, $b_{23} = \sigma_3^2$, $b_{33} = \left(\frac{\partial h}{\partial z_1}\sigma_1(z_1, t)\right)^2 + \sigma_3^2$. Consider the matrix $\| r_{\alpha\beta} \| = \| b_{\alpha\beta} - b_{\alpha\sigma}c_{\sigma\rho}b_{\rho\beta} \|$, which was introduced in (15): This is the variance-covariance matrix for the conditional probability density of the increments $\Delta z_1 = z_1(t + \Delta t) - z_1(t), \Delta z_2 = z_2(t + \Delta t) - z_2(t)$, given $z(t)$ and the increment $\Delta z_3 = z_3(t + \Delta t) - z_3(t)$. For the system (59) we find:

$$r_{11} = \sigma_1^2\left[1 - \frac{1}{1+\sigma_3^2/\left(\frac{\partial h}{\partial z_1}\sigma_1\right)^2}\right],$$

$$r_{12} = -\sigma_3^2\sigma_1^2\frac{\partial h}{\partial z_1}\left[\left(\frac{\partial h}{\partial z_1}\sigma_1\right)^2 + \sigma_3^2\right]^{-1}, r_{22} = \sigma_3^2\left[1 - \sigma_3^2\left[\left(\frac{\partial h}{\partial z_1}\sigma_1\right)^2 + \sigma_3^2\right]^{-1}\right], \text{and Det } \| r_{\alpha\beta} \| \equiv 0.$$

The case when σ_3 is small corresponds to the small errors of measurements. All the entries of the matrix $\| r_{\alpha\beta} \|$ (shown above) is small if σ_3 is small. That means that most of the samples $(z_1(t), z_2(t))|_0^n$ (given $z_3(t)|_0^n$) simulated with the use of the algorithm (17) continue to be localized in the domain where the a posteriori probability density is concentrated. Besides, the test (48) can be implemented (unless the function $h(x, t)$ is linear with respect to $x$and $\sigma_1 \equiv Const$) in order to discard a posteriori improbable samples with the use of "observed quadratic variation" (for the observations made in discrete time) of the observed process $z_3(t)$.

Thus, although the dimensionality of the unobservable process increased for the model of description (59) in comparison with the model (55), (56), the problem of nonlinear filtering (with observations made in discrete time) will be solved effectively with the use of the new Monte Carlo estimates (23)–(27) and with the use of the new algorithm of simulation (17), derived in the present paper.

Note that the model (59) is a hidden Markov model where unobservable process $(z_1(t), z_2(t))$ is a Markov process in itself. But if the samples $(z_1(t), z_2(t))|_0^n$ were simulated as trajectories of the Markov process $(z_1(t), z_2(t))$ in itself (as it is suggested in particle filters, developed and studied in the literature), then, in case of the low "level of intensity of noise" (i.e. if σ_3 is small) the large part of generated samples would be localized in the domain of low a posteriori probability (and that part is growing when σ_3 decreases). Then, in order to obtain some feasible accuracy of filtering, the number N should be increased significantly if the value of σ_3 decreased. Meanwhile, the large amount of calculations would be wasted in vain for processing of the samples of low a posteriori probability. Thus, this example (the system (59)) also proves the advantage of the new algorithms, developed in the present paper, in comparison with the known algorithms for hidden Markov models.

3. Conclusion

In the Conclusion, the new results obtained and presented by this author are shortly summarized and underscored.

(1) It was analytically proved by this author at first in the book (Khazen, 2009) that the increment of a multidimensional continuous diffusion Markov process $z(t)$, $\Delta z = z(t + \Delta t) - z(t)$, in small time interval Δt, given $z(t)$, obeys asymptotically to the multidimensional Gaussian probability distribution (with the first and second moments that are determined by its drift and diffusion coefficients; see expression (13)). As a corollary, the analytical expressions (14)–(16) are obtained, which describe Gaussian conditional probability density for the increment Δx of the unobservable components given observation of Δy and given the value $z(t)$; here $\Delta z = (\Delta x, \Delta y)$. The Gaussian probability density for Δy is also obtained (see (16)). In the present paper the new, precise algorithm (in closed analytical form) is obtained for simulation of samples $x(t_k)$ when $x(t)|_0^{k-1}$ and $y(t)|_0^k$ are given (see (17)). It is important that the influence of the current "next" observation $y(t_k)$ is taken into account when samples $x(t_k)$ are being simulated.

(2) The new Monte Carlo estimates of functions $f(x(t_n))$ given $y(t)|_0^n$ are obtained and presented in explicit, precise, closed analytical form, under the condition that the samples $x(t)|_0^n$ are being simulated with the use of the new proposed algorithm (17) (see (26), (25), (27)). The convergence of the Monte Carlo estimates (as the number N of samples increases, $N \to \infty$) is guaranteed by the Strong Law of Large Number (the Theorem of Kolmogorov). For the first time, the estimates, (26), (25), (27), are obtained for *the general case* of multidimensional diffusion Markov process $(x(t), y(t))$. In the particular case when $x(t)$ is a Markov process in itself and the process $(x(t), y(t))$ represents the simple Hidden Markov Model described by (28), the estimate (26) takes the form that is in agreement with the estimate that was obtained earlier in this particular case.

(3) For the first time, some tests are developed in order to discard at once the sample sequences $x(t)|_0^n$ that are highly improbable given the observation $y(t)|_0^n$ (see Section 2.4). This is important in order to prevent the "degeneration" of the set of considered samples. These tests are being performed independently for each sample $x(t)|_0^n$, and that is important also for the implementation with the use of parallel computing.

(4) Some branching sampling procedures are developed. The standard Sequential Importance Resampling (SIR) procedure is generalized for the general case of partially observed multidimensional diffusion Markov process, when the new algorithm of simulating of samples (17) should be used. Some new versions of branching resampling procedures are proposed, which can be easier implemented with the use of parallel computing.

(5) Analytical investigation of observed quadratic variations is developed (in case when diffusion coefficients of observed components, $y(t)$, depend on unobservable components $x(t)$) (see Section 2.5). For the first time, the analytical formulae (in explicit and closed form) that determine the dispersions of deviations of observed quadratic variations are obtained (see (38), (39)), and it is also proved that the deviations obey asymptotically (when $\Delta t \to 0$) to the Gaussian distribution.

(6) The important particular case of nonlinear filtering of a signal is considered in Section 2.6. Significant advantage of the new algorithms and estimates obtained in the present paper in comparison with the particle filters suggested earlier in the literature for Hidden Markov Models is demonstrated.

(7) In Section 2.5 it is demonstrated that the new filtering algorithms, developed in the present paper, provide opportunity to incorporate the observed quadratic variations into the set of observed data. They also provide opportunity to use additional precise observations in case if such observations are available. These results are new, and they also confirm the advantage of the proposed new solution of the nonlinear filtering problem.

The obtained new algorithms and estimates extend the range of applications of particle filtering beyond Hidden Markov Models and improve performance.

Funding
The author received no direct funding for this research.

Author details
Ellida M. Khazen[1]
E-mail: ellida_khazen@yahoo.com
[1] 13395 Coppermine Rd. Apartment 410, Herndon, VA 20171, USA.

References

Bernstein, S. N. (1934/1964). *Principes de la theorie des equations differentielles stochastiques*. Proceeding of the Phys.-Mat. Steklov Institute, 5, 95–124. Republished in Collected works of S.N. Bernstein (Vol. 4). Nauka: Moscow Publishing House (in Russian).

Bernstein, S. N. (1938/1964). Equations differentielles stochastiques. *Actualités Scientifiques et Industrielles, 738*, 5- 31. Conference International Sci. math. Univ. Geneve. Theorie des probabilites, V. Les fonctions aleatoires. Republished in Collected works of S.N. Bernstein, vol. 4. Nauka: Moscow Publishing House. (in Russian).

Carvalho, H., Del Moral, P., Monin, A., & Salut, G. (1997). Optimal nonlinear filtering in GPS/INS integration. *IEEE Transactions on Aerospace and Electronic Systems, 33*, 835–850. http://dx.doi.org/10.1109/7.599254

Crisan, D., & Doucet, A. (2002). A survey of convergence results on particle filtering methods for practitioners. *IEEE Transactions on Signal Processing, 50*, 736–746. http://dx.doi.org/10.1109/78.984773

Del Moral, P. (1998). Measure-valued processes and interacting particle systems. Application to nonlinear filtering problems. *Annals of Applied Probability, 8*, 438–495.

Del Moral, P., & Doucet, A. (2014). Particle methods: An introduction with applications. *ESAIM: Proceedings, 44*, 1–46. http://dx.doi.org/10.1051/proc/201444001

Doob, J. L. (1953). *Stochastic processes*. John Wiley & Sons, Chapman & Hall, New York, NY, London.

Doucet, A., & Johansen, A. (2011). A tutorial on particle filtering and smoothing: Fifteen years later. In D. Crisan & B. Rozovsky (Eds.), *The Oxford handbook of nonlinear filtering* (39 pp.). Oxford: Oxford University Press. Retrieved from http://www.stats.ox.ac.uk/~doucet/doucet_johansen_tutorialPF2011.pdf

Doucet, A., Godsill, S., & Andrieu, C. (2000). On sequential Monte Carlo sampling methods for Bayesian filtering. *Statistics and Computing, 10*, 197–208. http://dx.doi.org/10.1023/A:1008935410038

Doucet, A., Freitas, N., & Gordon, N. (Eds.). (2001). *Sequential Monte Carlo methods in practice*. Information science and statistics. New York, NY: Springer Verlag.

Gnedenko, B. V. (1976). *The theory of probability*. Moscow: Mir.

Hendeby, G., Karlsson, R., Gustafsson, F. (2010). Particle Filtering: The Need for Speed. *EURASIP Journal of Advances in Signal Processing, 2010*, Article ID 181403. doi:10.1155/2010/181403

Hull, J. C. (2000). *Options, Futures, & Other derivatives* (4th ed.). Upper Saddle River, NJ: Prentice Hall.

Khazen, E. M. (1968). *Methods of optimal statistical decisions and optimal control problems* (in Russian). Moscow: Soviet Radio.

Khazen, E. M. (1971). On stochastic differential equations for the a posteriori probability distribution in problems of adaptive filtering and signal detection. *Automation and Remote Control, 32*, 1776–1782.

Khazen, E. M. (1977). Chapters 2, 3, 5 by Khazen, E.M., pp. 47-66, 67-102, 193-243, in the book: Petrov, B.N., Ulanov, G.M., Ulyanov, S.V., Khazen, E.M. *Information theory in processes of optimal control and organization*. Nauka: Moscow Publishing House. (in Russian).

Khazen, E. M. (2009). *Methods of optimal statistical decisions, optimal control, and stochastic differential equations*. Bloomington, IN: Xlibris.

Kolmogorov, A.N. (1931/1986). Uber die analytischen Methoden in der Wahrscheinlichkeitsrechnung. Math. Ann., Bd. 104, S. 413-458. Russian translation in "Uspekhi Mat. Nauk", 1938, issue 5, pages 5-41. Republished in *Collected works of A.N. Kolmogorov, The Theory of probability and mathematical statistics*. Nauka: Moscow Publishing House. (in Russian).

Kolmogorov, A.N. (1933/1950). *Foundations of the Theory of Probability*. Published at first in German as "Grundbegriffe der Wahrscheinlichkeitsrechnung" in (1933). Russian editions in 1936, 1974 (English ed.). New York, NY: Chelsea.

Kolmogorov, A. N. (1933/1986). Zur Theorie der stetigen zufalligen Prozesse. Math. Ann., Bd. 108, S. 149–160. Republished in *Collected works of A.N. Kolmogorov, The Theory of Probability and Mathematical Statistics*. Nauka: Moscow Publishing House. (in Russian).

Kushner, H. (1971). *Introduction to stochastic control*. New York,NY: Holt, Rinchart, and Winston.

Míguez, J., Crisan, D., & Djurić, P. (2013). On the convergence of two sequential Monte Carlo methods for maximum a posteriori sequence estimation and stochastic global optimization. *Statistics and Computing, 23*, 91–107. http://dx.doi.org/10.1007/s11222-011-9294-4

Thrun, S., Fox, D., Burgard, W., & Dellaert, F. (2001). Robust Monte Carlo localization for mobile robots. *Artificial Intelligence, 128*, 99–141. http://dx.doi.org/10.1016/S0004-3702(01)00069-8

14

The iteratile: A new measure of central tendency

Blane Hollingsworth[1*]

*Corresponding author: Blane Hollingsworth, Scholar, Mathematics Department, Indiana University, Bloomington, IN, USA
E-mail: blane.hollingsworth@gmail. com.

Reviewing editor: Nengxiang Ling, Hefei University of Technology, China

Abstract: This article defines a new measure of central tendency (called the "iteratile") by iterating a function which maps a triple of data into a triple of the median, mean, and trimean (sorted). An explicit formula is given with proof, along with brief discussion of potential applications.

Subjects: Applied Mathematics; Mathematics & Statistics; Science

Keywords: mean; median; outlier; central tendency

The ideas in DiMarco and Savitz (2013, 2012) take N values of data and study all their m-tiles, which separate the data in m sections, for $2 \leq m \leq N+1$ (where the 2-tile is the median, the 4-tiles are quartiles, and the $(N+1)$-tile (the sample mean)). Observing that N values of data have N-many m-tiles, the author wondered if mapping all the data to their m-tiles and iterating would have a limit, and if so, what that limit would be.

In the case N = 3, we can get a formula. More precisely, given data (a, b, c), for $a \leq b \leq c$, *define* $f^*(a,b,c) = (b, \frac{a+2b+c}{4}, \frac{a+b+c}{3})$, *and then call f(a, b, c) the sorted version of* f^*, *where the components are in increasing order. Denote f composed with itself n times by* $f^{(n)}$.

THEOREM 1 $\lim_{n\to\infty} f^{(n)}(a,b,c)$ *exists and is of the form (I,I,I), where* $I = \frac{3a+8b+3c}{14}$. *(We call the limit the "iteratile" in what follows.)*

Proof If $a = b = c$, we are done, so assume $a < b < c$.

Note first that the limit exists; f is a contraction mapping since each of b, $\frac{a+2b+c}{4}$, *and* $\frac{a+b+c}{3}$ *must be greater than a and less than c.*

We find the general formula with the aid of two lemmas. The first allows us to focus only on the behavior of the middle term, and the second shows by induction how the middle term exhibits geometric series behavior.

LEMMA 1 *f(a, b, c) takes either the form:* $\left(b, \frac{a+2b+c}{4}, \frac{a+b+c}{3}\right)$ or $\left(\frac{a+b+c}{3}, \frac{a+2b+c}{4}, b\right)$.

Proof By cases: Case 1: $b - a = c - b$.

The assumption is equivalent to $2b = a + c$, so $\frac{a+2b+c}{4} = \frac{4b}{4} = b$, and a similar argument yields $\frac{a+b+c}{3} = b$.

ABOUT THE AUTHOR

Blane Hollingsworth received his PhD from Auburn University in stochastic differential equations, and is currently a visiting scholar at Indiana University. Currently, he is studying alternatives to means and medians as measures of central tendency, especially for small sets of data with outliers/ errors.

PUBLIC INTEREST STATEMENT

Often, small amounts of data and outliers make it difficult to do statistics. To measure the "typical" behavior, we usually use means or medians, but sometimes even those have problems. We offer an alternative way of measuring the center here.

Thus, $f(a,b,c) = (b,b,b)$, so $I = b$.

Case 2: $b - a > c - b$.

The assumption is equivalent to $2b > a + c$. Thus, $\frac{a+2b+c}{4} < \frac{2b+2b}{4} = b$ and $\frac{a+b+c}{3} < \frac{3b}{b} = b$. Further, $\frac{a+b+c}{3} - \frac{a+2b+c}{4} = \frac{a-2b+c}{12}$ which is negative by assumption. Thus,

$$b > \frac{a+2b+c}{4} > \frac{a+b+c}{3}.$$

Case 3: $b - a < c - b$. Similar arguments to Case 2 yield: □

$$b < \frac{a+2b+c}{4} < \frac{a+b+c}{3}.$$

Thus, it remains to study the difference from one iteration to the next in the middle term. Now, call $f^{(n)}(a,b,c) = (a_n, b_n, c_n)$ for convenience (so $a_0 = a$, etc.). We now give a formula for $b_{n+1} - b_n$ which will quickly complete the proof.

LEMMA 2 *For any* $n \in \mathbb{N}$, $b_{n+1} - b_n = \frac{a+c-2b}{4(-6)^n}$.

Proof By induction, we see that for $n = 0$:

$$b_1 - b_0 = \frac{a+2b+c}{4} - b = \frac{1}{4}(-a-c+2b) = \frac{1}{4}\frac{a+c-2b}{(-6)^0}.$$

For $n = 1$, first observe by Lemma 1 that

$$b_2 = \frac{1}{4}\frac{5a+14b+5c}{6}, \qquad (1)$$

since (without loss of generality) by Lemma 1,

$$b_2 = \frac{a_1+2b_1+c_1}{4} = \frac{1}{4}\left[\frac{a+b+c}{3} + \frac{a+2b-c}{4} + b\right] = \frac{1}{4}\frac{5a+14b+5c}{6}.$$

So,

$$b_2 - b_1 = \frac{1}{4}\frac{5a+14b+5c}{6} - \frac{a+2b+c}{4} = \frac{1}{4}\frac{a+c-2b}{(-6)^1}. \qquad (2)$$

Now, suppose that $b_{n+1} - b_n = \frac{a+c-2b}{4(-6)^n}$ for an arbitrary $n \in \mathbb{N}$. We want to show that it holds for $n+1$, so we study $b_{n+2} - b_{n+1}$.

An argument similar to (1) gives the relation:

$$b_{n+2} - b_{n+1} = \frac{1}{4}\frac{a_n + c_n - 2b_n}{(-6)}. \qquad (3)$$

So, Lemma 1 again gives

$$b_{n+2} - b_{n+1} = \frac{1}{4}[a_{n+1} - a_n + 2(b_{n+1} - b_n) + c_{n+1} - c_n], \qquad (4)$$

and it seems we need to study $a_{n+1} - a_n$ and $c_{n+1} - c_n$ to continue.

Again, Lemma 1 tells us that $a_{n+1} - a_n$ will either be of the form $b_n - a_n$ or $\frac{a_n+b_n+c_n}{3} - a_n$, and similarly for $c_{n+1} - c_n$. Luckily, we need to only consider the sum of $a_{n+1} - a_n$ and $c_{n+1} - c_n$, so without loss of generality we know,

$$a_{n+1} - a_n + c_{n+1} - c_n = \frac{a_n + b_n + c_n}{3} + b_n - a_n - c_n = \frac{1}{6}[8b_n - 4a_n - 4c_n]. \tag{5}$$

Now we have what we need to prove the claim. By (4),

$$b_{n+2} - b_{n+1} = \frac{1}{4}[a_{n+1} - a_n + 2(b_{n+1} - b_n) + c_{n+1} - c_n],$$

and by (5) and induction hypothesis,

$$= \frac{1}{4}\left[\frac{1}{6}(8b_n - 4a_n - 4c_n) + 2\left(\frac{1}{4}\frac{a+c-2b}{(-6)^n}\right)\right]$$
$$= \left[\frac{1}{6}(2b_n - a_n - c_n) + 2\left(\frac{1}{4}\frac{a+c-2b}{(-6)^n}\right)\right].$$

Using (3), we get the equation

$$b_{n+2} - b_{n+1} = 4(b_{n+2} - b_{n+1}) + \frac{1}{2}\frac{a+c-2b}{(-6)^n}.$$

We can write this as:

$$-3(b_{n+2} - b_{n+1}) = \frac{1}{2}\frac{a+c-2b}{(-6)^n},$$

which reduces to

$$b_{n+2} - b_{n+1} = \frac{a+c-2b}{(-6)^{n+1}},$$

and this proves the claim.

The upshot of Lemma 2 is that each subsequent iteration "steals $\frac{1}{6}$ from a (and from c) and gives it to b," which means we can set up the equation

$$\sum_{n=0}^{\infty} b_{n+1} - b_n = \sum_{n=0}^{\infty} \frac{a+c-2b}{4(-6)^n}.$$

The left-hand side is a telescoping series, and we know that $\lim_{n\to\infty} b_n = I$. Also, the right-hand side is a simple geometric series, thus the above becomes

$$I - b = \frac{3}{14}(a + c - 2b),$$

or

$$I = \frac{1}{14}(3a + 8b + 3c),$$

as claimed. □

Can this be extended to higher n? Unfortunately, there does not exist a universally agreed-upon notion of "percentile"; different definitions will yield different iteratiles. Thus, without specifying exactly which notion is desired, one cannot expect a general result. It seems that one gets a "symmetric convex combination" in any case.

How "good" or "useful" is the iteratile? One potential use is when there are errors in the data, so the data-set changes. What measure of central tendency minimizes the difference? Results along these lines are discussed in DiMarco, Hollingsworth, and Savitz (2015) in the context of z-scores and normal distributions.

For example, if the data-set (0, 0, 24) were changed to (0, 3, 16), then the difference in the iteratiles would be zero. But of course, one is not allowed to see the data change before selecting the appropriate measure. How does one pick a central tendency measure a priori to minimize the difference in measure for an arbitrary data change? There is a general trend: when data in the middle is (expected to be) altered, use the mean, and when data at the extremes are altered, use the median. When there is an "inbetween" for what to use? At least, relative to the trimean, the iteratile should be used if values toward the extreme are altered more, and the trimean when values toward the middle are altered more. Also, it seems that skewed alterations suggest use of iteratile.

In general, it is quite crude to have to resort to the median, as so much information is unaccounted for. The iteratile gives a bit more of that information and reduces the effects of outliers. As one piece of data gets arbitrarily large, the non-median measures will go to infinity, but at least, with different speeds (mean: 1/3, trimean: 1/4, iteratile: 3/14). As a result, perhaps using the iteratile is a decent alternative to the median. But still, is there anything special about the iteratile? One could, for example, merely make a measure of central tendency like $\frac{3a+10b+3c}{16}$, which would do even better ... but would it be optimal? Is there really something coming from the limit? These are complicated, longstanding, and potentially unanswerable questions, and even if the iteratile itself turns out to be inferior, perhaps the idea of iteratile may lead to other more useful alternatives and generalizations. At the very least, those who appreciate calculus may simply enjoy the result at a purely esthetic level.

Funding
The author received no direct funding for this reserach.

Author details
Blane Hollingsworth[1]
E-mail: blane.hollingsworth@gmail.com
[1] Mathematics Department, Indiana University, Bloomington, IN, USA.

References

DiMarco, D., & Savitz, R. (2012). The M-tile means, A new class of measures of central tendency: Theory and applications. *JIMS, 12*, 48–56.

DiMarco, D., & Savitz, R. (2013). The M-tile deviation: A new class of measures of dispersion. *International Journal of Business Research, 13*, 117–124.

DiMarco, D., Hollingsworth, B., & Savitz, R. (2015). On resistant versions of the standard score. *European Journal of Marketing, 15*, 7–16.

15

On the Bayesianity of minimum risk equivariant estimator for location or scale parameters under a general convex and invariant loss function

Amir T. Payandeh Najafabadi[1*]

*Corresponding author: Amir T. Payandeh Najafabadi, Department of Mathematical Sciences, Shahid Beheshti University, G.C. Evin, 1983963113 Tehran, Iran

E-mail: amirtpayandeh@sbu.ac.ir

Reviewing editor: Zudi Lu, University of Southampton, UK

Abstract: The Minimum Risk Equivariant (MRE), estimator is a widely used estimator which has several well-known theoretical and practical properties. It is well known that for the square error and absolute error loss functions, the MRE estimator is a generalized Bayes estimator. This article investigates the potential Bayesianity (or generalized Bayesianity) of the MRE estimator under a general convex and invariant loss function, $\rho(\cdot)$, for estimating the location and scale parameters of an unimodal density function.

Subjects: Mathematical Statistics; Mathematics & Statistics; Statistics; Science; Statistics & Probability

Keywords: Bayes estimator; the Fourier transform; the Minimum Risk Equivariant (MRE) estimator; convex and invariant loss functions

AMS subject classifications: 62F10; 62F30; 62C10; 62C15; 35Q15; 45B05; 42A99

1. Introduction

Compare with the uniform minimum variance unbiased estimator, the Minimum Risk Equivariant (MRE) estimator: (1) typically exists not only for convex loss function but also even for non-restricted loss functions and (2) does not need to consider randomized estimators (Lehmann & Casella, 1998, p. 156). Moreover, the MRE estimator is a widely used estimator which has several well-known theoretical (such as minimaxity and admissibility under some certain conditions) and practical properties. The MRE estimator has a wide range of applications in Finite sampling framework (Chandrasekar & Sajesh, 2013; Ledoit & Wolf, 2013), Reliability (Chandrasekar & Sajesh, 2013; Wei, Song, Yan, & Mao, 2000), Regression and non-linear models (Grafarend, 2006; Hallin & Jurečková, 2012), Contingency tables (Lehmann & Casella, 1998), Economic Forecasting (Elliott & Timmermann, 2013), etc.

ABOUT THE AUTHOR

Amir T. Payandeh Najafabadi is an Associate Professor in Department of Mathematics Sciences at Shahid Behashti University, Tehran, Evin. He was born on September 3, 1973. He received his PhD from University of New Brunswick, Canada in 2006. He has published 28 papers and was co-author of two books. His major research interests are: Statistical Decision Theory, Lévy processes, Risk theory, Riemann-Hilbert problem, & integral equations.

PUBLIC INTEREST STATEMENT

This article investigates the potential Bayesianity (or generalized Bayesianity) of the MRE estimator under a general convex and invariant loss function, $\rho(\cdot)$, for estimating the location and scale parameters of an unimodal density function.

It is well known that for the square error and the absolute error loss functions, the MRE estimator is a generalized Bayes estimator. This article extends this fact to a class of convex and invariant loss functions for unimodal location (or scale) density functions. This fact has been proven remarkably useful in solving a variety of problems in statistics. Subjects for which the problem is applicable range from truncated data (Manrique-Vallier & Reiter, 2014), imputation omitted data due to undercounting (Rubin, Gelman, & Meng, 2004, §12), Capture–recapture Estimation (Mitchell, 2014), etc.

The problem of finding a prior distribution that its corresponding Bayes estimator under a given loss function coincides with a given estimator started by Lehmann (1951). In his seminal paper, he considered a situation that a Bayes estimator under the squared error loss function is an unbiased estimator. His work has been followed and extended by several authors. For instance, Noorbaloochi and Meeden (1983, 2000) expanded Lehmann's (1951) finding for a general class of prior distributions. Kass and Wasserman (1996) reviewed the problem of selecting prior distributions that their corresponding Bayes estimators are invariance under some sort of transforms. Meng and Zaslavsky (2002) considered a class of single observation unbiased priors (i.e. such priors produced unbiased Bayes estimator under squared error loss function). They showed that under mild regularity conditions, such class of priors must be "noninformative" for estimating either *location* or *scale* parameters. Gelman (2006) constructed a non-central *t* student family of conditionally conjugate priors for hierarchical standard deviation parameters. For restricted parameter space, Kucerovsky, Marchand, Payandeh, and Strawderman (2009) provided a class of prior distributions which their corresponding Bayes estimator under absolute value error loss equal to the maximum likelihood estimator. Ma and Leijon (2011) found a conjugate beta mixture prior such that its corresponding Bayes estimator under the variational inference framework retains some given properties.

This paper provides a class of prior distributions that their corresponding Bayes estimator under general convex and invariant loss function coincides with the MRE estimator for location or scale family of distributions.

Section 2 collects some required elements for other sections. The problem of finding such prior distribution for location and scale parameters have been studied in Sections 3 and 4, respectively.

2. Preliminaries

Bayes estimator for an unknown parameter θ under a general loss function ρ has been evaluated from the posterior distribution $\pi(\theta|x)$. Therefore, to study some specific properties of a Bayes estimator, one has to study posterior distribution $\pi(\theta|x)$. The following provides a condition which leads to equivalent Bayes estimator under two prior distributions.

LEMMA 1 *Suppose X is a continuous random variable with density function f. Moreover, suppose that π_1 and π_2 are two priors distributions which lead to Bayes estimators δ_{π_1} and δ_{π_2}, under a general loss function ρ, respectively. Then, two Bayes estimators δ_{π_1} and δ_{π_2}, are equivalent estimator (i.e. $\delta_{\pi_1}(x) \equiv \delta_{\pi_2}(x)$) if and only if $\pi_1(\theta) = c\pi_2(\theta)$, for all $\theta \in \Theta$.*

Proof Bayes estimators with respect to π_1 and π_2 are equivalent if and only if the posterior distribution $\theta|x$ under these priors are equivalent, i.e.

$$\frac{\pi_1(\theta)f(x,\theta)}{\int_\Theta \pi_1(\theta)f(x,\theta)d\theta} = \frac{\pi_2(\theta)f(x,\theta)}{\int_\Theta \pi_2(\theta)f(x,\theta)d\theta}$$
$$\Leftrightarrow \frac{\pi_1(\theta)}{\pi_2(\theta)} = \frac{\int_\Theta \pi_1(\theta)f(x,\theta)d\theta}{\int_\Theta \pi_2(\theta)f(x,\theta)d\theta}$$

The rest of proof arrives from the fact that left hand side of the above equation is a function of θ while the right hand side is a function of x.

The following from Marchand and Payandeh (2011) recalls the Bayes estimator under generalized loss function ρ for location parameter μ.

LEMMA 2 *(Marchand & Payandeh, 2011) Suppose random variable X sampled from a location density function g_0. Moreover, suppose that $\delta_\pi(x)$ stands for the Bayes estimator under generalized loss function ρ_1 and prior distribution $\pi(\mu)$ for location parameter μ. Then, $\delta_\pi(x)$ satisfies*

$$\int_{-\infty}^{\infty} \rho_1'(\delta_\pi(x) - \mu) g_0(x - \mu)\pi(\mu) d\mu = 0, \quad x \in R \tag{1}$$

Now, we extend the above result for the problem of finding a Bayes estimator for a scale parameter θ, under general loss function ρ_2 and prior distribution $\tau(\theta)$.

LEMMA 3 *Suppose random variable Y sampled from a scale density function f_1. Moreover, suppose $\delta_\pi^*(y)$ stands for the Bayes estimator under generalized loss function ρ_2 and prior distribution $\tau(\theta)$ for a scale parameter θ. Then, $\delta_\pi^*(y)$ satisfies*

$$\int_0^{\infty} \rho_2'\left(\frac{\delta_\pi^*(y)}{\theta}\right) f_1\left(\frac{y}{\theta}\right) \frac{\pi(\ln(\theta))}{\theta^3} d\theta = 0, \quad y \geq 0 \tag{2}$$

where $\tau(\theta) = \pi(\ln(\theta))/\theta$.

Proof Marchand and Strawderman (2005) provided a connection between a scale parameter estimation problem, with elements (Y, θ, f_1, ρ_2) and a location parameter estimation problem with elements (Y, μ, g_0, ρ_1). They showed that by choosing $Y = ln(X)$, $\mu = \ln(\theta)$, $g_0(z) = e^z f_1(e^z)$, and $\rho_1(z) = \rho_2(e^z)$ transformations the problem of finding a Bayes estimator for a scale parameter θ under general loss function ρ_2 and prior distribution $\tau(\theta)$ can be restated as a problem of finding a Bayes estimator for location parameter μ under general loss function ρ_1 and prior distribution $\pi(\mu)$. Moreover, they showed that the Bayes estimator $\delta_\tau(x)$ of a location parameter μ with respect to prior $\tau(\mu)$ and the Bayes estimator $\delta_\pi^*(y)$ of a scale parameter θ with respect to prior $\pi(\theta)$ satisfies $\delta_\pi^*(y) = \exp\{\delta_\tau(\ln(x))\}$ and $\pi(\theta) = \tau(\ln(\theta))/\theta$. The desired proof arrives from the above observations.

The Fourier transform is an integral transform which defines for an integrable and real-valued $f: \mathbb{R} \to \mathbb{C}$ by

$$\mathcal{F}(f(t); t; \omega) := \int_{\mathbb{R}} f(t) e^{-i\omega t} dt, \quad \omega \in \mathbb{C} \tag{3}$$

The convolution theorem for the Fourier transforms states that

$$\mathcal{F}\left(\int_{\mathbb{R}} f(x-t) g(t) dt; x; \omega\right) = \mathcal{F}\left(f(x); x; \omega\right) \mathcal{F}\left(g(x); x; \omega\right) \tag{4}$$

(see Dym & Mckean, 1972 for more details).

Now, we recall definition and some properties of exponential type functions which are used later in proof of Theorem 4.

Definition 1 A function f in $L_1(\mathbb{R}) \cap L_2(\mathbb{R})$ is said to be an exponential type T function on the domain $D \subseteq \mathbb{C}$ if there are positive constants M and T such that $|f(\omega)| \leq M \exp\{T|\omega|\}$, for $\omega \in D$.

The Paley-Wiener theorem states that the Fourier (or inverse Fourier) transform of an $L_2(\mathbb{R})$ function vanishes outside of an interval $[-T, T]$, if and only if the function is an exponential type T (Dym & McKean, 1972, p. 158). An exponential type function is a continuous function which is infinitely differentiable everywhere and has a Taylor series expansion over every interval (see Champeney, 1987, p. 77; Walnut, 2002, p. 81). The exponential type functions are also called band-limited functions (see Bracewell, 2000, p. 119 for more details).

3. Bayesianity of the MRE estimator for location family of distributions

Suppose X has distribution P_μ with respect to Lebesgue density function $g_0(x-\mu)$ where $g_0(\cdot)$ is known and unknown location parameter $\mu \in \mathbb{R}$ is to be estimated by decision rule a under equivalent loss function $L(\mu, a) := \rho_1(a-\mu)$. An estimator (decision rule) T of μ is location invariant if and only if $T(X+c) = T(X)+c$, for all $c \in \mathbb{R}$. The MRE estimator is an estimator which has the smallest risk among all invariant estimators (Shao, 2003). It is well known that the MRE estimator for μ based upon one observation is $X+r$ where the value of r will minimize $E_0[l(X+r)]$, or for smooth $l(\cdot)$ satisfies $E_0[l'(X+r)] = 0$ (Lehmann & Casella, 1998).

Now, we consider the problem of finding prior distribution $\pi(\mu)$ that its corresponding Bayes estimator $\delta^{\rho_1}_{\text{Bayes}}(\cdot)$ under general convex and invariant loss function ρ_1, coincides with the MRE estimator $X+r$, i.e.

$$\delta^{\rho_1}_{\text{Bayes}}(X) \equiv X + r \tag{5}$$

The following studies possible solution of Equation 5 under the absolute value loss function.

THEOREM 1 *Suppose X is a continuous random variable with location density function g_0. Then, the Bayes estimator under the absolute value loss function and prior μ, coincides with the MRE estimator $X+r$, whenever $\pi(\mu) \equiv$ constant, for $\mu \in \mathbb{R}$.*

Proof Under the absolute value loss function and prior $\pi, \delta_\pi(x) = x$ is a Bayes estimator if and only if

$$p(\mu \le x + r | X = x) = \frac{1}{2}$$

$$\int_{-\infty}^{x+r} \pi(\mu) g_0(x-\mu) d\mu = \int_{x+r}^{\infty} \pi(\mu) g_0(x-\mu) d\mu$$

Letting $\pi(\mu) = c + \pi^*(\mu)$, $k(x) = g_0(x) I_{(-\infty,r)}(x) - f_0(x) I_{(r,\infty)}(x)$, and $a = -c\int_{-\infty}^{\infty} k(x)dx$. The above equation can be restated as

$$\int_{-\infty}^{\infty} \pi^*(\mu) k(x-\theta) d\theta = a, \quad x \in R$$

An application of the Fourier transform along with the convolution theorem (Equation 4), then, taking the Fourier transform back lead to

$$\pi(\mu) = c + \mathcal{F}^{-1}\left(\frac{2a\pi \text{Direct}(\omega)}{\mathcal{F}(k(x); x; \omega)}; \omega; \mu\right)$$
$$= \text{constant}$$

where Direct, $\mathcal{F}$, and $\mathcal{F}^{-1}$ stand for delta direct function and the Fourier and inverse Fourier transforms, respectively. The desired proof arrives from an application of Lemma 1.

The following theorem using Lemma 2 extends results of Theorem 1 to a general convex and invariant loss function ρ_1.

THEOREM 2 *Suppose X is a continuous random variable with location density function g_0. Moreover, suppose that $\rho_1(\cdot)$ stands for a convex and invariant loss function. Then, under prior $\pi(\mu) \equiv$ constant, the MRE estimator $X+r$ is a Bayes estimator for a location parameter μ.*

Proof Lemma 2 states that the Bayes estimate $\delta_\pi(x)$ satisfies the equation $\int_{-\infty}^{\infty} \rho'(\delta_\pi(x)-\mu)\, g_0(x-\mu)\,\pi(\theta)d\mu = 0$, for all x. Setting $\pi(\mu) = c + \pi^*(\mu)$, one can reduce solving equation $\delta_\pi(x) \equiv x$, in π, to

$$
\begin{aligned}
&\int_{-\infty}^{\infty} \rho'(\delta_\pi(x)-\mu) g_0(x-\mu)\pi^*(\mu)d\mu = -cb \\
&\Leftrightarrow \mathcal{F}(\pi^*(\mu); \sim \mu; \sim \omega) = \frac{-2cb\pi \mathrm{Direct}(\omega)}{\mathcal{F}(\rho'(-\mu-r)g_0(-\mu); \mu; \omega)} \\
&\Leftrightarrow \pi(\mu) = c - 2cb\pi\mathcal{F}^{-1}\left(\frac{\mathrm{Direct}(\omega)}{\mathcal{F}(\rho'(-\mu-r)g_0(-\mu); \mu; \omega)}; \omega; \mu\right) \\
&\pi(\mu) \equiv \text{constant}
\end{aligned}
$$

where $b = \int_{-\infty}^{\infty} \rho'(t)g_0(t)dt$. The last equation arrives from that fact that there exist a delta direct function in the nominator of the function insides of the inverse Fourier transform. An application of Lemma 1 warranties that, if there is another prior distribution $\pi_1(\mu)$ such that its corresponding Bayes estimator $\delta_{\pi_1}(x) \equiv x$. Then, $\pi_1(\mu) = c\pi(\mu)$, for all $\mu \in \Theta$.

The following two propositions verify findings of Theorem 2 for two class of loss functions.

The MRE estimator for normal distribution under the LINEX loss function $\rho_{\mathrm{LINEX}}(\delta,\mu) := exp\{a(\delta-\mu)\} - a(\delta-\mu) - 1$ is $x - a/2$. The following proposition studies Bayes ianity of such MRE estimator.

Proposition 1 Suppose X is a random variable which distributed according to a normal distribution with mean μ and variance 1. Then, the Bayes estimator for μ with respect to prior distribution $\pi(\mu) \equiv \text{constant}$, and under the LINEX loss function $\rho_{\mathrm{LINEX}}(\delta,\mu) := \exp\{a(\delta-\mu)\} - a(\delta-\mu) - 1$ coincides with the MRE estimator $X - a/2$.

Proof The Bayes estimate, say $\delta_\pi(X)$, with respect to prior π and under loss function ρ_{LINEX} finds out by $\delta_\pi(x) = -\ln(E_\pi(\exp\{-a\mu\}|X=x))/a$. Setting $\pi(\mu) = c + \pi^*(\mu)$, Theorem 2 reduces solving equation $\delta_\pi(x) \equiv x - a/2$, in π, to

$$
\begin{aligned}
\pi(\mu) &= c - \frac{ce^{a^2/2}\sqrt{8\pi^3}}{a}\mathcal{F}^{-1}\left(\frac{\mathrm{Direct}(\omega)}{\mathcal{F}((e^{a(\mu-a/2)}-1)e^{-\mu^2/2}; \mu; \omega)}; \omega; \mu\right) \\
&= c - \frac{ce^{a^2/2}\sqrt{8\pi^3}}{a}\left(-\frac{\sqrt{2}}{8\sqrt{\pi^3}}\right) \\
&= c\left(1 + \frac{1}{2a}e^{a^2/2}\right) \\
&= \text{constant}
\end{aligned}
$$

where constance c should be chosen such that $\pi(\mu) > 0$, for all $\mu \in \mathbb{R}$.

The following extends the above result to a convex combination of two LINEX loss functions, say ρ_{CLINEX}.

Proposition 2 Suppose X is a random variable which distributed according to a normal distribution with mean μ and variance 1. Then, the Bayes estimator for μ with respect to prior distribution $\pi(\mu) \equiv \text{constant}$, and under convex combination of two LINEX loss functions $\rho_{\mathrm{CLINEX}}(\delta,\mu) := \alpha(\exp\{a(\delta-\mu)\} - a(\delta-\mu) - 1) + (1-\alpha)(\exp\{-a(\delta-\mu)\} + a(\delta-\mu) - 1)$ coincides with the MRE estimator $X + r$.

Proof Setting $\pi(\mu) = c + \pi^*(\mu)$, along with result of Theorem 2, one may reduce solving equation $\delta_\pi(x) \equiv x + r$, in π, to

$$\pi(\mu) = c - \frac{cb\sqrt{8\pi^3}}{a}\mathcal{F}^{-1}\left(\frac{\text{Direct}(\omega)}{\mathcal{F}((\alpha(e^{a(\mu+r)}-1)-(1-\alpha)(e^{-a(\mu+r)}+1))e^{-\mu^2/2};\mu;\omega)};\omega;\mu\right)$$
$$= c - \frac{cb\sqrt{8\pi^3}}{2\pi a}\left(\frac{2+2e^{a^2/2-ra}+4\alpha e^{a^2/2}\sinh(ra)-4\alpha}{1+e^{a^2/2-ra}+2\alpha e^{a^2/2}\sinh(ra)-2\alpha}\right)$$
$$= \text{constant}$$

where $b := \int_{-\infty}^{\infty}(\alpha(e^{at}-1)-(1-\alpha)(e^{-at}+1))e^{-t^2/2}/\sqrt{2\pi}dt$ and constance c should be chosen such that $\pi(\mu) > 0$, for all $\mu \in \mathbb{R}$. The second equality arrives from the fact that $\mathcal{F}(e^{-\mu^2/2};\mu;\omega) = \sqrt{2\pi}e^{-\omega^2/2}$, $\mathcal{F}\left(e^{\pm a(\mu+r)-\mu^2/2};\mu;\omega\right) = \sqrt{2\pi}e^{\pm ar-(\omega\pm ai)^2/2}$, and $\mathcal{F}^{-1}(\text{Direct}(\omega)h(\omega);\omega;\mu) = h(0)$.

4. Bayesianity of the MRE estimator for scale family of distributions

Suppose X has distribution P_θ with respect to Lebesgue density function $f_1(x/\theta)$ where $f_1(\cdot)$ is known and unknown scale parameter $\theta \in \mathbb{R}^+$ is to be estimated by decision rule a under equivalent loss function $L(\theta, a) := \rho_2(a/\theta)$. An estimator (decision rule) T of θ is scale invariant if and only if $T(cX) = cT(X)$, for all $c \in \mathbb{R}$. The MRE estimator is an estimator which has the smallest risk among all invariant estimators (Shao, 2003). It is well known that the MRE estimator for θ based upon one observation is rX where the value of r will minimize $E_1[l(rX)]$, for smooth $l(\cdot)$ satisfies $E_1[l'(rX)] = 0$ (Lehmann & Casella, 1998).

Now, we consider the problem of finding prior distribution $\pi(\theta)$ that its corresponding Bayes estimator $\delta^{\rho_2}_{\text{Bayes}}(\cdot)$ under general convex and invariant loss function ρ_2, coincides with the MRE estimator rX, i.e.

$$\delta^{\rho_2}_{\text{Bayes}}(X) \equiv rX \tag{6}$$

The following studies possible solution of Equation 6 under the absolute value loss function for symmetric-scale distribution functions (see Jafarpour & Farnoosh, 2005 for more details on symmetric-scale distribution functions).

THEOREM 3 *Suppose non-negative and continuous random variable X distributed according to a scale density function f_1. Moreover, suppose that X/θ is symmetric about $1/r$. Then, the Bayes estimator, under the absolute value loss function and prior distribution $\tau(\theta) = 1/\theta$, coincides with the MRE estimator rX.*

Proof Under the absolute value loss function, solving equation $\delta_\tau(x) \equiv rx$, in τ, can be restated as

$$\begin{aligned}
median(\theta|X=x) = rx &\Leftrightarrow \int_0^{rx}\frac{\tau(\theta)}{\theta}f_1\left(\frac{x}{\theta}\right)d\theta = \int_{rx}^{\infty}\frac{\tau(\theta)}{\theta}f_1\left(\frac{x}{\theta}\right)d\theta\\
(\text{let } y = x/\theta) &\Leftrightarrow \int_0^{1/r}\frac{\tau(x/y)}{y}f_1(y)dy = \int_{1/r}^{\infty}\frac{\tau(x/y)}{y}f_1(y)dy\\
&\Leftrightarrow 2E_{f_1}\left(\frac{\tau(x/Y)}{Y}I_{[0,1/r]}(Y)|X=x\right) - E_{f_1}\left(\frac{\tau(x/Y)}{Y}|X=x\right) = 0\\
&\Leftrightarrow Cov_{f_1}\left(I_{[0,1/r]}(Y), \frac{\tau(x/Y)}{Y}|X=x\right) = 0
\end{aligned}$$

From the above equation, one may conclude that, a trivial solution of τ is to let $\tau(x/Y)/Y$ be free of Y. This observation along with an application of Lemma 1 complete the desired proof.

$X \sim Unif(0, 2\theta/r), \theta, r > 0$, is an obvious example for such symmetric-scale distribution satisfies Theorem 3's conditions. Since $rX/\theta \sim Unif(0,2)$ an expression $E(\rho_2(rX/\theta))$ does not depend on θ. Therefore, rX is a MRE estimator (see Rohatgi & Saleh, 2011, p. 446 for more details).

The following theorem using Lemma 3 extends results of Theorem 3 to a general convex and invariant loss function ρ_2.

THEOREM 4 *Suppose non-negative and continuous random variable X distributed according to a scale family distribution with density function f_1. Then, Bayes estimator, under convex and invariant loss function ρ_2 and prior distribution*

$$\tau(\theta) = \frac{c}{\theta}\left(1 - \mathcal{F}^{-1}\left(\frac{b\sqrt{\pi}e^{-(\omega-2)^2/4}}{\mathcal{F}(\rho_2'(re^{-y})f_1(e^{-y}); y; \omega-2)}; \omega; \ln(\theta)\right)\right)$$

coincides with the MRE estimator rX.

Proof Using Lemma 3, one may conclude that solving equation $\delta_\tau(x) \equiv rx$, in π^*, where $\tau(\theta) = (c + \pi^*(\ln(\theta)))/\theta$, can be reduced to

$$-c\frac{b}{x^2} = \int_0^\infty \rho_2'\left(\frac{rx}{\theta}\right) f_1\left(\frac{x}{\theta}\right)\frac{\pi^*(\ln(\theta))}{\theta^3} d\theta$$

$$-cbe^{-z^2} = \int_{-\infty}^{\infty} \rho_2'(e^{z-\gamma+\ln(r)})f_1(e^{z-\gamma})\pi^*(\gamma)e^{-2\gamma}d\gamma$$

where $b = \int_0^\infty t\rho_2'(rt)f_1(t)dt$ and in the second equality $z = \ln(x)$ and $\gamma = \ln(\theta)$. Taking the Fourier transform from both sides along with an application of the convolution theorem, the above equation can be restated as

$$\mathcal{F}(\pi^*(\gamma)e^{-2\gamma}; \gamma; \omega) = \frac{-cb\sqrt{\pi}e^{-\omega^2/4}}{\mathcal{F}(\rho_2'(re^{-y})f_1(e^{-y}); y; \omega)}$$

$$\Leftrightarrow \pi^*(\gamma) = -ce^{2\gamma}\mathcal{F}^{-1}\left(\frac{b\sqrt{\pi}e^{-\omega^2/4}}{\mathcal{F}(\rho_2'(re^{-y})f_1(e^{-y}); y; \omega)}; \omega; \gamma\right)$$

$$\Leftrightarrow \pi^*(\gamma) = -c\mathcal{F}^{-1}\left(\frac{b\sqrt{\pi}e^{-(\omega-2)^2/4}}{\mathcal{F}(\rho_2'(re^{-y})f_1(e^{-y}); y; \omega-2)}; \omega; \gamma\right)$$

$$\Leftrightarrow \pi^*(\ln(\theta)) = -c\mathcal{F}^{-1}\left(\frac{b\sqrt{\pi}e^{-(\omega-2)^2/4}}{\mathcal{F}(\rho_2'(re^{-y})f_1(e^{-y}); y; \omega-2)}; \omega; \ln(\theta)\right)$$

$$\Leftrightarrow \tau(\theta) = \frac{c}{\theta}\left(1 - \mathcal{F}^{-1}\left(\frac{b\sqrt{\pi}e^{-(\omega-2)^2/4}}{\mathcal{F}(\rho_2'(re^{-y})f_1(e^{-y}); y; \omega-2)}; \omega; \ln(\theta)\right)\right)$$

Positivity of $\tau(\cdot)$ arrives from the fact that $b\sqrt{\pi}e^{-(\omega-2)^2/4}/\mathcal{F}(\rho_2'(re^{-y})f_1(e^{-y}); y; \omega-2)$ is an exponential type 1 function. Now, the Paley-Wiener theorem warranties that its corresponding inverse Fourier transform takes its values inside of an interval $[-1, 1]$. The desired proof arrives by an application of Lemma 1.

It worthwhile mentioning that the above inverse Fourier transform may not evaluated analytically. Therefore, one has to employ some numerical approach, such as the fast Fourier transformation, to handel it.

5. Conclusion and suggestion

This paper provides a class of prior distributions which their corresponding Bayes estimator under general convex and invariant loss function coincides with the MRE estimator for location or scale family of distributions. This problem can be studied for scale-location family of distributions under general convex and invariant loss function.

Acknowledgements

Author would like to thank professor Xiao-Li Meng who introduced the problem and William Strawderman for his constructive comments. Referees' comments and suggestions are gratefully acknowledged by author.

Funding

The author has received no direct funding for this research.

Author details

Amir T. Payandeh Najafabadi[1]

E-mail: amirtpayandeh@sbu.ac.ir

[1] Department of Mathematical Sciences, Shahid Beheshti University, G.C. Evin, 1983963113 Tehran, Iran.

References

Bracewell, R. N. (2000). *The Fourier transform and its applications* (3rd ed.). New York, NY: McGraw-Hill.

Champeney, D. C. (1987). *A handbook of Fourier theorems*. New York, NY: Cambridge University Press.

Chandrasekar, B., & Sajesh, T. A. (2013). Reliability measures of systems with location-scale ACBVE components. *Theory & Applications, 28*, 7–15.

Dym, H., & Mckean, H. P. (1972). *Fourier series and integrals. Probability and mathematical statistics*. New York, NY: Academic Press.

Elliott, G., & Timmermann, A. (Eds.). (2013). *Handbook of economic forecasting* (Vol. 2). New York, NY: Newnes.

Gelman, A. (2006). Prior distributions for variance parameters in hierarchical models (comment on article by Browne and Draper). *Bayesian Analysis, 1*, 515–534.

Grafarend, E. W. (2006). *Linear and nonlinear models: Fixed effects, random effects, and mixed models*. New York, NY: Walter de Gruyter.

Hallin, M., & Jurečková, J. (2012). *Equivariant estimation. Encyclopedia of environmetrics*. New York, NY: Wiley.

Jafarpour, H., & Farnoosh, R. (2005). Comparing the kurtosis measures for symmetric-scale distribution functions considering a new kurtosis. In *Proceedings of the 8th WSEAS International Conference on Applied Mathematics* (pp. 90–94). Tenerife: World Scientific and Engineering Academy and Society (WSEAS).

Kass, R. E., & Wasserman, L. (1996). The selection of prior distributions by formal rules. *Journal of the American Statistical Association, 91*, 1343–1370.

Kucerovsky, D., Marchand, É., Payandeh, A. T., & Strawderman, W. E. (2009). On the Bayesianity of maximum likelihood estimators of restricted location parameters under absolute value error loss. *Statistics & Risk Modeling, 27*, 145–168.

Ledoit, O., & Wolf, M. (2013). *Optimal estimation of a large-dimensional covariance matrix under Stein's loss* (Working Paper No. 122). Zurich: University of Zurich Department of Economics.

Lehmann, E. L. (1951). A general concept of unbiasedness. *The Annals of Mathematical Statistics, 22*, 587–592.

Lehmann, E. L., & Casella, G. (1998). *Theory of point estimation*. New York, NY: Springer.

Ma, Z., & Leijon, A. (2011). Bayesian estimation of beta mixture models with variational inference. *IEEE Transactions on Pattern Analysis and Machine Intelligence, 33*, 2160–2173.

Manrique-Vallier, D., & Reiter, J. P. (2014). Bayesian estimation of discrete multivariate latent structure models with structural zeros. *Journal of Computational and Graphical Statistics, 23*, 1061–1079.

Marchand, É., & Payandeh, A. T. (2011). Bayesian improvements of a MRE estimator of a bounded location parameter. *Electronic Journal of Statistics, 5*, 1495–1502.

Marchand, É., & Strawderman, W. E. (2005). On improving on the minimum risk equivariant estimator of a scale parameter under a lower-bound constraint. *Journal of Statistical Planning and Inference, 134*, 90–101.

Meng, X., & Zaslavsky, A. M. (2002). Single observation unbiased priors. *The Annals of Statistics, 30*, 1345–1375.

Mitchell, S. A. (2014). *Capture--recapture estimation for conflict data and hierarchical models for program impact evaluation* (PhD thesis), Harvard University Cambridge, Cambridge, MA.

Noorbaloochi, S., & Meeden, G. (1983). Unbiasedness as the dual of being Bayes. *Journal of the American Statistical Association, 78*, 619–623.

Noorbaloochi, S., & Meeden, G. (2000). *Unbiasedness and Bayes estimators* (Technical Report No. 9971331). University of Minnesota. Retrieved from http://users.stat.umn.edu/gmeeden/papers/bayunb.pdf

Rohatgi, V. K., & Saleh, A. M. E. (2011). *An introduction to probability and statistics* (Vol. 910). New York, NY: Wiley.

Rubin, D. B., & Gelman, A. (Eds.). (2004). *Applied Bayesian modeling and causal inference from incomplete-data perspectives* (Vol. 561). New York, NY: Wiley.

Shao, J. (2003). *Mathematical statistics: Springer texts in statistics*. New York, NY: Springer.

Walnut, D. F. (2002). *An introduction to wavelet analysis* (2nd ed.). New York, NY: Birkhäuser Publisher.

Wei, J., Song, B., Yan, W., & Mao, Z. (2011, June). Reliability estimations of Burr-XII distribution under Entropy loss function. In *The 9th international Conference on IEEE Reliability, Maintainability and Safety (ICRMS)* (pp. 244–247). Guiyang, China.

16

A robust unbiased dual to product estimator for population mean through modified maximum likelihood in simple random sampling

Sanjay Kumar[1]* and Priyanka Chhaparwal[1]

*Corresponding author: Sanjay Kumar, Department of Statistics, Central University of Rajasthan, Bandarsindri, Kishangarh, Ajmer, Rajasthan 305817, India

E-mail: sanjay.kumar@curaj.ac.in

Reviewing editor: Guohua Zou, Chinese Academy of Sciences, China

Abstract: In simple random sampling setting, the ratio estimator is more efficient than the mean of a simple random sampling without replacement (SRSWOR) if $\rho_{yx} > \frac{1}{2}\frac{C_x}{C_y}$, provided $R > 0$, which is usually the case. This shows that if auxiliary information is such that $\rho_{yx} < -\frac{1}{2}\frac{C_x}{C_y}$, then we cannot use the ratio method of estimation to improve the sample mean as an estimator of population mean. So there is need for another type of estimator which also makes use of information on auxiliary variable x. Product method of estimation is an attempt in this direction. Product-type estimators are widely used for estimating population mean when the correlation between study and auxiliary variables is negatively high. This paper is developed to the study of the estimation of the population mean using of unbiased dual to product estimator by incorporating robust modified maximum likelihood estimators (MMLE's). Their properties have been obtained theoretically. For the support of the theoretical results, simulations studies under several super-population models have been made. We study the robustness properties of the modified estimators. We show that the utilization of MMLE's in estimating finite population mean results to robust estimates, which is very gainful when we have non-normality or common data anomalies such as outliers.

Subjects: Mathematics & Statistics; Science; Statistical Computing; Statistical Theory & Methods; Statistics; Statistics & Computing; Statistics & Probability

ABOUT THE AUTHORS

Sanjay Kumar obtained his PhD from Banaras Hindu University, Varanasi, India. He has been working as an assistant professor since 2011 in the Department of Statistics, Central University of Rajasthan, Ajmer, Rajasthan, India. His research interests include estimation, optimization problems, and robustness study in sampling theory.

Priyanka Chhaparwal is currently working as a research scholar at the Department of Statistics, Central University of Rajasthan, Ajmer, Rajasthan, India. Her research area includes estimating problems in sampling theory

PUBLIC INTEREST STATEMENT

In sampling theory, for obtaining the estimators of parameters of interest with more precision is an important objective in any statistical estimation procedure in the field of agriculture, medicine, and social sciences. For example, estimating quantity of fruits in a village. Supplementary information obtained from auxiliary variable helps in improving the efficiency of the estimators. For example, for estimating quantity of fruits in a village, size of plots can be used as supplementary information which will help in improving the estimators. Several authors have studied such problems under normality case. In this paper, we consider the case where the underlying distribution is not normal, which is a more realistic in real-life situations. We support the theoretical results with simulations under several super-population models and study the robustness property of the modified estimator.

Keywords: product estimator; unbiased dual to product estimator; auxiliary variable; simulation study; modified maximum likelihood; transformed auxiliary variable

Mathematics subject classification: 62D05

1. Introduction

The use of additional information supplied by auxiliary variables in sample survey have been considered mainly in the area of actuarial, medicine, agriculture, and social science at the stage of organization, designing, collection of units, and developing the estimation procedure. The use of such auxiliary information in sample surveys has been studied by Cochran (1940), who used it for estimating yields of agricultural crops in agricultural sciences. Product method of estimation is a popular estimation method in sampling theory. In case of negative correlation between study variable and auxiliary variable, Robson (1957) defined a product estimator for the estimation of population mean which was revisited by Murthy (1967). The product estimator performs better than the simple mean per unit estimator under certain conditions. The use of auxiliary information in sample surveys is widely studied in the books written by Yates (1960), Cochran (1977), and Sukhatme, Sukhatme, and Asok (1984). Further, Jhajj, Sharma, and Grover (2006), Bouza (2008, 2015), Swain (2013), and Chanu and Singh (2014) studied the use of auxiliary information under different sampling designs for improving several estimators.

Let $\bar{Y}\left(=\frac{1}{N}\sum_{i=1}^{N}y_i\right)$ and $\bar{X}\left(=\frac{1}{N}\sum_{i=1}^{N}x_i\right)$ be the population means of the study variable y and the auxiliary variable x, respectively, for the population $U:(U_1, U_2, \ldots U_N)$ of size N with coefficient of variations $C_y(=\frac{S_y}{\bar{Y}})$ and $C_x(=\frac{S_x}{\bar{X}})$ and correlation coefficient ρ_{yx}, where S_y and S_x are the population mean squares for the study variable (y) and the auxiliary variable (x) The traditional product estimator for population mean $\bar{Y}$ proposed by Murthy (1964) is given by

$$\bar{y}_p = \frac{\bar{y}}{\bar{X}}\bar{x}, \tag{1.1}$$

where $\bar{y} = \frac{1}{n}\sum_{i=1}^{n}y_i, \bar{x} = \frac{1}{n}\sum_{i=1}^{n}x_i$ and n is the size of the sample.

The bias and the mean square error (MSE) of the estimator $\bar{y}_p$ are given by

$$B(\bar{y}_p) = \left(\frac{1-f}{n}\right)\bar{Y}C_{yx} \tag{1.2}$$

and

$$MSE(\bar{y}_p) = \left(\frac{1-f}{n}\right)\bar{Y}^2\left(C_y^2 + C_x^2 + 2C_{yx}\right) \tag{1.3}$$

where $C_y^2 = \frac{S_y^2}{\bar{Y}^2}, C_x^2 = \frac{S_x^2}{\bar{X}^2}, C_{yx} = \frac{S_{yx}}{\bar{Y}\bar{X}}, S_y^2 = \frac{1}{N-1}\sum_{i=1}^{N}(y_i - \bar{Y})^2, S_x^2 = \frac{1}{N-1}\sum_{i=1}^{N}(x_i - \bar{X})^2$

$f = \frac{n}{N}$ and S_{yx} is the covariance between the study variable and auxiliary variable.

An unbiased estimator $\bar{y}_{pu}$ of the population mean $\bar{Y}$ after correcting the bias of $\bar{y}_p$ is given by

$$\bar{y}_{pu} \cong \bar{y}_p - B(\bar{y}_p) \tag{1.4}$$

To $O(1/n)$, $MSE(\bar{y}_{pu}) \cong MSE(\bar{y}_p) = \left(\frac{1}{n} - \frac{1}{N}\right)\bar{Y}^2\left(C_y^2 + C_x^2 + 2C_{yx}\right)$

By making transformation, $z_i = \frac{N\bar{X}-nx_i}{N-n}(i = 1, 2, \ldots, N)$, Bandopadhyay (1980) proposed a dual to product estimator, which is given by

$$t_1 = \frac{\bar{y}}{\bar{z}}\bar{X}, \tag{1.5}$$

where the sample mean of z is $\bar{z} = \frac{N\bar{X}-n\bar{x}}{N-n}$, the population mean of z is $\bar{Z} = \bar{X}$, y and x is negatively correlated and y is positively correlated with transformed variable z.

Further, $V(\bar{z}) = \left(\frac{1}{n} - \frac{1}{N}\right)\gamma^2 S_x^2$ and $Cov(\bar{y}, \bar{z}) = -\left(\frac{1}{n} - \frac{1}{N}\right)\gamma S_{yx}$,

where $\gamma = \frac{n}{N-n}$.

The bias and the MSE of the estimator t_1 are given by

$$B(t_1) = \left(\frac{1-f}{n}\right)\gamma(k+1)\bar{Y}C_x^2 \tag{1.6}$$

and

$$MSE(t_1) = \left(\frac{1-f}{n}\right)\bar{Y}^2\left(C_y^2 + \gamma^2 C_x^2 + 2\gamma\rho_{yx}C_y C_x\right) \tag{1.7}$$

where $\rho_{yx}(<0)$ is the correlation between y and x, $k = \frac{C_{yx}}{C_x^2} = \rho_{yx}\frac{C_y}{C_x}$.

The estimator t_1 is preferred to $\bar{y}_p$ when, $k > -\frac{1}{2}(1+\gamma)$, $(1-\gamma) > 0$, k being negative because $\rho_{yx} < 0$.

Further, using this transformation and applying the technique of Hartley and Ross (1954), we have an unbiased dual to product estimator (see Singh, 2003) given by

$$t_2 = \bar{r}_1\bar{Z} + \frac{n(N-1)}{N(n-1)}(\bar{y} - \bar{r}_1\bar{z}), \tag{1.8}$$

where $\bar{r}_1 = \frac{1}{n}\sum_{i=1}^{n}\frac{y_i(N-n)}{N\bar{X}-nx_i}$.

The variance of t_2 to $O(1/n)$ is given by

$$V(t_2) = E(\bar{y} - \bar{Y})^2 + \bar{R}_1^2\gamma^2 V(\bar{x}) + 2\bar{R}_1\gamma Cov(\bar{y}, \bar{x}) = \left(\frac{1}{n} - \frac{1}{N}\right)(S_y^2 + \bar{R}_1^2\gamma^2 S_x^2 + 2\bar{R}_1\gamma S_{yx}), \tag{1.9}$$

where $\bar{R}_1 = \frac{1}{N}\sum_{i=1}^{N}\frac{y_i(N-n)}{N\bar{X}-nx_i}$

However, in all of these studies mentioned above, the underlying distribution of y is assumed to be from a normal population. In this paper, we consider the case where the underlying distribution is not normal, which is a more realistic in real-life situations.

Zheng and Al-Saleh (2002) and Islam, Shaibur, and Hossain (2009) have studied the effectivity of modified maximum likelihood estimators (MMLE's) which plays a key role in increasing the efficiency of the estimators. Using modified maximum likelihood (MML) methodology (see Tiku, Tan, & Balakrishnan, 1986), we propose a new dual to product type estimator that is based on order statistics. We have shown that the proposed estimator has always smaller mean square error (MSE) with respect to the corresponding unbiased dual to product estimator (1.8), unless the underlying distribution is normal. When the underlying distribution is normal, both the estimators provide exactly the same mean square error. We support the theoretical result with simulations under several super-population models and study the robustness property of the modified dual to product estimator. We show that utilization of MMLE for estimating finite populations mean results to robust estimate, which is very gainful when we have non-normality or other common data anomalies such as outliers.

2. Long-tailed symmetric family

For the super-population linear regression model, $y_i = \theta x_i + e_i$; $i = 1, 2, \ldots, n$, let the underlying distribution of the study variable y follow the long-tailed symmetric family.

$$f(y) = LTS(p,\sigma) = \frac{\Gamma p}{\sigma\sqrt{K}\Gamma\left(\frac{1}{2}\right)\Gamma\left(p-\frac{1}{2}\right)}\left\{1+\frac{1}{K}\left(\frac{y-\mu}{\sigma}\right)^2\right\}^{-p}; -\infty < y < \infty, \tag{2.1}$$

where $K = 2p-3, p \geq 2$ is the shape parameter (p is known) with $E(y) = \mu$ and $Var(y) = \sigma^2$. Here, it can be obtained that the kurtosis of (2.1) is $\frac{\mu_4}{\mu_2^2} = 3K/(K-2)$.

The coefficients of kurtosis of the LTS family that we consider in this family are ∞, 6, 4.5, 4.0 for p = 2.5, 3.5, 4.5, 5.5, respectively.

We realize that when $p = \infty$ (2.1) reduces to a normal distribution. The likelihood function obtained from (2.1) is given by

$$LogL \propto -nlog\sigma - p\sum_{i=1}^{n}\log\left\{1+\frac{1}{K}z_i^2\right\}; z_i = \frac{y_i-\mu}{\sigma}. \tag{2.2}$$

The MLE of μ (assuming σ is known) is the solution of the likelihood equation

$$\frac{dLogL}{d\mu} = \frac{2p}{K\sigma}\sum_{i=1}^{n}g(z_i) = 0, g(z_i) = z_i/\left\{1+\frac{1}{K}(z_i^2)\right\}, \tag{2.3}$$

which does not have explicit solutions.

Vaughan (1992a) showed that Equation (2.2) is known to have multiple roots for all $p < \infty$ but unknown and the number of roots increases as n increases.

The robust MMLE which is known to be asymptotically equivalent to the MLE are obtained in following three steps:

(1) The likelihood equations are expressed in terms of the ordered variates: $y_{(1)} \leq y_{(2)} \leq \cdots \leq y_{(n)}$,

(2) The function $g(z_i)$ are linearized using the first two terms of a Taylor series expansion around $t_{(i)} = E\left(z_{(i)}\right), z_{(i)} = \frac{y_{(i)}-\mu}{\sigma}; 1 \leq i \leq n,$

(3) The resulting equations are solved for the parameters which gives a unique solution (MMLE).

The values of $t_{(i)}; 1 \leq i \leq n$ are given in Tiku and Kumra (1981) for p = 2 (0.5)10 and Vaughan (1992b) for p = 1.5 when $n \leq 20$. For $n > 20$, the approximate values of $t_{(i)}$ can be used which are obtained from the equations

$$\frac{\Gamma p}{\sigma\sqrt{K}\Gamma\left(\frac{1}{2}\right)\Gamma\left(p-\frac{1}{2}\right)}\int_{-\infty}^{t_{(i)}}\{1+\frac{1}{K}z^2\}^{-p}dz = \frac{i}{n+1}; 1 \leq i \leq n. \tag{2.4}$$

We note that $t = \sqrt{\frac{v}{k}}z$ follows a Student's T-distribution with degrees of freedom $v = 2p-1$.

We have now

$$\frac{dLogL}{d\mu} = \frac{2p}{K\sigma}\sum_{i=1}^{n}g\left(z_{(i)}\right) = 0, since \sum_{i=1}^{n}y_i = \sum_{i=1}^{n}y_{(i)} \tag{2.5}$$

A Taylor series expansion of $g\left(z_{(i)}\right)$ around $t_{(i)}$ with first two terms of expansion gives

$$g(z_{(i)}) \cong g\left(t_{(i)}\right) + \left\{z_{(i)} - t_{(i)}\right\}\left\{\frac{d\{g(z)\}}{dz}\Big|_{z=t_{(i)}}\right\} = \alpha_i + \beta_i z_{(i)};\ 1 \le i \le n, \tag{2.6}$$

$$\text{where } \alpha_i = \left(\frac{2}{K}\right)\frac{t_{(i)}^3}{\{1+(1/K)t_{(i)}^2\}^2} \text{ and } \beta_i = \frac{1-(1/K)t_{(i)}^2}{\{1+(1/K)t_{(i)}^2\}^2}. \tag{2.7}$$

Further, for symmetric distributions, it may be noted that $t_{(i)} = -t_{(n-i+1)}$ and hence

$$\alpha_i = -\alpha_{(n-i+1)}, \sum_{i=1}^{n}\alpha_i = 0 \text{ and } \beta_i = \beta_{(n-i+1)}. \tag{2.8}$$

Now, using (2.6) and (2.7) in (2.5), we have the modified likelihood equation which is given by

$$\frac{dLogL}{d\mu} \cong \frac{dLogL^*}{d\mu} = \frac{2p}{K\sigma}\sum_{i=1}^{n}(\alpha_i + \beta_i z_{(i)}) = 0. \tag{2.9}$$

Hence, the solution of (2.9) is the MMLE $\hat{\mu}$ is given by

$$\hat{\mu} = \frac{\sum_{i=1}^{n}\beta_i y_{(i)}}{m} \tag{2.10}$$

where $m = \sum_{i=1}^{n}\beta_i$

Tiku and Vellaisamy (1996) showed that

$$E(\hat{\mu} - \overline{Y}) = 0 \tag{2.11}$$

and

$$E(\hat{\mu} - \overline{Y})^2 = V(\hat{\mu}) - \frac{2n}{N}Cov(\hat{\mu}, \bar{y}) + \frac{\sigma^2}{N}. \tag{2.12}$$

The exact variance of $\hat{\mu}$ is given by $V(\hat{\mu}) = (\boldsymbol{\beta}'\Omega\boldsymbol{\beta})\sigma^2/m^2$, where $\boldsymbol{\beta}' = (\beta_1, \beta_2, \beta_3, \ldots, \beta_n)$ and $\boldsymbol{\Omega}$ is the variance-covariance matrix of the standard variates $z_{(i)} = \frac{y_{(i)}-\mu}{\sigma}$; $1 \le i \le n$. The term $Cov(\hat{\mu}, \bar{y})$ in (2.12) can be evaluated as

$Cov(\hat{\mu}, \bar{y}) = (\boldsymbol{\beta}'\Omega\boldsymbol{\omega})\sigma^2/m$, where $\boldsymbol{\omega}'$ is the $1 \times n$ row vector with elements $1/n$. The elements of $\boldsymbol{\Omega}$ are tabulated in Tiku and Kumra (1981) and Vaughan (1992b).

When σ is not known, the MMLE $\hat{\sigma}$ can be obtained as given by Tiku and Suresh (1992) and Tiku and Vellaisamy (1996), i.e.

$$\hat{\sigma} = \frac{F + \sqrt{F^2 + 4nC}}{2\sqrt{n(n-1)}}, \tag{2.13}$$

where $F = \frac{2p}{K}\sum_{i=1}^{n}\alpha_i y_{(i)}$ and $C = \frac{2p}{K}\sum_{i=1}^{n}\beta_i(y_{(i)} - \hat{\mu})^2$

The methodology of MML is employed in those situations where maximum likelihood (ML) estimation is intractable as widely used by Puthenpura and Sinha (1986), Tiku and Suresh (1992), and Oral (2006). Under some regularity conditions, MMLEs have exactly the same asymptotic properties as ML estimators (MLEs) as discussed in Vaughan and Tiku (2000), and for small n values they are known to be essentially as efficient as MLEs.

3. The proposed dual to product estimator and its variance

In the context of sampling theory, Tiku and Bhasln (1982) and Tiku and Vellaisamy (1996) used the MMLE (2.10) and showed that utilizing the MMLEs leads to improvements in efficiencies in estimating the finite population mean.

Motivated from such approach, we propose a new unbiased dual to product estimator which is given by

$$T_1 = \bar{r}_1\bar{Z} + \frac{n(N-1)}{N(n-1)}\left(\hat{\mu} - \bar{r}_1\bar{Z}\right) \qquad (3.1)$$

assuming the population mean of the auxiliary variable $\overline{X}$ is known.

The expression for the variance of the proposed estimator T_1, up to the terms of order n^{-1} is given as follows:

Let $\hat{\mu} = \overline{Y}(1+\in_0), \bar{z} = \overline{Z}(1+\in_1)$ such that $E(\in_0) = 0 = E(\in_1)$

Using simple random sampling without replacement method of sampling, we have,

$$E(\in_0^2) = \frac{1}{\overline{Y}^2}E(\hat{\mu}-\overline{Y})^2 = \frac{1}{\overline{Y}^2}\left\{V(\hat{\mu}) - \frac{2n}{N}Cov(\hat{\mu},\bar{y}) + \frac{\sigma^2}{N}\right\},$$

$$E\left(\in_1^2\right) = \tfrac{1}{\overline{Z}^2}V\left(\overline{Z}\right) = \tfrac{1}{\overline{X}^2}\left(\tfrac{n}{N-n}\right)^2 V(\overline{x}) = \tfrac{1}{\overline{X}^2}\left(\tfrac{n}{N-n}\right)^2\left(\tfrac{1}{n} - \tfrac{1}{N}\right)S_x^2$$

$$= \tfrac{1}{\overline{X}^2}\left(\tfrac{n}{N-n}\right)^2\left(\tfrac{1}{n} - \tfrac{1}{N}\right)\tfrac{1}{N-1}\sum_{i=1}^{N}\left(x_i - \overline{X}\right)^2$$

$$= \tfrac{1}{\overline{X}^2}\tfrac{n}{(N-n)N(N-1)}\sum_{i=1}^{N}\left(x_i - \overline{X}\right)^2$$

$$E(\in_0,\in_1) = \frac{1}{\overline{Y}\overline{X}}Cov(\hat{\mu},\overline{Z}) = -\frac{1}{\overline{Y},\overline{X}}\gamma Cov(\hat{\mu},\bar{x})$$

Now, we have

$B\left(T_1\right) = 0,$

$$V(T_1) = E(\hat{\mu}-\overline{Y})^2 + \bar{R}_1^2\gamma^2 V(\bar{x}) + 2\bar{R}_1\gamma Cov(\hat{\mu},\bar{x}) \qquad (3.2)$$

where $Cov(\hat{\mu},\bar{x}) = \left(\frac{1}{\theta}\right)\left\{Cov(\hat{\mu},\bar{y}-\bar{e})\right\} = (1/\theta)\left\{Cov(\hat{\mu},\bar{y}) - Cov(\theta\bar{x}_{[\cdot]} + \bar{e}_{[\cdot]},\bar{e})\right\}$

$\bar{x}_{[\cdot]} = \sum_{i=1}^{n}\beta_i\bar{x}_{[\cdot]}/m, \bar{e}_{[\cdot]} = \sum_{i=1}^{n}\beta_i\bar{e}_{[\cdot]}/m, \bar{e}_{[\cdot]} = y_{(i)} - \theta x_{[i]}$, and $x_{[i]}$ is the concomitant of $y_{(i)}$, i.e. $x_{[i]}$ is that observation x_i which is coupled with $y_{(i)}$, when (y_i, x_i) are ordered with respect to y_i; $i \le i \le n$. Here, we realize that x is assumed to be non-stochastic in nature in the super-population linear regression model $y = \theta x + e$, $Cov\left(x_i, e_j\right)$ is not affected by the ordering of the y values for $1 \le i \le n$ and $1 \le j \le n$; hence

$$Cov(\hat{\mu},\bar{x}) = (1/\theta)\left\{Cov(\hat{\mu},\bar{y}) - Cov(\bar{e}_{[\cdot]},\bar{e})\right\},$$

where $Cov(\bar{e}_{[\cdot]},\bar{e}) = (\beta'\Omega\omega)\frac{\sigma_e^2}{m}$,

Note that if the sampling fraction n/N exceeds 5%, the finite population correction $(N - n)/N$ can be introduced as

$$Cov(\hat{\mu},\bar{x}) = \{(N-n)/N\theta\}\left\{Cov(\hat{\mu},\bar{y})\right\} - Cov(\bar{e}_{[\cdot]},\bar{e})\}$$

4. Monte Carlo simulation study

In this study for the simulation, we have used R-programming software. In the super-population models generated, we use the model

$$y_i = \theta x_i + e_i, i = 1, 2, \ldots, N, \quad (4.1)$$

where we generate e_i and x_i independently and calculate y_i for i = 1, 2, ..., N. Let the errors $e_1, e_2, \ldots, e_N$ be the random observations from a super-population from (2.1) with $E(e) = 0$ and $V(e) = \sigma_e^2$. Let U_N denotes the corresponding finite population consists of N pairs (x_1, y_1), (x_2, y_2), ..., (x_N, y_N). To calculate the MSE of the proposed estimator in (3.1), we calculate T_1 for all possible simple random samples $\binom{N}{n}$ of size n (= 5, 11, 15) from U_N. Since $\binom{N}{n}$ is extremely large, so we conduct all Monte Carlo studies as follows.

We take N = 500 in each simulation and generate U_{500} pairs from an assumed super-population. From the generated finite population U_{500}, we have selected a sample of size $n(= 5, 11, 15)$ by simple random sampling without replacement. Now, we choose at random S = 10,000 samples for all the possible $\binom{500}{n}$ samples of size n (= 5, 11, 15), which gives 10,000 values of T_1. To compare the efficiency of the proposed estimator under different models for a given n, we calculate the values of mean square errors as follows:

$MSE(T_1) = \frac{1}{s}\sum_{j=1}^{s}(T_{1j} - \bar{Y})^2$, $MSE(t_1) = \frac{1}{s}\sum_{j=1}^{s}(t_{1j} - \bar{Y})^2$, $MSE(\bar{y}_{pu}) = \frac{1}{s}\sum_{j=1}^{s}(\bar{y}_{pj} - \bar{Y})^2$ and $MSE(t_2) = \frac{1}{s}\sum_{j=1}^{s}(t_{2j} - \bar{Y})^2$.

For setting the population correlation ρ_{yx} sufficiently high, we choose the value of parameter θ in the model $y = \theta x + e$, such that the correlation coefficient between study variable (y) and auxiliary variable (x) is ρ_{yx}. To determine the value of θ that satisfies this condition, we follow a similar way given by Rao and Beegle (1967) and write the population correlation between the study variable(y) and the auxiliary variable (x). For example if $X \sim U$ (0,1), the value of θ for which the population correlation between y and x becomes $\theta^2 = \frac{12\sigma^2\rho_{yx}^2}{1-\rho_{yx}^2}$ for the LTS family. Similarly, if x is generated from Exp (1), the value of θ for the population correlation becomes $\theta^2 = \frac{\sigma^2\rho_{yx}^2}{1-\rho_{yx}^2}$ for the symmetric family. In the same way, we can have $x \sim exp(0.5), x \sim N(0, 1), x \sim U(-1, 2.5)$ etc. and the corresponding values of θ can be calculated accordingly. Here, we take $\sigma^2 = 1$, in all situations without loss of generality and calculate the required parameter θ for which $\rho_{yx} = -0.45$.

5. Comparison of efficiencies of the proposed estimator

The conditions under which the proposed estimator T_1 is more efficient than the corresponding estimators $\bar{y}_{pu}$, t_1 and t_2 are given as follows:

$MSE(T_1) \leq MSE(t_2)$ if

$$Cov(\bar{y}, \bar{x}) \geq \frac{1}{2\bar{R}_1\gamma}\{E(\hat{\mu} - \bar{Y})^2 - E(\bar{y} - \bar{Y})^2\} + Cov(\hat{\mu}, \bar{x}), \quad (5.1)$$

$MSE(T_1) \leq MSE(t_1) if$

$$Cov(\bar{y}, \bar{x}) \geq \frac{1}{2R\gamma}\{E(\hat{\mu} - \bar{Y})^2 - E(\bar{y} - \bar{Y})^2\} + \frac{\gamma}{2R}V(\bar{x})\{\bar{R}_1^2 - R^2\} + \frac{\bar{R}_1}{R}Cov(\hat{\mu}, \bar{x}) \quad (5.2)$$

$MSE(T_1) \leq MSE(\bar{y}_{pu}) if$

$$Cov(\bar{y},\bar{x}) \geq \frac{1}{2R}\{E(\hat{\mu}-\bar{Y})^2 - E(\bar{y}-\bar{Y})^2\} + \frac{1}{2R}V(\bar{x})\{\bar{R}_1^2\gamma^2 - R^2\} + \frac{\bar{R}_1\gamma^2}{R}Cov(\hat{\mu},\bar{x}), \tag{5.3}$$

$MSE(T_1) \leq MSE(t_2) \geq MSE(t_1) if$

$$\frac{1}{2\bar{R}_1\gamma}\{E\left(\hat{\mu}-\bar{Y}\right)^2 - E\left(\bar{Y}-\bar{Y}\right)^2\} + Cov(\hat{\mu},\bar{x}) \leq Cov(\bar{y},\bar{x}) \leq \frac{\gamma}{2\bar{R}_1}V(\bar{x})\{R^2 - \bar{R}_1^2\} + \frac{R}{\bar{R}_1}Cov(\bar{y},\bar{x}) \tag{5.4}$$

$MSE(T_1) \leq MSE(t_2) \leq MSE(\bar{y}_{pu})$ if

$$\frac{1}{2\bar{R}_1\gamma}\{E\left(\hat{\mu}-\bar{Y}\right)^2 - E\left(\bar{y}-\bar{Y}\right)^2\} + Cov(\hat{\mu},\bar{x}) \leq Cov(\bar{y},\bar{x}) \leq \frac{1}{2\bar{R}_1\gamma}V(\bar{x})\left\{R^2 - \bar{R}_1^2\gamma^2\right\} + \frac{R}{\bar{R}_1\gamma}Cov(\bar{y},\bar{x}), \tag{5.5}$$

and

$MSE(T_1) \leq MSE(t_2) \leq MSE(t_1) \leq MSE(\bar{y}_{pu})$ if

$$\frac{1}{2\bar{R}_1\gamma}\{E\left(\hat{\mu}-\bar{Y}\right)^2 - E\left(\bar{y}-\bar{Y}\right)^2\} + Cov(\hat{\mu},\bar{x}) \leq Cov(\bar{y},\bar{x}) \leq \frac{\gamma}{2\bar{R}_1}V(\bar{x})\left\{R^2 - \bar{R}_1^2\right\} + \frac{R}{\bar{R}_1}Cov(\bar{y},\bar{x})$$
$$\leq \frac{1}{2\bar{R}_1\gamma}V(\bar{x})\{R^2 - \bar{R}_1^2\gamma^2\} + \frac{R}{\bar{R}_1\gamma}Cov(\bar{y},\bar{x}), \tag{5.6}$$

where $Cov(\bar{y},\bar{x}) = \left(\frac{1}{n} - \frac{1}{N}\right) S_{yx}$.

We assume two different super-population models given below to see how much efficiency we gain with the proposed modified estimator, when the conditions given in Section 5 are satisfied under non-normality:

(1) $x \sim U(-1, 2.5)$ and $e \sim LTS(p, 1)$.

(2) $x \sim exp\ (1)$ and $e \sim LTS(p, 1)$.

For the models (1) and (2), the values θ which makes the population correlation $\rho_{yx} = -0.45$ are given in Table 1.

Here, we note that for the LTS family (2.1), the value of θ does not depend on the shape parameter p.

To verify that the super-populations are generated appropriately, we provide a scatter graph and the underlying distribution of model for $p = 3.5$ for model (2) in Figures 1 and 2.

Relative efficiencies are calculated by $RE = \frac{MSE(\bar{y}_{pu})}{MSE(\cdot)} * 100$,

where MSE (.) and relative efficiency (RE) are given in Table 2 for the model (1) and (2).

Table 1. Parameter values of θ used in models (1)–(2) that give $\rho_{yx} = -0.45$

Population	p		
	2.5	4.5	5.5
Model (1)	−1.746	−1.746	−1.746
Model (2)	−0.504	−0.504	−0.504

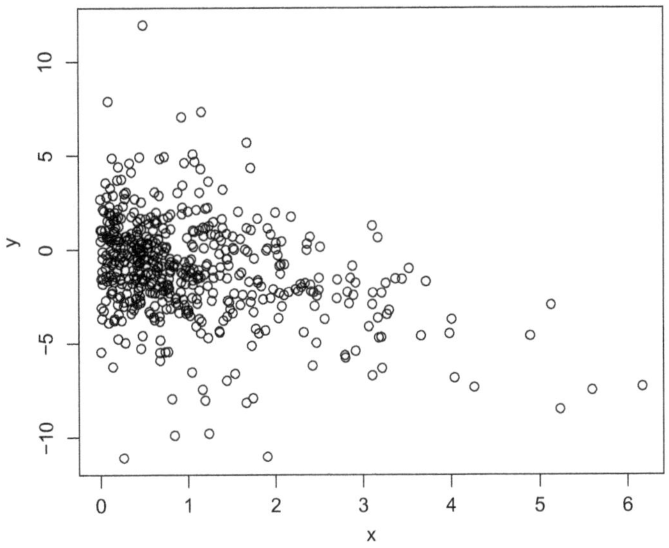

Figure 1. A scatter graph of the study variable and auxiliary variable obtained from model (2) for p = 3.5.

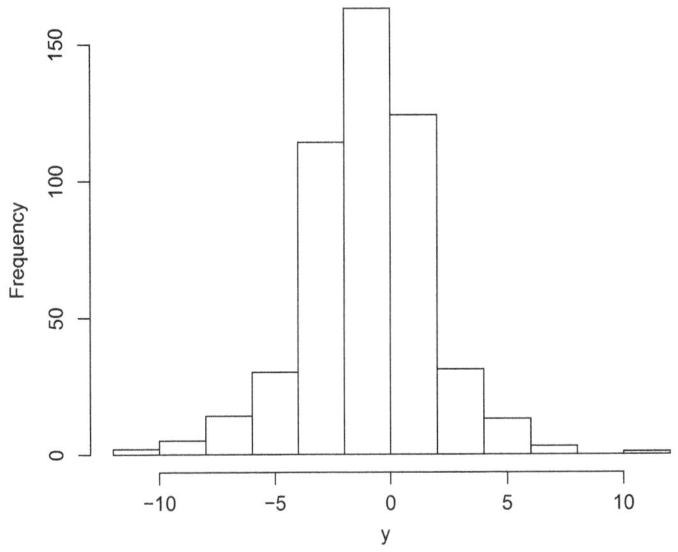

Figure 2. Underlying distribution of the study variable obtained from model (2) for p = 3.5.

From Table 2, we see that the proposed estimator T_1 is more efficient than the corresponding estimators $\bar{y}_{pu}$, t_1 and t_2 because the theoretical conditions given in Section 5 are satisfied. We also observe that when sample size increases, mean square error decreases.

6. Robustness of the proposed estimator

The outliers in sample data are normally a focused problem for survey statistician. In practice, the shape parameters p in $LTS(p, \sigma)$ might be mis-specified. Therefore, it is very important for estimators to have efficiencies of robustness estimates such as an estimator is full efficient or nearly so for an assumed model and maintains high efficiencies for plausible to the assumed model.

Here, we take $N = 500$ and $\sigma^2 = 1$ without loss of generality and we study the robustness property of proposed estimator under different outlier models as follows.

We assume $x \sim U(-1, 2.5)$ as well as $x \sim Exp(1)$ and $y \sim LTS(p = 3.5, \sigma^2 = 1)$. We determine our super-population model as follow:

Table 2. Mean square error and efficiencies of the estimators under super-populations (1–2)

	Est.	x ~ *U*(0,1) and e ~ *LTS* (*p*,1)			x ~ exp(1) and e ~ *LTS* (*p*,1)		
		n			*n*		
		5	11	15	5	11	15
$p = 2.5$	T_1	229.74 (0.2333)	215.50 (0.0961)	195.85 (0.0747)	214.54 (0.2152)	202.80 (0.0963)	213.27 (0.0678)
	t_2	197.20 (0.2718)	187.08 (0.1107)	163.46 (0.0225)	191.34 (0.2413)	171.91 (0.1136)	170.12 (0.0850)
	t_1	197.13 (0.2719)	186.91 (0.1108)	163.28 (0.0896)	191.18 (0.2415)	171.62 (0.1138)	169.62 (0.0851)
	$\bar{y}_{pu}$	100.00 (0.536)	100.00 (0.2071)	100.00 (0.1463)	100.00 (0.4617)	100.00 (0.1953)	100.00 (0.1446)
$p = 4.5$	T_1	226.97 (0.2403)	177.83 (0.1087)	180.72 (0.0773)	179.04 (0.2447)	168.95 (0.1124)	172.99 (0.0822)
	t_2	224.35 (0.2431)	174.14 (0.1110)	178.42 (0.0783)	172.28 (0.2543)	161.21 (0.1178)	164.58 (0.0864)
	t_1	224.26 (0.2432)	173.99 (0.1111)	178.19 (0.0784)	172.01 (0.2547)	160.93 (0.1180)	164.39 (0.0865)
	$\bar{y}_{pu}$	100.00 (0.5484)	100.00 (0.1933)	100.00 (0.1397)	100.00 (0.4381))	100.00 (0.1899)	100.00 (0.1422)
$p = 5.5$	T_1	209.83 (0.2513)	180.92 (0.1174)	191.68 (0.0793)	192.57 (0.2490)	176.75 (0.1161)	158.31 (0.0842)
	t_2	208.25 (0.2532)	178.64 (0.1189)	190.00 (0.0800)	188.71 (0.2541)	171.29 (0.1198)	152.87 (0.0872)
	t_1	208.17 (0.2533)	178.49 (0.1190)	189.76 (0.0801)	188.48 (0.2544)	171.00 (0.1200)	152.69 (0.0873)
	$\bar{y}_{pu}$	100.00 (0.5273)	100.00 (0.2124)	100.00 (0.1520)	100.00 (0.4795)	100.00 (0.2052)	100.00 (0.1333)

Note: Mean square errors are in the parenthesis.

(5) True model: $LTS(p = 3.5, \sigma^2 = 1)$

(6) Dixon's outliers model: $N–N_o$ observations from *LTS* (3.5, 1)and N_o (we do not know which) form *LTS* (3.5, 2.0)

(7) Mis-specified model: *LTS* (4.0, 1)

Here, we realize that the model (5), the assumed super-population model is given for the purpose of comparison and the models (6) and (7) are taken as its plausible alternatives. Here, we have assumed the super-population model *LTS* (3.5, 1). The coefficients (α_i, β_i) from (2.7) are calculated with $p = 3.5$ and are used in models (5) and (6). N_o in the model (6) is calculated from the formula $(|0.5 + 0.1 * N| = 50)$ for $N = 500$. We standardized the generated $e_i^{'}s, (i = 1, 2, \ldots, N)$ in all the models to have the same variance as that of *LTS* (3.5, 1) i.e. it should be equal to 1. The simulated values of MSE and the relative efficiency are given in Table 3. Here, theoretical conditions are satisfied for the models.

From the Table 3, we see that the proposed estimator T_1 is more efficient than the corresponding estimators $\bar{y}_{pu}$, t_1 and t_2 because the theoretical conditions are satisfied. We also observe that when sample size increases, mean square error decreases.

7. Determination of the shape parameter

It may be possible that the shape parameter *p* is unknown, then in such a case in order to determine whether a particular density is appropriate for the underlying distribution of the study variable *y*, a Q–Q plot is made by plotting the population quantiles for the density against the ordered values of *y*.

Table 3. Mean square errors and efficiencies under super-populations (5)–(7) for LTS family

Est.	*n*			*n*		
	5	11	15	5	11	15
	True Model (5): $x \sim Uni\ (0, 1)$			Dixon outlier Model (6): $x \sim Uni\ (0, 1)$		
T_1	220.58 (0.2473)	200.19 (0.1037)	191.73 (0.0810)	275.37 (0.9490)	233.97 (0.3441)	194.52 (0.2374)
t_2	213.25 (0.2558)	188.04 (0.1104)	177.48 (0.0875)	248.53 (1.0515)	210.15 (0.3831)	182.39 (0.2532)
t_1	213.17 (0.2559)	187.87 (0.1105)	177.28 (0.0876)	248.25 (1.0527)	209.93 (0.3835)	182.24 (0.2534)
$\bar{y}_{pu}$	100.00 (0.5455)	100.00 (0.2076)	100.00 (0.1553)	100.00 (2.6133))	100.00 (0.8051))	100.00 (0.4618)
	Mis - specified Model (7): $x \sim Uni\ (0, 1)$			True Model (5): $x \sim Exp\ (1)$		
T_1	231.94 (0.2683)	183.21 (0.1102)	185.40 (0.0822)	207.97 (0.2308)	184.50 (0.1026)	174.59 (0.0724)
t_2	226.62 (0.2746)	179.15 (0.1127)	180.57 (0.0844)	201.34 (0.2384)	167.97 (0.1127)	159.80 (0.0791)
t_1	226.62 (0.2747)	178.99 (0.1128)	180.36 (0.0845)	201.01 (0.2388)	167.82 (0.1128)	159.60 (0.0792)
$\bar{y}_{pu}$	100.00 (0.6223)	100.00 (0.2019)	100.00 (0.1524)	100.00 (0.4800)	100.00 (0.1893)	100.00 (0.1264)
	Dixon outlier Model (6): $x \sim Exp\ (1)$			Mis - specified Model (7): $x \sim Exp\ (1)$		
T_1	246.62 (0.3029)	198.58 (0.1354)	211.18 (0.0966)	188.42 (0.2349)	174.13 (0.1090)	174.04 (0.0782)
t_2	243.64 (0.3066)	188.68 (0.1404)	197.87 (0.1031)	183.19 (0.2416)	163.90 (0.1158)	169.28 (0.0804)
t_1	242.77 (0.3077)	188.14 (0.1408)	197.30 (0.1034)	182.82 (0.2421)	163.76 (0.1159)	169.07 (0.0805)
$\bar{y}_{pu}$	100.00 (0.7470)	100.00 (0.2649)	100.00 (0.2040)	100.00 (0.4426)	100.00 (0.1898)	100.00 (0.1361)

Note: Mean square errors are in the parenthesis.

The population quantiles $t_{(i)}$ are determined from the equation

$$\int_{-\infty}^{t_{(i)}} t(u)du = \frac{i}{n+1};\ 1 \le i \le n, \text{ where } n \text{ is the sample size.}$$

The Q-Q plot that closely approximates a straight line would be assumed to be the most appropriate. Using such procedure, we can also obtain a plausible value for the shape parameter simply.

8. Conclusions

In this study, we show that when the underlying distribution of the study variable is not normal (e.g. Pareto distribution etc.), which is applicable in most of areas, MML integrated estimators can improve the efficiency of the estimators. In the paper, we show when the underlying distribution of the study variable is a long-tailed symmetric distribution, MML integrated dual to product estimator (T_1) can improve the efficiency of the unbiased dual to product estimator t_2. The proposed estimator is also more efficient than the product estimators $\bar{y}_{pu}$ and t_1. We also show that the MML integrated dual to product estimator (T_1) is robust to outliers as well as other data anomalies.

Acknowledgment
The authors are grateful to the Editors and referees for their valuable suggestions which led to improvements in the article.

Funding
The authors received no direct funding for this research.

Author details
Sanjay Kumar[1]
E-mail: sanjay.kumar@curaj.ac.in
Priyanka Chhaparwal[1]
E-mail: priyankachhaparwal4@gmail.com
[1] Department of Statistics, Central University of Rajasthan, Bandarsindri, Kishangarh, Ajmer, Rajasthan 305817, India.

References

Bandopadhyay, S. (1980). Improved ratio and product estimators. *Sankhya, 42*, 45–49.

Bouza, C. N. (2008). Ranked set sampling for the product estimator. *Revista Investigación Operacional, 29*, 201–206.

Bouza, C. N. (2015). A family of ratio estimators of the mean containing primals and duals for simple random sampling with replacement and ranked set sampling designs. *Journal of Basic and Applied Research International, 8*, 245–253.

Chanu, W. W., & Singh, B. K. (2014). Improved class of ratio -cum-product estimators of finite population mean in two phase sampling. *Global Journal of Science Frontier Research: Mathematics and Decision Sciences, 14*, 69–81.

Cochran, W. G. (1940). The estimation of the yields of cereal experiments by sampling for the ratio of grain to total produce. *The Journal of Agricultural Science, 30*, 262–275. http://dx.doi.org/10.1017/S0021859600048012

Cochran, W. G. (1977). *Sampling techniques*. New York, NY: Wiley.

Hartley, H. O., & Ross, A. (1954). Unbiased ratio estimators. *Nature, 174*, 270–271. http://dx.doi.org/10.1038/174270a0

Islam, T., Shaibur, M. R., & Hossain, S. S. (2009). Effectivity of modified maximum likelihhod estimators using selected ranked set sampling data. *Austrian Journal of Statistics, 38*, 109–120.

Jhajj, H. S., Sharma, M. K., & Grover, L. K. (2006). Dual of ratio estimators of finite population mean obtained on using linear transformation to auxiliary variable. *JOURNAL OF THE JAPAN STATISTICAL SOCIETY, 36*, 107–119. http://dx.doi.org/10.14490/jjss.36.107

Murthy, M. N. (1964). Product method of estimation. *Sankhya A, 26*, 69–74.

Murthy, M. N. (1967). *Sampling theory and methods*. Calcutta: Statistical Publishing Society.

Oral, E. (2006). Binary regression with stochastic covariates. *Communications in Statistics: Theory and Methods, 35*, 1429–1447. http://dx.doi.org/10.1080/03610920600637123

Puthenpura, S., & Sinha, N. K. (1986). Modified maximum likelihood method for the robust estimation of system parameters from very noisy data. *Automatica, 22*, 231–235. http://dx.doi.org/10.1016/0005-1098(86)90085-3

Rao, J. N. K., & Beegle, L. D. (1967). A Monte Carlo study of some ratio estimators. *Sankhy¯a: Series B, 29*, 47–56.

Robson, D. (1957). Applications of multivariate polykays to the theory of unbiased ratio-type estimation. *Journal of the American Statistical Association, 52*, 511–522. http://dx.doi.org/10.1080/01621459.1957.10501407

Singh, S. (2003). *Advanced sampling theory with applications* (Vol. 1). Kluwer Academic Publishers, the Netherlands. http://dx.doi.org/10.1007/978-94-007-0789-4

Sukhatme, P. V., Sukhatme, B. V., & Asok, C. (1984). *Sampling theory of surveys with applications*. New Delhi: Indian Society Agricultural Statistics.

Swain, A. K. P. (2013). On some modified ratio and product type estimators-revisited. *Revista Investigación Operacional, 34*, 35–57.

Tiku, M. L., & Bhasln, P. (1982). Usefulness of robust estimators in sample survey. *Communications in Statistics: Theory and Methods, 11*, 2597–2610. http://dx.doi.org/10.1080/03610918208828409

Tiku, M. L., & Kumra, S. (1981). Expected values and variances and covariances of order statistics for a family of symmetric distributions (student's t). *Selected Tables in Mathematical Statistics, American Mathematical Society, 8*, 141–270.

Tiku, M. L., & Suresh, R. P. (1992). A new method of estimation for location and scale parameters. *Journal of Statistical Planning and Inference, 30*, 281–292. http://dx.doi.org/10.1016/0378-3758(92)90088-A

Tiku, M. L., & Vellaisamy, P. (1996). Improving efficiency of survey sample procedures through order statistics. *Journal of Indian Society Agricultural Statistics, 49*, 363–385.

Tiku, M. L., Tan, M. Y., & Balakrishnan, N. (1986). *Robust inference*. New York, NY: Marcel Dekker.

Vaughan, D. C. (1992a). On the tiku-suresh method of estimation. *Communications in Statistics-Theory and Methods, 21*, 451–469. http://dx.doi.org/10.1080/03610929208830788

Vaughan, D. C. (1992b). Expected values, variances and covariances of order statistics for student's t-distribution with two degrees of freedom. *Communications in Statistics-Simulation and Computation, 21*, 391–404. http://dx.doi.org/10.1080/03610919208813025

Vaughan, D. C., & Tiku, M. L. (2000). Estimation and hypothesis testing for a nonnormal bivariate distribution with applications. *Mathematical and Computer Modelling, 32*, 53–67. http://dx.doi.org/10.1016/S0895-7177(00)00119-9

Yates, F. (1960). *Sampling methods in censuses and surveys*. London: Charles Griffin.

Zheng, G., & Al-Saleh, M. F. (2002). Modified maximum likelihood estimators based on ranked set samples. *Annals of the Institute of Statistical Mathematics, 54*, 641–658. http://dx.doi.org/10.1023/A:1022475413950

17

Different types of Bernstein operators in inference of Gaussian graphical model

Melih Ağraz[1] and Vilda Purutçuoğlu[1]*

*Corresponding author: Vilda Purutçuoğlu, Department of Statistics, Middle East Technical University, Ankara 06800, Turkey
E-mail: vpurutcu@metu.edu.tr

Reviewing editor: Yong Hong Wu, Curtin University of Technology, Australia

Abstract: The Gaussian graphical model (GGM) is a powerful tool to describe the relationship between the nodes via the inverse of the covariance matrix in a complex biological system. But the inference of this matrix is problematic because of its high dimension and sparsity. From previous analyses, it has been shown that the Bernstein and Szasz polynomials can improve the accuracy of the estimate if they are used in advance of the inference as a processing step of the data. Hereby in this study, we consider whether any type of the Bernstein operators such as the Bleiman Butzer Hahn, Meyer-König, and Zeller operators can be performed for the improvement of the accuracy or only the Bernstein and the Szasz polynomials can satisfy this condition. From the findings of the Monte Carlo runs, we detect that the highest accuracies in GGM can be obtained under the Bernstein and Szasz polynomials, rather than all other types of the Bernstein polynomials, from small to high-dimensional biological networks.

Subjects: Bioinformatics; Biology; Bioscience; Mathematics & Statistics; Multivariate Statistics; Science; Statistical Theory & Methods; Statistics; Statistics & Probability; Statistics for the Biological Sciences

Keywords: Gaussian graphical model; Bernstein operators; Bleiman Butzer Hahn operators; Meyer-König and Zeller operators; systems biology; bioinformatics; statistics

ABOUT THE AUTHOR

Vilda Purutçuoğlu is the associate professor in the Department of Statistics at Middle East Technical University (METU) and also affiliated faculty in the Informatics Institute, Institute of Applied Mathematics and Department of Biomedical Engineering at METU. She has completed her BSc and MSc in statistics and minor degree in economics. She received her doctorate at the Lancaster University. Purutçuoğlu's current researches are in the field of bioinformatics, systems biology and biostatistics. She has a research group with seven doctorates and four master students who have been working on deterministic and stochastic modelings of biological networks and their inferences via Bayesian and frequentist theories. Hereby, the findings in this study are the part of the researches about the improvement of probabilistic inferences of complex networks and these findings can be helpful for eliminating batch effects in their observations before any statistical analyses.

PUBLIC INTEREST STATEMENT

The Gaussian graphical model (GGM) is one of the well-known probabilistic modeling approaches for the complex biological systems. Briefly, this model expresses the biological interactions between components of systems, which are genes or proteins, by using the property of the conditional dependency under multivariate normally distributed data. On the other hand, the Bernstein Polynomials are one of the famous Bernstein-type of operators in the approximation theory to smoothen the data via statistical distributions within the range of [0, 1]. From previous studies, it has been shown that these polynomials are successful in getting high accuracy in inference of interactions between genes in the system when they are used as the processing step in advance of GGM. In this study, we consider all Bernstein-type of operators, as the alternatives of the Bernstein polynomials, for the pre-processing step of GGM to discard the bath effects in the original observations.

1. Introduction

The approximation theory is concerned with the study of how well given functions can be proximated by basic functions. In this theory, it is usual to apply the approximating functions in the form of linear positive operators, such as the Bernstein, Szasz–Mirakyan polynomials, the Bleiman Butzer Hahn (BBH) operator, Meyer-König and Zeller (MKZ) operator. From previous works, it is known that the Bernstein and Szasz polynomials can significantly improve the accuracy of estimates from the point of view of their inference via the Gaussian graphical model (GGM) (Purutçuoğlu, Ağraz & Wit, 2015). Therefore, here we investigate whether the strong alternatives of the Bernstein polynomials, i.e. the Bleiman and Hahn operator and the Meyer-König and Zeller operator, are as successful as the Bernstein polynomials and can be used alternately.

The Gaussian graphical model is a probabilistic and undirected statistical model for the complex biological networks under the normally distributed random multivariate variables whose dependency structure is represented by a graph (Dempster, 1972). In Figure 1, we represent the structure of the filtered yeast interactome (FYI) network (Han et al., 2004) as an example of realistically complex biological systems that GGM can handle with. Hereby, in GGM, we assume that the observations follow a multivariate normal distribution with a mean vector μ and the variance–covariance matrix Σ, and the conditional independent structure of the observations can be described via a set of nodes and edges constructing an undirected network by means of the inverse of the covariance matrix, also called the precision $\Theta = \Sigma^{-1}$. In Θ, the non-zero entries indicate the interactions between nodes, i.e. genes, and zero entries imply no interaction between the selected pair of nodes.

On the other hand, the Bernstein operators which cover the Bernstein and Szasz polynomials are the approximations that are based on the binomial and the Poisson distribution, respectively. But in the literature, different Bernstein-type operators are presented as well. Specifically, the Bleiman-Butzer Hahn (BBH) and the Meyer-König and Zeller (MKZ) operators which are derived from the Bernstein operators are the most well-known ones. The theoretical properties of the BBH operator (Agratini, 1996) and MKZ operator (Abel, 1995; Becker, 1977) have been studied extensively.

In biological networks, the Bernstein polynomials enable us to transform the data in a new range (Lorentz, 1953; Bernstein, 1912). Accordingly, in this study, we consider to implement the Bernstein polynomials, the BBH operator and the MKZ operator in different dimensional biological systems before estimating the model parameters of GGM.

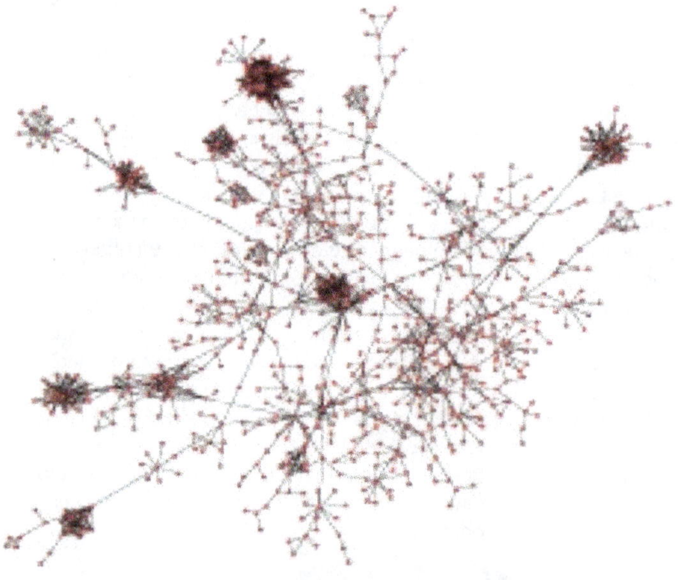

Figure 1. The main component of the FYI network (Han et al., 2004).

Accordingly, in the organization of this paper, we introduce GGM and different types of operators in Section 2. In Section 3, we report our outputs based on the Monte Carlo studies. Finally, in Section 4, we summarize our results and discuss our future works.

2. Methods

2.1. Gaussian graphical model

The biological system is an expression between components of the complex biological networks and the interactions between components in this system can be probabilistically represented by the Gaussian graphical model. In this modeling approach, the nodes are indicated as $X = (X_1, \dots, X_p)$ under the assumption that the state vector X has a multivariate Gaussian (normal) distribution via

$$X \sim N(\mu, \Sigma) \tag{1}$$

in which μ is a p-dimensional mean vector and Σ shows the $(p \times p)$ variance-covariance matrix for the p-dimensional normal density of the random variable X with n samples per nodes, i.e. genes. Hereby, the p-dimensional normal density of X shown in Equation (1) can be denoted by the following probability density function.

$$f(X) = \frac{1}{(2\pi)^{n/2}|\Sigma|^{1/2}} \exp\left[-\frac{1}{2}(X-\mu)'\Sigma^{-1}(X-\mu)\right]. \tag{2}$$

In Equation (2), Σ is symmetric and invertible matrix, and its inverse called the precision, Θ, indicates the dependency structure between two nodes. Accordingly, in GGM, if there exists an edge between two nodes, i.e. $\Theta_{ij} = \Theta_{ji} \neq 0$, these two nodes are not conditionally independent given other nodes (Whittaker, 1990). In other words, the non-zero entries of the off-diagonal entries of Θ imply a physical or functional interaction between the nodes (Dempster, 1972; Wermuth, 1976; Whittaker, 1990). In Figure 2, we draw a small network having five nodes in which X_1 is conditionally independent on X_3 for given X_2. Similarly, X_3 is conditionally independent on X_5 for given X_2 and X_4 in the same figure.

There are lots of techniques to estimate entries of the precision matrix in GGM. The neighborhood selection method with the lasso regression (Meinshaussen & Bühlmann, 2006) is one of the computationally efficient approaches for sparse and high-dimensional graphs. This method belongs to the class of the covariance selection approaches that is based on non-parametric calculation. In this model, given that the set of nodes and the number of nodes in the graph are denoted by Φ and $|\Phi(n)|$, respectively, the neighborhood of the node p in Φ is the smallest subset of the node $\Phi \setminus p$. Since each state of the node, i.e. gene, p, X_p, is defined as conditionally independent on the state of all remaining nodes, X_{-p}, an optimal prediction of the vector of the regression coefficient of X_p in the lasso regression (Tibshirani, 1996), θ^p, can be found by the following expression.

$$\theta^p = \arg\min_{\theta_p=0} E\left(X_p - \sum_{k\in\Phi(n)} \theta_k X_k\right)^2. \tag{3}$$

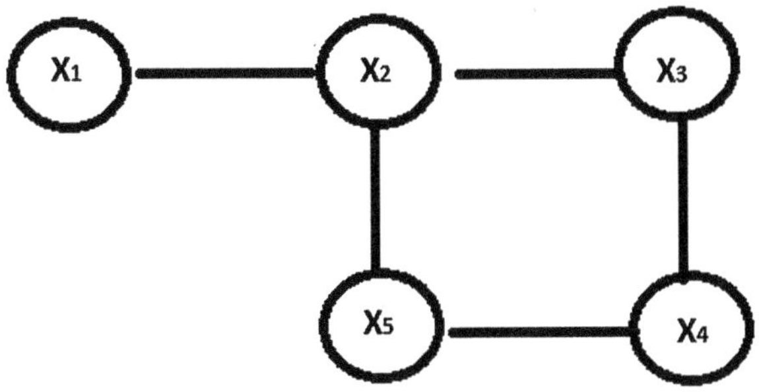

Figure 2. Simple representation of the relationship between five nodes in a system.

In Equation (3), Φ denotes the set of nodes with n number of observations each, $E(\cdot)$ shows the expectation of the given random variable, and Θ_k^p stands for $\Theta_k^p = -\Sigma_{kp}/\Sigma_{pp}$. Here, the best predictor for X_p is detected as a linear function of variables in the set of neighbors of the node (Meinshaussen & Bühlmann, 2006).

Furthermore, the graphical lasso, also named as glasso, approach (Friedman, Hastie, & Tibshiran, 2008), is another common technique in inference of biological networks. Briefly, this method controls the type I and type II errors by detecting the probability under very low levels when the penalized likelihood expression is defined as below.

$$\max_{\Theta}[\log(|\Theta|) - \mathrm{Tr}(S\Theta) - \lambda \|\Theta\|_1]. \tag{4}$$

In Equation (4), $\|\cdot\|_1$ shows the l_1-norm, which is the sum of the absolute value of the given element, and $\mathrm{Tr}(\cdot)$ presents the trace. S indicates the sample covariance matrix $S = \Theta^{-1}$. Accordingly, in inference of S in Equation (4) as an unbiased estimator of Σ in Equation (2), the sample covariance matrix is applied, i.e. $\hat{\Theta} = S$ via

$$s_{ij} = \frac{1}{n}\sum_{k=1}^{n}(x_k^i - \bar{x}^{(i)})(x_k^j - \bar{x}^{(j)}). \tag{5}$$

Here, s_{ij} denotes the ith row and the jth entry of S and n is the total sample size per gene as stated previously. Finally, x_k^i stands for the kth sample of the ith gene and $\bar{x}^{(j)}$ refers to the mean term for the ith gene having totally n observations.

On the other hand, different researchers have also investigated other parametric solutions of this problem. But the main challenge under these calculations is the high dimension of genes (nodes) with respect to the observations per gene, denoted by n, i.e. $n << p$. This inequality causes a singular covariance matrix, resulting in multiple of solutions if we perform the standard maximum likelihood technique in inference of model parameters. Hereby, in order to unravel this problem, different types of penalized methods are also suggested besides the neighborhood selection and glasso approaches as described above. These methods are based on distinct optimization approaches of the lasso regression in the estimation of Θ. We can also list grouped lasso (Yuan & Lin, citeyuan), elastic net method (Zou & Hastie, 2005), fused lasso (Tibshirani, Saunders, Rosset, Zhu, & Knight, 2005), adaptive lasso (Zou, 2006), coordinate-wise descent algorithm within the l_1-penalized lasso (Friedman, Hastie, & Tibshiran, 2007), and multivariate methods with scout algorithm (Witten & Tibshirani, 2009) approaches as the other most well-known approaches used in GGM. All these methods mainly suggest similar model construction for the penalized maximum likelihood function defined for the lasso regression by either the l_1 norm or l_2-norm as the penalizing term. On the other side, recently, certain semi-parametric approaches such as the non-paranormal SKEPTIC algorithm (Liu, Han, Yuan, Lafferty, & Wasserman, 2012) and fully parametric approaches via Bayesian techniques such as the birth-death Markov chain Monte Carlo (MCMC) (Muhammadi & Wit, 2014) approach are proposed in order to estimate Θ without using any optimization methods. In the literature of GGM, different types of Monte Carlo approaches have been also performed. Atay-Kayis and Massam (2005) apply Monte Carlo methods in the calculation of Θ when the network graph in GGM is non-decomposable. Dobra, Lenkoski, and Rodriguez (2011) and Wang and Li (2012) introduce the reversible jump MCMC and Dauwels, Yu, Xu, and Wang (2013) implement the Monte Carlo expectation maximization method within copula GGM. Whereas among all these alternatives, the neighborhood selection approach, suggested by Meinshaussen and Bühlmann (2006), is one of the most common techniques due to its simplicity in inference. Hence, in our analyses, we prefer this method and perform it in junction with Monte Carlo simulations in order to evaluate the performance of different estimates.

2.2. Bernstein polynomials

The Bernstein polynomial was introduced by Sergey Natanovich Bernstein as a new proof of the Weierstrass approximation theorem. This polynomial privileges in the approximation theory since it applies the constructive new polynomials to prove the Weierstrass approximation instead of polynomials that are already known as mathematicians.

For the mathematical expression of these polynomials, let N be the set of positive integers for every function $f:[0,1] \to R$. Thus, the Bernstein polynomial for the nth degree is defined by

$$B_n(f;t) = \sum_{k=0}^{n} f(\frac{k}{n}) b_{k,n}(t), \tag{6}$$

where $b_{k,n}$ is called the Bernstein basis polynomial and can shown as

$$b_{k,n}(t) = \binom{n}{k}(1-t)^{n-k}t^k \tag{7}$$

for $k = 0, 1, \dots, n$ and $\binom{n}{k}$ binomial coefficient.

If $f:[0,1] \to R$ is a continuous function, then the sequence of the Bernstein polynomials B_n converges uniformly to f on [0, 1],

The Bernstein polynomial (Lorentz, 1953) has certain properties such as the positivity, symmetry, and the degree raising. The first characteristic means $b_{k:n} = b_{n-k:n}$ while $i = 1, \dots, n$. On the other hand, the second feature presents $b_{k:n}(x) > 0$ and finally, the third one implies that any lower degree Bernstein polynomial is written as a linear combination of the nth Bernstein polynomials.

The generalized version of the Bernstein operator is called the Szas-Mirakyan operator and is defined as

$$S_n(f;x) = e^{-nx} \sum_{k=0}^{\infty} f\left(\frac{k}{n}\right)\frac{(nx)^k}{k!}, \tag{8}$$

where $x \in [0,1]$ and the function f is defined on an infinite interval $R^+ = [0,\infty)$.

2.3. Butzer Hahn operator

Bleiman, Butzer and Hahn (BBH) operator (Bleiman, Butzer, & Hahn, 1980) which is defined by the Bernstein-type can be represented as below for the nth degree with the k basis polynomials for the value x.

$$L_n(f;x) = \frac{1}{(1+x)^n}\sum_{k=0}^{n}\binom{n}{k}x^k f\left(\frac{k}{n+1-k}\right). \tag{9}$$

For Equation (9), the following inequality is satisfied.

$$|L_n(f;x)| \le ||f||_{C_B} \quad (f \in C_B[0,\infty)). \tag{10}$$

Equation (10) implies that the BBH operator is linear and bounded for $x \in [0,\infty)$ when $(1+x)^{-n}\sum_{k=0}^{n}\binom{n}{k}x^k = 1$. Here, $C_B[0,\infty)$ is the class of the real-valued function f defined within the interval $[0,\infty)$ and for all functions of f in this interval, $\lim L_n(f;x) = f(x)$ for each $x \in [0,\infty)$ when $n \to \infty$.

Additionally, the property of the uniform approximation of the BBH operator has been studied (Totik, 1984) when f belongs to $C[0,\infty]$ for the continuous function on $[0,\infty)$. Also Mercer (1989) has independently derived the Voronovskaya-type theorem which gives an asymptotic error term for the Bernstein polynomials for the functions which are twice differentiable as follows.

$$\lim_{n\to\infty} n((L_n(f;x) - f(x)) = \frac{x(1+x)^2}{2} f''(x) \tag{11}$$

for all $f \in C^2[0,\infty)$ with $f(x) = O(x)$ when $x \to \infty$, and $f''(x)$ is the second derivative of the function.

Abel (1996) has extended this study by giving the complete asymptotic expansion for the BBH operator as the following form.

$$L_n(f;x) = f(x) + \sum_{k=1}^{\infty} c_k(f;x)(n+1)^{-k} \tag{12}$$

$n \to \infty$. Here, c_k represents all the coefficients from $k = 0$ to $k = n$.

The Szasz operator is the limiting operator of BBH (Khan, 1988) and $L_n(f;x) \geq L_{n+1}(f;x) \geq \ldots f(x)$ if f is convex (Khan, Nevai, & Pinkus, 1991). $L_n(f;x)$ is convex itself if f is a non-increasing convex function.

Jayasri and Sitaraman (1993) have determined that L_n is a pointwise approximation process in which the largest subclass of $C[0,\infty)$ for the Bernstein-type of operator.

Then, the following function class is introduced in the study of Hermann, Szbados, and Tandori (1991).

$$\mathcal{H} = f \in C[0,\infty]: \log(|f(x)| + 1) = o(x). \tag{13}$$

He has proved that if f belongs to $\mathcal{H}$, then for each $x > 0$, the pointwise convergence is $\lim_{n\to\infty} L_n f = f$ on $[0,\infty)$. Moreover, for some $a > 0$, $f(x) = e^{ax}$, then $\lim_{x\to\infty} L_n(f;x) = \infty$. Also the operator L_n is arisen from the random variable with n observations, X_n, which has the Bernoulli distribution as below.

$$P(\{X_n = k|(n-k+1)\}) = \binom{n}{k} p^k q^{n-k} \tag{14}$$

for the parameters $p = x/(1+x)$, $q = 1/(1+x)$ and $k = 0, 1, \ldots, n$.

2.4. *Meyer-König and Zeller operator*

The operator

$$M_n(f;x) = \sum_{k=0}^{\infty} f\left(\frac{k}{n+k+1}\right) m_{n+1,k}(x) \tag{15}$$

is known as the Meyer-König and Zeller (MKZ) operator when

$$m_{n,k}(x) = \binom{n+k}{k} x^k (1-x)^{n+1} \tag{16}$$

for the nth degree and the kth basis polynomial for the value of x as stated previously.

These operators are known as the Bernstein-type of operators. Cheney and Sharma (1996) have defined this operator as a power series of the Bernstein operators.

The MKZ operator can also be obtained from the negative binomial distribution via

$$M_n(f;x) = (1-x)^{n+1} \sum_{k=0}^{\infty} \binom{n+k}{k} f\left(\frac{k}{n+k}\right) x^k. \tag{17}$$

3. Application

In the assessment of the polynomials' results, we consider four different scenarios. In the first scenario, we estimate the precision matrix Θ from the simulated data-sets under different kinds of the Bernstein polynomials and the Bernstein-type of operators, which are the MKZ operator and the BBH operator. For the analyses of the model, we generate 50, 100, and 500 dimensional data-sets in which each gene has 20 observations. Then, we set all the off-diagonal entries of Θ to 0.9 arbitrarily and generate scale-free networks by using the huge package in the R programme language. Then, in order to evaluate whether the entry of the off-diagonal terms has any effect in inference, we change it via moderately small and small values too. Hereby, in the second and the third scenarios, the off-diagonal elements of the precision matrix set to 0.7 and 0.5, respectively, when the networks are scale-free. Finally, as the fourth plan, we change the sparsity of the system from the scale-freeness to the hubs property since the former typically indicates a high sparsity level around 90% or above and the latter implies relatively a lower sparsity level at around 80–90%. On the other hand, in our analyses, we have not controlled the performance of methods lower than these sparsities levels as the high sparsity is one of the main features of biological systems (Barabasi & Otiva, Barabasi04).

Accordingly, in all scenarios, we initially generate a data-set for the true network and keep its true path for the best model selection in further steps. Later, we transform this actual data-set via the Bernstein polynomial, the Szasz polynomial, the MKZ operator, and the BBH operator. Finally, all these non-transformed and transformed data are used to estimate the precision matrices by using the neighborhood selection (NB) method. In the application, we choose NB among alternatives due to its computational gain. For the assessment, we calculate the F-measure and the precision for each run as shown below.

$$\text{Precision} = \frac{TP}{TP+FP} \quad \text{and} \quad \text{F-measure} = 2\frac{\text{Precision} \times \text{Recall}}{\text{Precision} + \text{Recall}}. \tag{18}$$

In Equation (18), TP (true positive) indicates the number of correctly classified objects that have positive labels and FN (false negative) shows the number of misclassified objects that have negative labels and finally, Recall is calculated as $\text{Recall} = TP/(TP+FN)$.

Then, we repeat the calculation of the underlying statistics for 1000 Monte Carlo runs and their means are computed. The results are presented in Table 1.

From Table 1, it is seen that for low dimensions, the Bernstein polynomials give better results than others, in particular, the Szasz polynomials have the highest F-measure. We obtain the same results for the precision values as well. Moreover, we detect that when the dimension of the matrix increases, the F-measure and the precision value decrease. Furthermore, as shown in Tables 2 and 3, we observe similar findings in the sense the Szasz polynomials typically produce better F-measure and precision even though we decrease the correlation between genes (by off-diagonal entries 0.7 and 0.5). Whereas under these scenarios, it is found that the MKZ operator is as good as the Szasz operator in terms of the accuracy of the estimates under certain conditions. Finally when we evaluate the outputs of Tables 2–4, we see that the results of both the Szasz polynomial and the MKZ operator are very close to each other and overperform with respect to the remaining operators.

Table 1. Comparison of the F-measure and the precision values computed with/without operators in inference of Θ with 0.9 off-diagonal entries under scale-free networks

Measure	Dimension of Θ	No operator	Bernstein polynomial	Szasz polynomial	BBH operator	MKZ operator
F-measure	(50 × 50)	0.0014	0.1714	0.1590	0.1132	0.1492
	(100 × 100)	0.0001	0.0904	0.0863	0.0594	0.0812
	(500 × 500)	0.0000	0.0171	0.0171	0.0094	0.0163
Precision	(50 × 50)	Not Computable	0.4852	0.4789	0.3794	0.4729
	(100 × 100)	Not Computable	0.4769	0.4749	0.4601	0.4699
	(500 × 500)	Not Computable	0.4663	0.4682	0.4466	0.4624

Table 2. Comparison of the F-measure and the precision values computed with/without operators in inference of Θ with 0.7 off-diagonal entries under scale-free networks

Measure	Dimension of Θ	No operator	Bernstein polynomial	Szasz polynomial	BBH operator	MKZ operator
F-measure	(50 × 50)	Not Computable	0.0699	0.0711	0.0704	0.0712
	(100 × 100)	Not Computable	0.0361	0.0366	0.0362	0.0368
	(500 × 500)	Not Computable	0.0077	0.0083	0.0078	0.0080
Precision	(50 × 50)	Not Computable	0.0403	0.0403	0.0404	0.0401
	(100 × 100)	Not Computable	0.0201	0.0200	0.0200	0.0200
	(500 × 500)	Not Computable	0.0040	0.0040	0.0040	0.0040

Table 3. Comparison of the F-measure and the precision values computed with/without operators in inference of Θ with 0.5 off-diagonal entries under scale-free networks

Measure	Dimension of Θ	No operator	Bernstein polynomial	Szasz polynomial	BBH operator	MKZ operator
F-measure	(50 × 50)	Not Computable	0.0703	0.0710	0.0702	0.0716
	(100 × 100)	Not Computable	0.0356	0.0365	0.0362	0.0367
	(500 × 500)	Not Computable	0.0077	0.0080	0.0077	0.0080
Precision	(50 × 50)	Not Computable	0.0404	0.0402	0.0402	0.0403
	(100 × 100)	Not Computable	0.0198	0.0199	0.0200	0.0199
	(500 × 500)	Not Computable	0.0040	0.0040	0.0039	0.0040

On conclusion, all these outputs imply that when the sparsity of networks decreases, all Bernstein-type of operators compute similar results and the Szasz polynomial as well as the MKZ operator are slightly better than others. On the contrary, if the sparsity level raises as mostly observed in

Table 4. Comparison of the F-measure and the precision values computed with/without operators in inference of Θ with 0.9 off-diagonal entries under hubs networks

Measure	Dimension of Θ	No operator	Bernstein polynomial	Szasz polynomial	BBH operator	MKZ operator
F-measure	(50 × 50)	Not Computable	0.0668	0.0687	0.0677	0.0691
	(100 × 100)	Not Computable	0.0352	0.0365	0.0358	0.0366
	(500 × 500)	Not Computable	0.0074	0.0076	0.0075	0.0076
Precision	(50 × 50)	Not Computable	0.0383	0.0387	0.0386	0.0387
	(100 × 100)	Not Computable	0.0196	0.0198	0.0198	0.0198
	(500 × 500)	Not Computable	0.0038	0.0038	0.0038	0.0038

biological networks, the operators have different accuracy values and the performance of the Bernstein polynomials, especially, the Szasz polynomial, becomes better.

4. Discussion

In this study, we have compared all well-known Bernstein-type of operators to detect which alternative produces the highest accuracy if it is applied with GGM. For this purpose, we have analyzed the Monte Carlo results of the Bernstein polynomials, BBH and MKZ operators. The results have indicated that the Bernstein polynomials have the highest accuracies if they are performed in advance of the inference of the model parameters of GGM and the MKZ operator can be another good alternative if the sparsity level of the system decreases. But for all choices of operators, we have found that these operators can improve the accuracies of estimates for different dimensional biochemical systems if they are implemented as the pre-processing step before the inference of the precision matrix. As the extension of this study, we consider other approximation methods such as the Fourier transformations in such a way that the distributional feature of the observations can be embedded to the new transformed data in order to smooth the original data-set smartly and get estimates with higher accuracy.

Acknowledgements
The authors would like to thank the anonymous referees, editors, and guest editors of this Special Issue of the journal for their valuable contributions which improve the quality of the paper.

Funding
Vilda Purutçuoğlu would like to thank to SysPatho EU FP-7 Collaborative Project [project number 260429] for their support.

Author details
Melih Ağraz[1]
E-mail: agraz@metu.edu.tr
Vilda Purutçuoğlu[1]
E-mail: vpurutcu@metu.edu.tr
[1] Department of Statistics, Middle East Technical University, Ankara 06800, Turkey.

References

Atay-Kayis, A., & Massam, H. (2005). A Monte Carlo method for computing the marginal likelihood in nondecomposable gaussian graphical models. *Biometrika, 92*, 317–335.
Abel, U. (1995). The moments for the Meyer--Knig and Zeller operators. *Journal of Approximation Theory, 82*, 352–361.
Abel, U. (1996). On the asymptotic approximation with operators of Bleimann. *Butzer and Hahn, Indagationes Mathematicae, 7*(1), 1–9.
Agratini, O. (1996). A class of Bleimann, Butzer and Hahn type operators. *Analele Universitătii Din Timişoara, 34*, 173–180.
Barabasi, A. L., & Otiva, Z. N. (2004). Understanding the cell's functional organization. *Nature Reviews Genetics, 5*, 101–113.
Bernstein, N. (1912). Démonstration du théréme de Weierstrass fondée sur le calcul de probabilités. *Community of Kharkov Mathematical Society, 13*, 1–2.
Becker, M., Kucharski, D., & Nessel, R. J. (1977). Global approximation theorems for the Szasz–Mirakyanoperators in exponential weight spaces, in linear spaces and approximation. *Proceeding of Conference of Oberwolfach, Basel ISNM, 40*, 319–333.
Bleiman, G., Butzer, P. L., & Hahn, L. (1980). A Bernstein-type operator approximating continuous functions on the semi-axis. *Indagationes Mathematicae, 42*, 255–262.
Cheney, E. W., & Sharma, A. (1996). Bernstein power series. *Canadian Journal of Mathematics, 16*, 241–253.
Dauwels, J., Yu,, H., Xu, S., & Wang, X. (2013). Copula Gaussian graphical model for discrete data. *IEEE International Conference on Acoustics, Speech and Signal Processing (ICASSP)*. Vancouver.
Dempster, A. (1972). Covariance selection. *Biometrics, 28*, 157–175.
Dobra, A., Lenkoski, A., & Rodriguez, A. (2011). Bayesian inference for general Gaussian graphical models with application to multivariate lattice data. *Journal of the American Statistical Association, 106*, 1418–1433.
Friedman, J. H., Hastie, T., & Tibshiran, R. (2007). Pathwise coordinate optimization. *Annals of Applied Statistics, 2*, 302–332.
Friedman, J. H., Hastie, T., & Tibshiran, R. (2008). Sparse inverse covariance estimation with the graphical lasso. *Biostatistics, 9*, 432–441.
Han, J.-D., Bertin, N., Hao, T., Goldberg, D. S., Berriz, G. F., Zhang, L. V., ... Vidal, M. (2004). Evidence for dynamically organized modularity in the yeast protein-protein interaction network. *Letters to Nature, 430*, 88–93.
Hermann, T., Szbados, J., & Tandori, K. (Eds.). (1991). *Approximation Theory, Proceedings Conference Kecskemet/Hung., 1990, On the operator of Bleimann, Butzer and Hahn*(58, pp. 355–360). North-Holland.
Jayasri, C., & Sitaraman, Y. (1993). On a Bernstein-type operator of Bleimann-Butzer and Hahn. *Journal of Computational and Applied Mathematics, 47*, 267–272.
Khan, K. A., Nevai, P., & Pinkus, A. (1991). *Some properties of a Bernstein type pf operator of Bleimann, Butzer and Hahn*, In P. Nevai & A. Pinkus (Eds.), Progress in approximation theory (pp. 497–504). New York, NY: Academic Press.
Khan, R. A. (1988). A note on a Bernstein-type operator of Bleimann. *Butzer and Hahn, Journal of Approximation Theory, 53*, 295–303.
Liu, H., Han, F., Yuan, M., Lafferty, J., & Wasserman, L. (2012). High dimensional semiparametric gaussian copula graphical models. *The Annals of Statistics, 40*, 2293–2326.
Lorentz, G. G. (1953). *Bernstein polynomilas* (Mathematical Exposition No. 8). Toronto: University of Toronto Press.
Meinshaussen, N., & Bühlmann, P. (2006). High dimensional graphs and variable selection with the lasso. *The Annals of Statistics, 34*, 1436–1462.
Mercer, A. (1989). A Bernstein-type operator approximating continuous functions on the half-line. *Bulletin of Calcutta Mathematical Societiy, 31*, 133–137.

Muhammadi, A., & Wit, E. C. (2014). Bayesian structure learning in sparse Gaussian graphical models. *Bayesian Analysis, 10,* 109–138.

Purutçuoğlu, V., Ağraz, M., & Wit, E. (2015). *Bernstein approximations in glasso-based estimation of biological networks.*

Tibshirani, R. (1996). Regression shrinkage and selection via the lasso. *Journal of the Royal Statistical Society, Series B, 58,* 267–288.

Tibshirani, R., Saunders, M., Rosset, S., Zhu, J., & Knight, K. (2005). Sparsity and smoothness via the fused lasso. *Journal of the Royal Statistical Society, Series B, 67,* 99–108.

Totik, V. (1984). Uniform approximation by Bernstein-type operators. *Nederlandse Akademie van Wetenschappen (Indagationes Mathematicae), 50,* 87–93.

Wang, H., & Li, S. Z. (2012). Efficient Gaussian graphical model determination under G-Wishart prior distributions. *Electronic Journal of Statistics, 6,* 168–198.

Witten, D. M., Frieman, J. H., & Simon, N. (2011). New insights and faster computations for the graphical lasso. *Journal of Computational and Graphical Statistics, 20,* 892–900.

Whittaker, J. (1990). *Graphical models in applied multivariate statistics.* New York, NY: Wiley.

Yuan, M., & Lin, Y. (2007). Model selection and estimation in the Gaussian graphical model. *Biometrika, 94,* 19–35.

Zou, H. (2006). The adaptive lasso and its oracle properties. *Journal of the American Statistical Association, 101,* 1418–1429.

Zou, H., & Hastie, T. (2005). Regularization and variable selection via the elastic net. *Journal of the Royal Statistical Society, Series B, 67,* 91–108.

18

Multilevel multinomial logistic regression model for identifying factors associated with anemia in children 6–59 months in northeastern states of India

Sanku Dey[1§] and Enayetur Raheem[2*§]

*Corresponding author: Enayetur Raheem, Department of Statistics, St. Anthony's College, Shillong, Meghalaya 793001, India
E-mail: enayetur.raheem@unco.edu
Reviewing editor: Zudi Lu, University of Southampton, UK

Abstract: Objective: To examine the factors influencing the occurrence of childhood anemia in northeast India. Method: A nationally representative systematic multi-stage stratified cross-sectional sample of singleton children aged 6–59 months from all states of India. Data consist of 10,136 children in the age group 6–59 months in eight northeastern states. The level of anemia was the outcome variable with four ordinal categories (severe, moderate, mild, and non-anemic). A two-level random intercept multivariate logistic regression model was considered with state of residence as the level-2 variable. Results: About 53% of the children are anemic in the northeastern states of India. Tripura has the highest prevalence of anemia cases (74%), whereas the lowest percentage of anemia cases was in Manipur (42%). Multivariate analysis suggests that age at marriage (OR = 1.13, 95% CI: 1.05, 1.21) and the number of children ever born (OR = 1.09, 95% CI: 1.03, 1.15) have significant effects on being at or below a hemoglobin level (severely anemic). Furthermore, age

ABOUT THE AUTHORS

Sanku Dey's, PhD, research focuses on the development and application of statistical approaches to scientific and medical issues. Another area of his research encompasses both frequentist and Bayesian inference of reliability theory and general distribution theory. This area of research has real-life applications in engineering and medical sciences. Another important research in this area is to work on distribution theory using records. Record values and record times have been of interest to humans throughout history. Meteorologists frequently deal with upper and lower record temperatures and precipitation levels. Record values often appear in sporting events. I have developed several mathematical models based on distributions like Rayleigh, generalized Rayleigh, generalized exponential, and generalized inverted exponential distributions.

Enayetur Raheem's, PhD, research is focused on building better predictive models through improved estimation of regression parameters. He is interested in data/text mining/feature extraction, and machine learning. Other interests include application of statistical methods in medical/health, biological and environmental sciences.

PUBLIC INTEREST STATEMENT

A multilevel multinomial logistic regression model was considered to identify potential risk factors of anemia in children aged 6–59 months in northeastern India. Level of anemia was considered as the outcome variable with four ordinal categories (severe, moderate, mild, and non-anemic) based on hemoglobin concentration in blood. A two-level random intercept model was used with state of residence as the level-2 variable. Considering the hierarchical nature of the data, we did not find place of residence (urban vs. rural), sex of child, or household living standard (low, medium, and high SES) to be significant. Rather, mother's age at marriage and the number of children ever born were significantly associated with anemia. Our findings suggest that as the child becomes 48 months or older, likelihood of severe anemia decreases. This study throws light on the fact that anemia intervention needs to focus more at state level followed by individual level.

of the child (OR = 0.92, 95% CI: 0.86 – 1.00) was a significant predictor, indicating that odds of severe anemia decrease if the child is 48 months or older. Conclusions: The high prevalence of mild and moderate anemia demands due emphasis in the programs and policies of the government so that the overall prevalence of anemia among children aged 6–59 months can be reduced. Comprehensive strategies need to focus more at state level followed by individual level for combating anemia.

Subjects: Mathematics & Statistics; Multivariate Statistics; Science; Statistical Theory & Methods; Statistics; Statistics & Probability

Keywords: multilevel multinomial logistic regression; child anemia; odds ratio

1. Introduction

Anemia is considered the most prevalent nutritional deficiency globally (McLean, Cogswell, Egli, Wojdyla, & de Benoist, 2009). About one-third of the global population (over two billion) is anemic (de Benoist, McLean, Egll, & Cogswell, 2008) while the prevalence of anemia among preschool children is 47.4% (IIPS, 2000). According to WHO estimates, India has the highest prevalence of anemia among the South Asian countries in all age groups. The Third National Family Health Survey (IIPS, 2007) of India revealed that 70% of the children are anemic in the age group of 6–59 months, including 3% severely anemic, 40% moderately anemic, and 26% mildly anemic (IIPS, 2000). As per five major surveys (National Family Health Survey (NFHS Assunção, Santos, de Barros, Gigante, & Victora, 2007; de Benoist et al., 2008), District Level Household Survey 2 (DLHS), Indian Council of Medical Research (ICMR) Micronutrient Survey (Togeja & Singh, 2004) and Micronutrient Survey (NNMB, 2004), over 70% of preschool children are anemic in the country.

Iron deficiency (ID) is listed as one of the "top ten risk factors contributing to death" (Dubey, 1994). Iron deficiency anemia (IDA) is prevalent in South Asia, predominantly India, Bangladesh, and Pakistan. However, the prevalence of IDA has declined substantially in neighboring countries such as Bangladesh and Pakistan (Gillespie, 1997), whereas it has plummeted from 20 to 8% within a decade in China (Lokeshwar, Manglani, Rao, Patel, & Kulkarni, 1990). Studies carried out in Egypt (Seshadri, Gopaldas, Walter, & Heywood, 1989), India (Soemantri, Gopaldas, Seshadri, & Pollitt, 1989), Thailand (Pollitt, Hathirat, Kotchabhakdi, Missell, & Valyasevi, 1989), and the USA (Pollitt, 1991) revealed that iron and folate deficiency anemia reduce the learning capacity, attentiveness, and intelligence of children aged below five years. Several studies have been carried out on anemia in India since 1980. However, very few studies focused on child anemia at the national and regional levels (Bharati, Pal, Chakrabarty, & Bharati, 2015; Dey, Goswami, & Dey, 2013; Singh & Patra, 2014). The risk factors of anemia most often cited in the literature are: low family income and low maternal level of education, lack of access to health care services, inadequate sanitary conditions, and a diet with poor quantities of iron (Oliveira, Osório, & Raposo, 2007; Osório, Lira, & Ashworth, 2004) .

Nutritional problem is very common among children below five years in all the states of India, particularly, in those states whose performances are very poor in respect of demographic and socioeconomic indicators. There are several factors responsible for IDA, for example: low dietary intake, inadequate iron (less than 20 mg/day) and folic acid intake (less than 70 mg/day), and chronic blood loss due to infection such as malaria and hookworm infestations, among others (NNMB, 2004; Togeja & Singh, 2004). A recent study on the determination of socioeconomic and demographic determinants of anemia among Indian children aged 6–59 months found an increasing trend of prevalence of anemia up to two years of age and then decreased thereafter (Goswmai & Das, 2015). Therefore, it is imperative to find out the factors that contributed to anemia and to examine the contribution of existing programs in combating child anemia, especially in the less developed areas such as northeastern states of India. In view of the above, our objective is to determine the prevalence of anemia among children (6–59 months) from the northeastern states of India. To comprehend the prevalence of anemia and to identify the significant predictors, socioeconomic differentials have been taken into consideration in our study.

2. Methods

The present paper uses the data-set from the 2005–2006 National Family Health Survey (released in 2008) for the northeastern states of India. Respondents were selected through a systematic multi-stage stratified sample survey conducted in all 29 states of India (IIPS, 2007). In each state, populations were stratified by urban and rural area of residence, and the sample size at the state level was proportional to the size of the state's urban and rural population. A uniform sample design was adopted in all states. In each state, rural sample was selected in two stages: at the first stage, primary sampling units (PSUs), which are villages, were selected with probability proportional to population size (PPS). At the second stage, households within each PSU were selected randomly. In urban areas, a three-stage sampling procedure was adopted. At first, wards were selected with PPS sampling. In the second stage, one census enumeration block (CEB) was randomly selected from each sample ward. In the final stage, households were randomly selected within each sample CEB. The study population constitutes a nationally representative cross-sectional sample of singleton children aged 0–59 months and born after January 2000 or January 2001 (n = 50,750) to mothers aged 15–49 years from all 29 states of India. Information on children was obtained by a face-to-face interview with mothers, with a response rate of 94.5%. The survey also provided district-level information on the prevalence of under-nutrition [weight-for-age, using the standard deviation (SD) classification] among children in the age group of 0–59 month(s); prevalence of anemia (Hb estimation by HemoCue Hb201 + analyzer, Angelholm, Sweden) for the children aged 6–59 months, adolescent girls aged 10–19 years, and pregnant women; household availability of iodized salt; and the coverage of vitamin A program, with appropriate dosage. To meet the objectives of the study, we have produced a data-set that pertains to the northeastern states of India. Our study comprises 10,136 children within 6–59 months of age.

3. Variables

The variables of the study are briefly described in the following.

3.1. Outcome/response variable

We create the response variable using the variables that measures hemoglobin level in blood. We split the variable into four categories as per the specification of WHO to define level of anemia. In particular, we define a new variable called hglevel to indicate level of hemoglobin in blood (g/dL). We set hglevel = 1 if hemoglobin level is below 7 g/dL and call it severely low level (indicating severe anemia), hglevel = 2 if hemoglobin level is 7–9.9 g/dL (indicating moderate anemia), hglevel = 3 if hemoglobin level is 10–10.9 g/dL (indicating mild anemia), and hglevel = 4 if hemoglobin level is above 11 g/dL indicating no anemia.

3.2. Independent variables/predictors

The study includes a set of independent variables to understand the extent and differentials in the level of anemia among children aged 6–59 months.

4. Objective of the study

Our objective is to study the relationship between some household-level predictors and some maternal variables on the probability of being at or below a hemoglobin level. Based on our classification of anemic children, we calculate odds ratio and the predictive probability of being at or below a hemoglobin level based on a polytomous logistic regression model.

In particular, we like to answer the following research questions:

(1) What is the likelihood of being at or below each level of hemoglobin in the blood (g/dL) for children at a typical household?

(2) Does the likelihood of being at or below each level of hemoglobin (i.e. anemia level) vary across state of residence?

(3) What are the relationships between the household and maternal variables, and the likelihood of a child being at or below a given hemoglobin level?

5. Statistical analysis

In the present study, we employ bivariate as well as multivariate techniques to identify the factors that are associated with anemia in children of 6–59 months of age. In the bivariate analysis, cross-tabulation was made between the potential risk factors and the presence of anemia. Pearson's chi-square test and *p*-values were used to test for the significance of each of the potential risk factor in bivariate analysis. Further analyses were carried out to study the relationship of the potential risk factors with the hemoglobin level in a multivariate setup. We use multinomial multilevel logistic regression model to predict the level of anemia as a function of mother's age at marriage, number of children ever born to mother, religion, literacy of mother, household living standard, place of residence, sex of the child, and age of the child (in months). The wealth index was constructed using household asset data. Each household asset was assigned a score generated through principal components analysis and the resulting scores were standardized and summed up for each household (IIPS, 2007).

Due to the stratified nature of data in NFHS, the children are naturally nested into mothers, mothers are nested into households, households into PSUs, and PSUs into the states. Hence, keeping in view of the hierarchically clustered nature of the survey data, we use multilevel regression model to avoid possible under-estimation of parameters from a single-level model (Griffiths, Matthews, & Hinde, 2002). The advantage of multilevel model is that they properly account for the correlation structure of the data that frequently occurs in social sciences and in multistage survey sampling.

To properly account for the hierarchical nature of the data, we consider state of residence as the level-2 variable under which the respondents are nested. Thus, respondent-level variables are the level-1 variables used in this study to predict the level of anemia in children. Our outcome variable is ordinal polytomous with four levels for anemia. The model for such data uses a multinomial distribution and cumulative logit link function to compute the cumulative odds for each category of the response (O'Connnell, Goldstein, Rogers, & Peng, 2008).

For the polytomous response variable, let us denote η_{kij} as the log odds of a child in the *i*th household in the *j*th state being at or below *k*th level of hemoglobin in blood. The model for level-1 may be written as:

$$\eta_{kij} = \log\left(\frac{P(R_{ij} \le k)}{1 - P\left(R_{ij} \le k\right)}\right) = \beta_{0j} + \beta_{1j} X_{ij} + \delta_k \tag{1}$$

Here, β_{0j} is the intercept, assumed to be random, represents the average log odds of being at or below severely low hemoglobin level; $P\left(R_{ij} \le k\right)$ represents the probability of responding at or below *k*th level of the outcome variable; X_{ij} represents level-1 predictorβ_{1j}, the slope coefficient corresponding to X_{ij} which measures the change in the probability of being at a given hemoglobin level per unit change in the level-1 predictor; and δ_k epresents the difference between the *k*th category and the preceding one. Essentially, for $k = 1$, we have $\delta_1 = 0$.

We consider the intercept to be random having the following model:

$$\beta_{0j} = \gamma_{00} + u_{0j},$$

where γ_{00} is the log odds of severe anemia relative to no anemia when the predictor variables in the model are evaluated at zero, and u_{0j} is the random error term with $u_{0j} \sim N\left(0, \tau_{00}\right)$ where $u\tau_{00}$ is the level-2 error variance.

Now, substituting the expression for β_{0j} in equation (McLean et al., 2009), we get

$$\eta_{kij} = \log\left(\frac{P(R_{ij} \leq k)}{1 - P\left(R_{ij} \leq k\right)}\right) = \gamma_{00} + \beta_{1j}X_{ij} + \delta_k + u_{0j} \tag{2}$$

This model represents a scenario with no level-2 predictor (as in our case), and one level-1 predictor. However, in our study, we have more than one level-1 predictor, and so the model is meant to be augmented by those additional covariates as they are considered in the analysis.

Model (2) gives log odds when fitted to data. Alternatively, predicted probability of the event of interest (e.g. being at or below a hemoglobin level) can be calculated using the following formula as discussed in (Ene, Leighton, Blue, & Bell, 2014):

$$\text{Predicted probability (PP)}\ \theta_{kij} = \frac{e^{\eta_{kij}}}{1 + e^{\eta_{kij}}} \tag{3}$$

This is simply a conversion of log odds of the event of interest to probability of the event of interest in order to provide a meaningful interpretation of the numbers. In this expression, θ_{kij} is the probability of the event (e.g. being at or below a hemoglobin level), $1 - \theta_{kij}$ is the corresponding probability of being above a given hemoglobin level, η_{kij} represents the log odds of the event of interest, and e is approximately 2.72.

We used PROC GLIMMIX in SAS 9.4 for multinomial distribution, and CLOGIT link function. To the best of our knowledge, there is no model diagnostic procedure available under PROC GLIMMIX for multinomial model. We considered several candidate models, and the best model was selected based on −2 log likelihood as a criterion. Details are discussed later in the paper.

5.1. Model-building strategies for multilevel logistic regression

A multilevel multinomial logistic regression model was considered to predict the probability of being at or below a hemoglobin level using the available predictors. Since the outcome variable is ordinal, we consider cumulative logit link function. The steps of the model-building process are outlined in Table 3. We consider random intercept model where the model is fitted with only the intercept (Model 1). Then, in Model 2, we add maternal characteristics such as number of children ever born (2 or less, 3–4, and 5 or more) and age at marriage (below 18, 18–26, and above 26 years of age). In Model 3, we add child's age in months (<48 months, 48, or more months). To get Model 4, we add religion and literacy of mother to Model 3. Finally, we augment Model 4 by adding the sex of the child, household living standard (low, medium, high), and place of residence (urban vs. rural). We call it Model 5.

6. Results

6.1. Bivariate analysis

Of the 10,136 children included in this study, 51.6% were male and 48.4% were female. Fifty-two percent of the children were anemic (53% male and 52% female). Table 1 presents cross-tabulation of prevalence of anemia (frequency and percentages) by level of anemia for northeastern states of India. The last column shows total number of children in the study sample from each state along with the percentages of children with anemia. The last column of Table 1 shows crude prevalence rate of each type of anemia per hundred children in the sample. Overall, 12% of the children have severe anemia, 34% have moderate anemia, 6% have mild anemia, and 48% have no anemia. We observe that Tripura has the highest percentage (74%) of anemia cases among the children in the sample. Of the 361 children in Tripura, about 19% are severely anemic, 50% moderate, and 6% have mild anemia. Sikkim falls next to Tripura with about 71% anemia cases (22% severe, 43% moderate,

Table 1. Prevalence of anemia by state of residence

State	Anemia level				
	Severely anemic	Moderately anemic	Mildly anemic	Non-anemic	Total (% anemic)
Sikkim	159	301	42	206	708
	22.46%	42.51%	5.93%	29.10%	71%
Arunachal Pradesh	317	724	261	1,611	2,913
	10.88%	24.85%	8.96%	55.30%	45%
Nagaland	33	180	35	277	525
	6.29%	34.29%	6.67%	52.76%	47%
Manipur	241	526	118	1,198	2,083
	11.57%	25.25%	5.66%	57.51%	42%
Mizoram	51	258	57	296	662
	7.70%	38.97%	8.61%	44.71%	55%
Tripura	67	181	20	93	361
	18.56%	50.14%	5.54%	25.76%	74%
Meghalaya	22	55	17	105	199
	11.06%	27.64%	8.54%	52.76%	47%
Assam	334	1,189	136	1,026	2,685
	12.44%	44.28%	5.07%	38.21%	62%
Total (crude %)	1,224	3,414	686	4,812	10,136
	(12%)	(34%)	(6%)	(48%)	

Note: Overall Pearson's chi-square statistics value is 622.84 with a *p*-value of <0.0001. We found statistically significant association between state of residence and anemia level.

and 6% mild). Nagaland and Meghalaya each have 47% anemia cases, while Arunachal Pradesh has 45% followed by Manipur which has the lowest percentage of anemia cases (42%). Sikkim has the highest percentage of severely anemic children (about 23%) followed by Tripura (about 19%). Mizoram has the lowest rate of severely anemic children with only 8% prevalence rate.

Pearson's chi-square test for association shows that the place of residence is associated with the level of anemia ($\chi^2 = 622.84$, $p < 0.0001$). In other words, prevalence of anemia among children varies significantly among the states.

Anemia is slightly more prevalent among males (53%) than females (52%). Among the severely anemic children, 85% are in rural areas compared to 15% in urban areas. Similarly, 79% of moderately anemic children and 75% of mildly anemic children live in rural areas. Muslim children have the highest prevalence rate of any form of anemia (62%) followed by 56% among Hindus. Christians have the lowest prevalence of anemia (48%). However, among the severely anemic child, 44% are Hindu, 7% Muslim, 29% are Christian (Table not shown).

Of the severely anemic cases, 62% are from households with low living standard, 27% are from medium, and 10% are from high living standard households. Interestingly, prevalence of any form of anemia is low among the mothers who cannot read or write. However, the association is not statistically significant ($\chi^2 = 6.43$, $p = .09$). Among the mothers of severely anemic children, 46% have two or less children, 36% have three to four children, and 18% have more than five children. Majority of the mothers of severely anemic children (63%) were married between the ages 18 and 26 years, 32% were married below 18 years, while 5% were married when they were above 26 years of age.

Results of bivariate analysis are presented in Table 2. We observe that there is a statistically significant association between the level of anemia (severe, moderate, mild, and none) and religion of the respondents ($p < 0.0001$), and place of residence ($p < 0.0001$), and between the level of anemia and age of mother at marriage ($p = 0.0425$). There is no statistically significant association found between anemia level and household living standard, sex of child, literacy of mother, total children ever born to mother, and age of the child (months). Owing to this, we investigate the relationship between anemia level and the variables cited above using multilevel polytomous logistic regression.

6.2. Multivariate analysis

In multivariate analysis, the fitted models along with the estimated effects and their standard errors are presented in

Table 4. We used Laplace estimation method so that we can compare the models based on negative 2 log likelihood (−2LL) or the deviance test. The last row of Table 4 shows the *p*-values for the deviance test based on Chi-square statistic.

We consider five difference models and they are presented in Table 3. Notice that Model 1 is nested under Model 2, which in essence is nested under Model 3, and this in turn is nested under Model 4, and again Model 4 nested under Model 5. This allows us to compare these models based on −2LL. We found Model 3 to be better than Model 1 as the drop in −2LL is statistically significant ($p < 0.0001$), and Model 3 to be better than Model 2 ($p = 0.0452$) as well. Models 4 and 5 include more variables and thus reduce the −2LL values further. However, the gain in reduction of −2LL is not statistically significant enough to consider them for our analysis. We, therefore, resort to Model 3 and use this model to answer our research questions.

In regard to our first research question about the likelihood of being at or below a given hemoglobin level (indicative of anemia level) at a given household, we consider the fixed effects estimates provided under Model 1 of Table 4. The estimated effects in Table 4 (Model 1) are the log-odds of being at or below a hemoglobin level with typical background characteristics (no covariates in the model). We use these estimates to calculate the predictive probability (PP) of a child being at or below a hemoglobin level. For example, the probability of being at or below severely low hemoglobin level (i.e. severely anemic) for a given child with average background characteristics (with no covariate in the model) is

$$\text{P (being at or below severe anemic level)} = \frac{e^{\eta_{ij}}}{1+e^{\eta_{ij}}} = \frac{e^{-1.95}}{1+e^{-1.95}} = \frac{0.1423}{1+0.1423} = 0.1247$$

Similarly, the probability of being at or below moderately anemic level is 0.4825 and at or below mildly anemic is 0.5523. These are cumulative probabilities of being at or below a given level of anemia. In order to obtain the exact probability of being at a given level, we have to subtract the adjacent probabilities. For example, the predicted probability of being at or below severely anemic level is 0.1247, at the moderately anemic level is (0.4825 − 0.1247=) 0.3578, and at the mild anemic level is (0.5523 − 0.4825=) 0.0698. Also, probability of being non-anemic is 1 − 0.5523 = 0.4477.

In the second research question, we wanted to know if the likelihood of being at or below each level of hemoglobin (i.e. anemia level) varies across state of residence. To answer this question, we estimated the covariance parameter for the variable "state." The estimated intercept for "state" is 0.2015 ($z = 1.94$, $p = 0.0263$). Thus, there exists a statistically significant variation across states for the likelihood of being at or below a hemoglobin level. Additionally, the intra-class correlation coefficient (ICC) was computed using the covariance parameter estimate for the state (0.2015). The ICC indicates how much of the total variation in the probability of being at or below a hemoglobin level is due to the variation among the states. Following the procedure of (Snijders & Bosker, 1999), we calculate ICC as:

Table 2. Frequency distribution of anemia level by the predictors of anemia. The first column shows chi-square test for independence; the cells represent the frequency (first row), row percentage (second row). The last column shows crude odds ratio for overall anemia

Variables (Chi-square and *p*-value)		Severely anemic	Moderately anemic	Mildly anemic	Non anemic	Crude odds ratio
Place of residence Chi-square = 27.16 $p < 0.0001$	Rural	1,035	2,705	516	3,818	1.04
		12.82%	33.50%	6.39%	47.29%	
	Urban (ref)	189	709	170	994	
		9.17%	34.38%	8.24%	48.21%	
Religion Chi-square = 166.35 $p < 0.0001$	Hindu	537	1,330	238	1,667	1.23*
		14.24%	35.26%	6.31%	44.19%	
	Muslim	83	442	48	350	1.59*
		8.99%	47.89%	5.20%	37.92%	
	Christian	352	1,050	245	1,824	0.88
		10.14%	30.25%	7.06%	52.55%	
	Others (ref)	252	592	155	971	
		12.79%	30.05%	7.87%	49.29%	
Household living standard Chi-square = 8.33 $p = 0.2146$	Low (ref)	764	2,031	400	2,956	
		12.42%	33.02%	6.50%	48.06%	
	Medium	336	999	199	1,312	1.01
		11.81%	35.10%	6.99%	46.10%	
	High	124	384	87	544	1.08
		10.89%	33.71%	7.64%	47.76%	
Sex of child Chi-square = 2.08 $p = 0.5548$	Male	636	1,772	368	2,453	1.08
		12.16%	33.89%	7.04%	46.91%	
	Female (ref)	588	1,642	318	2,359	
		11.98%	33.46%	6.48%	48.07%	
Literacy of Mother Chi-square = 6.43 $p = 0.923$	Can read & write (ref)	692	2,024	385	2,771	
		11.78%	34.47%	6.56%	47.19%	
	Can't read & write	497	1,257	276	1,867	0.97
		12.75%	32.26%	7.08%	47.91%	
Total children ever born to mother Chi-square = 9.72 $p = 0.1366$	Up to two children)	548	1,444	276	2,014	0.91
		12.80%	33.72%	6.45%	47.03%	
	Three or four children	423	1,214	271	1,805	0.97
		11.39%	32.70%	7.30%	48.61%	
	Five or above children (ref)	218	623	114	819	
		12.29%	35.12%	6.43%	46.17%	
Age at marriage Chi-square = 13.03 $p = 0.0425$	Below 18 Years	384	1,016	183	1,556	1.03
		12.23%	32.37%	5.83%	49.57%	
	18 to 26 years	748	2,090	445	2,846	0.91
		12.20%	34.10%	7.26%	46.43%	
	Above 26 Years (ref)	57	175	33	236	
		11.38%	34.93%	6.59%	47.11%	
Age of child in months Chi-square = 5.07 $p = 0.1669$	>48 months	499	1,477	313	2,034	1.03
		11.54%	34.17%	7.24%	47.05%	
	<48 months	725	1,937	373	2,778	
		12.47%	33.32%	6.42%	47.79%	

*Significant results based on 95% confidence interval for the odds ratio.

Table 3. Model-building strategies

Model 1	Model 2	Model 3	Model 4	Model 5
No predictors but only ran-dom effects for the "state"	Model 1 + level-1 fixed effects that are related to maternal characteristics such as number of children ever born (2 or less, 3–4, and 5 or more) and age at mar-riage (below 18, 18–26, and above 26 years).	Model 2 + age of the child (months)	Model 3 + level-1 fixed ef-fects religion and literacy of mother	Model 4 + sex of child (male/female), household living standard (low, medium, and high), and place of residence (urban/rural)
Results used to determine the percentage variation in anemia level explained by the level-2 units (state of residence)	Results indicate the rela-tionship between level-1 (household level) predictors and the anemia level	Results indicate if addition of child's age improves the fit.	Results indicate whether addition of demographic variables improves the model fit	Results indicate whether addition of SES variables improves the model fit

Table 4. Estimates for two-level multinomial logistic regression models for predicting anemia level (*N* = 10,136)

	Model 1	Model 2	Model 3	Model 4	Model 5
Fixed effects					
Intercept 1 (se-verely anemic)	−1.95* (0.16)	−2.29* (0.19)	−2.35* (0.19)	−2.42* (0.20)	−2.23* (0.23)
Intercept 2 (moderately anemic)	−0.07 (0.16)	−0.42 (0.19)	−0.48* (0.19)	−0.55 (0.20)	−0.37 (0.23)
Intercept 3 (mildly anemic)	0.21 (0.16)	−0.14 (0.19)	−0.19 (0.19)	−0.27 (0.20)	−0.08 (0.23)
Age at marriage		0.12* (0.04)	0.12* (0.04)	0.12* (0.04)	0.12* (0.04)
Total children ever born		0.08* (0.03)	0.09* (0.03)	0.08* (0.03)	0.08* (0.03)
Age of child (<48 months)			−0.08* (0.04)	0.08* (0.04)	0.08* (0.04)
Religion				0.03 (0.02)	0.02 (0.02)
Literacy of mother				0.02 (0.04)	0.003 (0.04)
Household living standard					−0.02 (0.03)
Place of resi-dence					−0.06 (0.05)
Sex of child					−0.03 (0.04)
Error variance					
Intercept	0.2015* (0.10)				
Model fit					
−2LL	23,063.22	22,228.13** (p < 0.0001)	22,224.12** (p = 0.0452)	22,222.03 (p = 0.3517)	22,218.83 (p = 0.3618)

Note: Entries in the table are estimated effects while the standard errors are reported in the parenthesis.

*Level of significance at $p < 0.05$.

**Significant LR test; ICC = 0.0577. PROC GLIMMIX in SAS 9.4 with Laplace estimation method was used.

$$\mathrm{ICC} = \frac{\tau_{00}}{\tau_{00} + 3.29} = \frac{0.2015}{0.2015 + 3.29} = 0.0577$$

The ICC = 0.0577 indicates that approximately 6% of the total variation in the response is accounted for by the states. Thus, the remaining 94% variability is due to the variation within the respondents/households and other unknown factors.

Finally, to answer our last research question, we use the best fit model (Model 3) as discussed earlier. Model 3 contains three significant covariates (age at marriage, number of children ever born, and age of the child in months). We find significant effect of age at marriage ($\hat{\beta} = 0.12$, $p = 0.04$, number of children ever born ($\hat{\beta} = 0.09$, $p = 0.03$), and age of the child ($\hat{\beta} = 0.08$, $p = 0.04$. The estimates for mother's age at marriage and the number children ever born have positive effects on the log odds, whereas age of child has a negative effect. This implies if the age at marriage increases by one unit, the corresponding change in the log odds is 0.12. Similarly, a unit increase in the number of children ever born to mother increases the log odds by 0.09, and getting 48 months or older reduces the log odds by 0.08 units. In other words, age at marriage and number of children born to women increase the probability of being at the severely anemic level. On the other hand, increase in age of the child decreases the probability of being at a given anemia level. Since we have four levels for the outcome variables, we have three intercepts as shown in Table 4. We performed significance tests for these fixed intercepts and they were all found to be significant (result not shown). We calculated odds ratio for these predictors for easier interpretation. For age at marriage (OR = 1.13, 95% CI: 1.05, 1.21), one unit increase in age at marriage results in 1.13 times increase of having lower hemoglobin level (i.e. becoming more anemic). For number of children even born (OR = 1.09, 95% CI: 1.03, 1.15), one unit increase in the number of children born to women inflates the odds of having low hemoglobin level (indicating severely anemic) by 1.09. On the other hand, age of the child (OR = 0.92, 95% CI: 0.86 – 1.00) indicates odds of being at a lower hemoglobin level and decreases if the child is 48 months or older.

Using Model 3, the predicted probability of being at a given anemia level can be calculated following the formula given in Equation (3).

7. Discussion

Detection of the risk factors is detrimental to planning and implementation of programs to eradicate child anemia, especially in those groups where prevalence is very high. Our analysis suggests that childhood anemia is highly prevalent in the northeastern states of India. The overall prevalence of anemia in children in the northeastern states is 53%. There are important and significant relationships between anemia and some of the selected potential risk factors. Bivariate analysis suggests that there is no significant association between anemia level and household living standard, sex of child, literacy of mother, total children ever born to mother, and age of the child (months), whereas results from multivariate analysis show that age of the child, mother's age at marriage, and number of children ever born have significant effects on child anemia. We also found statistically significant variation across states for the chance of being at or below a certain hemoglobin level. The intra-class correlation coefficient, ICC = 0.0577, indicates approximately 6% of the total variation in the response variable explained by the state of residence. This is a sizeable contribution in explaining the total variability in the response variable.

In contrary to our findings, several other studies (le Cessie et al., 2002; Unsal, Bor, Tozun, Dinleyici, & Erenturk, 2007) indicated that male children were at greater risk of anemia than female children. Also, children living in rural areas were at greater risk of anemia as compared to their urban counterparts. Results also show that children belonging to households with low and medium standards of living index are more susceptible to anemia compared to their counterparts which corroborates the findings in Brazil and other countries (Assunção et al., 2007; Mamiro et al., 2005). We also notice that there is a declining tendency of the prevalence of anemia after two years of age. This may be due to the iron intake which improves with age as a result of a more varied diet, including the introduction of meat and other hemoglobin-containing diets.

Interestingly, the prevalence of any form of anemia is low among illiterate mothers. Results also indicate that the highest prevalence rate of any form of anemia was among the Muslim children (62%) followed by Hindus (56%) and Christians (48%). The reason might be that the children from Muslim and Hindu religions are from lowest quintiles and illiterate mothers. The fertility of the mother impacted anemia in children i.e. the higher number of children of the mother increases the requirement for childcare, demand for food, and inadequate supply of nutritional diet to all the children which ultimately make the children more vulnerable to the risk of anemia (Singh & Patra, 2014)

Mother's age at marriage had also a considerable effect on anemia in their children. Children of women who got married between 18 and 26 years were at greater risk of anemia (Agho, Dibley, D'Este, & Gibberd, 2008; Unsal et al., 2007).

The results from this study indicate that ignoring the hierarchical nature of the data could result in over statement of the significance of some of the variables included in the model. Most importantly, the standard errors would be biased downward. Anemia is a widespread health problem in India, especially in less developed regions like northeastern states of India. The strength of the paper is that it provides important information for policy-makers, program planners, and implementers that seek to reduce childhood anemia in northeastern states of India. The study findings have some important and relevant policy messages. High prevalence of mild and moderate anemia demands multiple interventions and strategies to tackle the burden of anemia. Special policies must be formulated and executed for prevention of anemia.

The present study demonstrates the endemicity of childhood anemia in northeastern states of India which may be due biological, social, and cultural factors. However, in this study, mother's age at marriage and the number of children ever born are the factors significantly associated with anemia in children. Results also suggest that age of the child is also a significant predictor, as the child ages to 48 months or more, the likelihood of severe anemia decreases. This study throws light on the fact that anemia intervention needs to focus more at state level followed by individual level. The limitations of the study are that indicators such as mother's anemia level, serum ferritin, dietary intake, worm infestations, malaria, and infectious diseases (Sinha, Deshmukh, & Garg, 2008) have not been considered.

Acknowledgments
All suggestions from the associated editors and reviewers will be acknowledged as appropriate.

Funding
The authors received no direct funding for this research.

Author details
Sanku Dey[1]
E-mail: sanku_dey2k2003@yahoo.co.in
Enayetur Raheem[2]
E-mail: enayetur.raheem@unco.edu
ORCID ID: http://orcid.org/0000-0001-8880-3649
[1] Department of Statistics, St. Anthony's College, Shillong, Meghalaya 793001, India.
[2] Applied Statistics and Research Methods, University of Northern Colorado, Greeley, CO 80639, USA.
[§] Sanku Dey identified the research problem. Both authors have developed the plan of analysis. Enayetur Raheem carried out the analysis. Both authors have contributed in writing the revising the manuscript.

References

Agho, K. E., Dibley, M. J., D'Este, C., & Gibberd, R. (2008). Factors associated with haemoglobin concentration among Timor-Leste children aged 6–59 months. *Journal of Health, Population and Nutrition, 26*, 200–209.

Assunção M. C. F., Santos, I., de Barros A. J. D, Gigante, D. P., & Victora, C. G. (2007). Anemia in children under six: Population-based study in Pelotas, Southern Brazil. *Revista de Saúde Pública, 41*, 328–335. http://dx.doi.org/10.1590/S0034-89102007000300002

Bharati, S., Pal, M., Chakrabarty, S., & Bharati, P. (2015). Socioeconomic determinants of iron-deficiency anemia among children aged 6 to 59 months in India. *Asia-Pacific Journal of Public Health, 27*, NP1432–NP1443. doi:10.1177/1010539513491417

de Benoist, B., McLean, E., Egll, I., & Cogswell, M. (2008). Worldwide prevalence of anaemia 1993–2005: WHO global database on anaemia, vi–41. Retrieved from http://apps.who.int/iris/bitstream/10665/43894/1/9789241596657_eng.pdf

Dey, S., Goswami, S., & Dey, T. (2013). Identifying predictors of childhood anaemia in North-East India. *Journal of Health, Population and Nutrition, 31*, 462–470.

Dubey A. (1994). Iron deficiency anaemia: Epidemiology, diagnosis and clinical profile. In H. Sachdev & P. Choudhury, (Eds.), *Nutrition in children: Developing country concerns* (pp. 217–235). New Delhi: B.I. Publications.

Ene, M., Leighton, E. A., Blue, G. L., & Bell, B. A. (2014). *Multilevel models for categorical data using SAS® GLIMMIX: The basics. SAS Global Forum*. Retrieved from http://analytics.ncsu.edu/sesug/2014/SD-13.pdf

Gillespie S. (Ed.) (1997). *Malnutrition in South Asia: A regional profile* (189 p). Kathmandu: United Nations Children's Fund.

Goswmai, S., & Das, K. K. (2015). Socio-economic and demographic determinants of childhood anemia. *Jornal de Pediatria, 91*, 471–477. http://dx.doi.org/10.1016/j.jped.2014.09.009

Griffiths, P., Matthews, Z., & Hinde, A. (2002). Gender, family, and the nutritional status of children in three culturally contrasting states of India. *Social Science & Medicine, 55*, 775–790. http://dx.doi.org/10.1016/S0277-9536(01)00202-7

IIPS. (2000). *National Family Health Survey (NFHS-2), 1998-99*. Retrieved from http://dhsprogram.com/pubs/pdf/FRIND2/FRIND2.pdf

IIPS. (2007). *National Family Health Survey (NFHS-3), 2005-06*, International Institute for Population Sciences (IIPS) and Macro International. Retrieved from http://pdf.usaid.gov/pdf_docs/PNADK385.pdf

le Cessie, S., Verhoeff, F. H., Mengistie, G., Kazembe, P., Broadhead, R., & Brabin, B. J. (2002). Changes in haemoglobin levels in infants in Malawi: Effect of low birth weight and fetal anaemia. *Archives of Disease in Childhood - Fetal and Neonatal Edition, 86*, F182–F187. Retrieved from http://www.ncbi.nlm.nih.gov/pubmed/11978749

Lokeshwar, M., Manglani, M., Rao, S., Patel, S., & Kulkarni, M. (1990). Iron deficiency—Clinical manifestation and management. In M. Mehta & M. Kulkarni, (Eds.) *Proceedings of national symposium cum work shop on child nutrition: The Indian scene* (pp. 269–275). Bombay: Division of Gastroenterology and Nutrition, LTM Medical College and Hospital.

Mamiro, P. S., Kolsteren, P., Roberfroid, D., Tatala, S., Opsomer, A. S., & Van Camp, J. H. (2005). Feeding practices and factors contributing to wasting, stunting, and iron-deficiency anaemia among 3-23-month old children in Kilosa district, rural Tanzania. *Journal of Health, Population and Nutrition, 23*, 222–230.

McLean, E., Cogswell, M., Egli, I., Wojdyla, D., & de Benoist, B. (2009). Worldwide prevalence of anaemia, WHO Vitamin and Mineral Nutrition Information System, 1993–2005. *Public Health Nutrition, 12*, 444–454. http://dx.doi.org/10.1017/S1368980008002401

NNMB. (2004). *Micronutrient survey*. Hyderabad. Retrieved from http://nnmbindia.org/NNMBMNDREPORT2004-Web.pdf

O'Connnell, A. A., Goldstein, J., Rogers, H. J., & Peng, C. Y. (2008). Multilevel logistic models for dichotomous and ordinal data. In A. A. O'Connell & D. B. McCoach (Eds.) *Multilevel modeling of educational data* (pp. 199–242). Charlotte, NC: Information Age Publishing,

Oliveira, M. A., Osório, M. M., & Raposo, M. C. (2007). Socioeconomic and dietary risk factors for anemia in children aged 6 to 59 months. *Jornal de Pediatria, 83*, 39–46. http://dx.doi.org/10.2223/JPED.1579

Osório, M. M., Lira, P. I., & Ashworth, A. (2004) Factors associated with Hb concentration in children aged 6-59 months in the State of Pernambuco, Brazil. *British Journal of Nutrition, 91*, 307–314. http://dx.doi.org/10.1079/BJN20031042

Pollitt E. (1991). Effects of a diet deficient in iron on the growth and development of preschool and school aged children. *Food and Nutrition Bulletin, 13*, 110–118.

Pollitt, E., Hathirat, P., Kotchabhakdi, N. J., Missell, L., & Valyasevi, A. (1989). Iron deficiency and educational achievement in Thailand. *American Journal of Clinical Nutrition, 50*, 687–697.

Seshadri, S., Gopaldas, T., Walter, T., & Heywood, A. (1989). Impact of iron supplementation on cognitive functions in preschool and school-aged children: The Indian experience. *American Journal of Clinical Nutrition, 50*, 675–686.

Singh, R. K., & Patra, S. (2014). Extent of anaemia among preschool children in EAG states, India: A challenge to policy makers. *Anemia, 2014*, 868752, 9 p. doi:10.1155/2014/868752

Sinha, N., Deshmukh, P. R., & Garg, B. S. (2008). Epidemiological correlates of nutritional anemia among children (6-35 months) in rural Wardha, Central India, *Indian Journal of Medical Sciences, 62*, 45–54.

Snijders, T. A. B, & Bosker, R. J. (1999). Multilevel analysis: An introduction to basic and advanced multilevel modeling. *Comparative and General Pharmacology*, viii, 266. Retrieved from http://www.amazon.com/Multilevel-Analysis-Introduction-Advanced-Modeling/dp/0761958908

Soemantri, A. G., Gopaldas, T., Seshadri, S., & Pollitt, E. (1989). Preliminary findings on iron supplementation and learning achievement of rural Indonesian children. *American Journal of Clinical Nutrition, 50*, 698–702.

Togeja, G., & Singh, P. (2004). *Micronutrient profile of Indian population*. New Delhi: Indian Council of Medical Research.

Unsal, A., Bor, O., Tozun, M., Dinleyici, E. C., & Erenturk, G. (2007). Prevalence of anemia and related risk factors among 2-11 months age infants in Eskisehir. *Turkish Journal of Medical Sciences, 7*, 1335–1339.

Permissions

All chapters in this book were first published in CM, by Cogent OA; hereby published with permission under the Creative Commons Attribution License or equivalent. Every chapter published in this book has been scrutinized by our experts. Their significance has been extensively debated. The topics covered herein carry significant findings which will fuel the growth of the discipline. They may even be implemented as practical applications or may be referred to as a beginning point for another development.

The contributors of this book come from diverse backgrounds, making this book a truly international effort. This book will bring forth new frontiers with its revolutionizing research information and detailed analysis of the nascent developments around the world.

We would like to thank all the contributing authors for lending their expertise to make the book truly unique. They have played a crucial role in the development of this book. Without their invaluable contributions this book wouldn't have been possible. They have made vital efforts to compile up to date information on the varied aspects of this subject to make this book a valuable addition to the collection of many professionals and students.

This book was conceptualized with the vision of imparting up-to-date information and advanced data in this field. To ensure the same, a matchless editorial board was set up. Every individual on the board went through rigorous rounds of assessment to prove their worth. After which they invested a large part of their time researching and compiling the most relevant data for our readers.

The editorial board has been involved in producing this book since its inception. They have spent rigorous hours researching and exploring the diverse topics which have resulted in the successful publishing of this book. They have passed on their knowledge of decades through this book. To expedite this challenging task, the publisher supported the team at every step. A small team of assistant editors was also appointed to further simplify the editing procedure and attain best results for the readers.

Apart from the editorial board, the designing team has also invested a significant amount of their time in understanding the subject and creating the most relevant covers. They scrutinized every image to scout for the most suitable representation of the subject and create an appropriate cover for the book.

The publishing team has been an ardent support to the editorial, designing and production team. Their endless efforts to recruit the best for this project, has resulted in the accomplishment of this book. They are a veteran in the field of academics and their pool of knowledge is as vast as their experience in printing. Their expertise and guidance has proved useful at every step. Their uncompromising quality standards have made this book an exceptional effort. Their encouragement from time to time has been an inspiration for everyone.

The publisher and the editorial board hope that this book will prove to be a valuable piece of knowledge for researchers, students, practitioners and scholars across the globe.

List of Contributors

David Trafimow
Department of Psychology, New Mexico State University, MSC 3452, P.O. Box 30001, Las Cruces, NM 88003-8001, USA

Yohei Kawasaki and Kazuki Ide
Department of Drug Evaluation & Informatics, School of Pharmaceutical Sciences, University of Shizuoka, Shizuoka, Japan

Masaaki Doi
Clinical Data Science & Quality Management Department, Toray Industries, Inc., Tokyo, Japan

Graduate School of Science and Engineering, Chuo University, Tokyo, Japan

Etsuo Miyaoka
Department of Mathematics, Tokyo University of Science, Tokyo, Japan

Stan Lipovetsky
GfK North America, 8401 Golden Valley Road, Minneapolis, MN 55427, USA

Ashish Kumar, Monika Saini and Kuntal Devi
Department of Mathematics, Manipal University Jaipur, Jaipur 303007, Rajasthan, India

Aamir Saghir
Department of Mathematics, Mirpur University of Science and Technology (MUST), Mirpur, Pakistan

Diarmuid O'Driscoll
Department of Mathematics and Computer Studies, Mary Immaculate College, Limerick, Ireland

Donald E. Ramirez
Department of Mathematics, University of Virginia, Charlottesville, VA, USA

Iryna Golichenko
Department of Mathematical Analysis and Probability Theory, National Technical University of Ukraine "Kyiv Polytechnic Institute", Kyiv 03056, Ukraine

Oleksandr Masyutka
Department of Mathematics and Theoretical Radiophysics, Taras Shevchenko National University of Kyiv, Kyiv 01601, Ukraine

Mikhail Moklyachuk
Department of Probability Theory, Statistics and Actuarial Mathematics, Taras Shevchenko National University of Kyiv, Kyiv 01601, Ukraine

Maksym Luz and Mikhail Moklyachuk
Department of Probability Theory, Statistics and Actuarial Mathematics, Taras Shevchenko National University of Kyiv, Kyiv, 01601, Ukraine

Maksym Luz and Mikhail Moklyachuk
Department of Probability Theory, Statistics and Actuarial Mathematics, Taras Shevchenko National University of Kyiv, Kyiv 01601, Ukraine

Christian H. Weiß
Department of Mathematics and Statistics, Helmut Schmidt University, Postfach 700822, 22008 Hamburg, Germany

Aamir Saghir
Department of Mathematics, Mirpur University of Science and Technology (MUST) Mirpur, Mirpur, Pakistan

Department of Mathematics, Zhejiang University, Hangzhou, P.R. China

Zhengyan Lin
Department of Mathematics, Zhejiang University, Hangzhou, P.R. China

Ching-Wen Chen
National Kaohsiung First University of Science and Technology, Kaohsiung, Taiwan

Diarmuid O'Driscoll
Department of Mathematics and Computer Studies, Mary Immaculate College, Limerick, Ireland

Donald E. Ramirez
Department of Mathematics, University of Virginia, Charlottesville, VA, 22904, USA

Ellida M. Khazen
13395 Coppermine Rd. Apartment 410, Herndon, VA 20171, USA

Blane Hollingsworth
Mathematics Department, Indiana University, Bloomington, IN, USA

Amir T. Payandeh Najafabadi
Department of Mathematical Sciences, Shahid Beheshti University, G.C. Evin, 1983963113 Tehran, Iran

Sanjay Kumar and Priyanka Chhaparwal
Department of Statistics, Central University of Rajasthan, Bandarsindri, Kishangarh, Ajmer, Rajasthan 305817, India

Melih Ağraz and Vilda Purutçuoğlu
Department of Statistics, Middle East Technical University, Ankara 06800, Turkey

Sanku Dey
Department of Statistics, St. Anthony's College, Shillong, Meghalaya 793001, India

Enayetur Raheem
Applied Statistics and Research Methods, University of Northern Colorado, Greeley, CO 80639, USA

Index

www.ingramcontent.com/pod-product-compliance
Lightning Source LLC
Chambersburg PA
CBHW082045190326
41458CB00010B/3462
* 9 7 8 1 6 3 2 3 8 5 6 0 4 *